MODELING AND CONTROL OF FUEL CELLS

BOOKS IN THE IEEE PRESS SERIES ON POWER ENGINEERING

Principles of Electric Machines with Power Electronic Applications, Second Edition
M.E. El-Hawary

Pulse Width Modulation for Power Converters: Principles and Practice
D. Grahame Holmes and Thomas Lipo

Analysis of Electric Machinery and Drive Systems, Second Edition
Paul C. Krause, Oleg Wasynczuk, and Scott D. Sudhoff

Risk Assessment for Power Systems: Models, Methods, and Applications
Wenyuan Li

Optimization Principles: Practical Applications to the Operations of Markets of the Electric Power Industry
Narayan S. Rau

Electric Economics: Regulation and Deregulation
Geoffrey Rothwell and Tomas Gomez

Electric Power Systems: Analysis and Control
Fabio Saccomanno

Electrical Insulation for Rotating Machines: Design, Evaluation, Aging, Testing, and Repair
Greg Stone, Edward A. Boulter, Ian Culbert, and Hussein Dhirani

Signal Processing of Power Quality Disturbances
Math H. J. Bollen and Irene Y. H. Gu

Instantaneous Power Theory and Applications to Power Conditioning
Hirofumi Akagi, Edson H. Watanabe and Mauricio Aredes

Maintaining Mission Critical Systems in a 24/7 Environment
Peter M. Curtis

Elements of Tidal-Electric Engineering
Robert H. Clark

Handbook of Large Turbo-Generator Operation and Maintenance, Second Edition
Geoff Klempner and Isidor Kerszenbaum

Introduction to Electrical Power Systems
Mohamed E. El-Hawary

Modeling and Control of Fuel Cells: Distributed Generation Applications
M. Hashem Nehrir and Caisheng Wang

MODELING AND CONTROL OF FUEL CELLS

Distributed Generation Applications

M. Hashem Nehrir
Caisheng Wang

IEEE PRESS SERIES ON POWER ENGINEERING

IEEE

IEEE Press

WILEY

A JOHN WILEY & SONS, INC., PUBLICATION

IEEE Press
445 Hoes Lane
Piscataway, NJ 08854

Library of Congress Cataloging-in-Publication Data is available.

ISBN 978-0-470-23328-3

CONTENTS

Preface **xiii**

Acknowledgments **xvii**

1 Introduction **1**

1.1 Background: A Brief History of U.S. Electric Utility Formation and Restructuring 1
1.2 Power Deregulation and Distributed Generation 3
1.3 DG Types 7
1.4 Fuel Cell DG 9
1.5 The Hydrogen Economy 13
1.5.1 Introduction 13
1.5.2 Challenges of Transition to a Hydrogen Economy 14
1.5.3 Hydrogen Production 15
1.5.3.1 Hydrogen Production by Reforming Natural Gas 16
15.3.2 Hydrogen Production from Coal 17
1.5.3.3 Hydrogen Production from Nuclear Energy 18
1.5.3.4 Hydrogen Production by Water Electrolysis 19

1.5.3.5 Solar Energy to Hydrogen 19
1.5.3.6 Wind Energy to Hydrogen 20
1.5.3.7 Biomass Energy to Hydrogen 20
1.5.4 Hydrogen Storage and Distribution 21
1.5.5 Department of Energy Hydrogen-Related Activities 22
1.5.5.1 Hydrogen Production 22
1.5.5.2 Hydrogen Basic Research 23
1.5.5.3 Hydrogen Delivery 23
1.5.5.4 Hydrogen Storage 24
1.5.5.5 Hydrogen Energy Conversion (Fuel Cells) 24
1.5.6 The Role of This Book 26
References 27

2 Principles of Operation of Fuel Cells 29

2.1 Introduction 29
2.2 Chemical and Thermal Energy of an Element 30
2.3 Fundamentals of Thermodynamics 31
2.3.1 The First Law of Thermodynamics 31
2.3.2 The Second Law of Thermodynamics 32
2.4 Fundamentals of Electrochemical Processes 34
2.4.1 The Gibbs Free Energy 34
2.5 Energy Balance in Chemical Reactions 35
2.6 The Nernst Equation 37
2.7 Fuel Cell Basics 38
2.8 Types of Fuel Cells 40
2.9 Fuel Cell Equivalent Circuit 53
2.10 Capacitance of Double-Layer Charge Effect 54
2.11 Summary 55
References 56

3 Dynamic Modeling and Simulation of PEM Fuel Cells 57

3.1 Introduction: Need for Fuel Cell Dynamic Models 57
3.2 Nomenclature (PEMFC) 58
3.3 PEMFC Dynamic Model Development 60
3.3.1 Gas Diffusion at the Electrodes 62
3.3.2 Material Conservation 64
3.3.3 PEMFC Output Voltage 65

3.3.4 PEMFC Voltage Drops 67
3.3.5 Thermodynamic Energy Balance for PEMFC 69
3.4 PEMFC Model Structure 71
3.5 Equivalent Electrical Circuit Model of PEMFC 72
3.6 PEMFC Model Validation 77
References 83

4 Dynamic Modeling and Simulation of Solid Oxide Fuel Cells 85

4.1 Introduction 85
4.2 Nomenclature (SOFC) 86
4.3 SOFC Dynamic Model Development 88
4.3.1 Effective Partial Pressures 89
4.3.2 Material Conservation 92
4.3.3 SOFC Output Voltage 94
4.3.3.1 Activation Voltage Drop 95
4.3.4 Thermodynamic Energy Balance for Tubular SOFC 98
4.3.4.1 The Fuel Cell Tube 99
4.3.4.2 Fuel 100
4.3.4.3 Air Between Cell and Air Supply Tube (AST) 100
4.3.4.4 Air Supply Tube 101
4.3.4.5 Air in AST 101
4.4 SOFC Dynamic Model Structure 102
4.5 SOFC Model Response—Constant Fuel Flow Operation 103
4.5.1 Steady-State Characteristics 103
4.5.2 Dynamic Response 106
4.5.2.1 Dynamics Due to the Double-Layer Charge Effect 106
4.5.2.2 Dynamics Due to the Effect of Pressure 108
4.5.2.3 Dynamics Due to the Effect of Temperature 109
4.6 SOFC Model Response—Constant Fuel Utilization Operation 111
4.6.1 Steady-State Characteristics 112
4.6.2 Dynamic Response 113
References 114

5 Principles of Operation and Modeling of Electrolyzers **116**

5.1 Principle of Operation of Electrolyzers 116
5.2 Dynamic Modeling of Electrolyzers 117
5.2.1 Electrolyzer Steady-State (V–I) Characteristics 119
5.2.2 Modeling Hydrogen Production Rate 120
5.2.3 Electrolyzer Thermal Model 122
5.3 Electrolyzer Model Implementation 123
References 125

6 Power Electronic Interfacing Circuits for Fuel Cell Applications **126**

6.1 Introduction 126
6.2 Overview of Basic Power Electronic Switches 128
6.2.1 Diode 128
6.2.2 Thyristor 129
6.2.3 Bipolar Junction Transistor (BJT) 130
6.2.4 Metal-Oxide Semiconductor Field Effect Transistor (MOSFET) 131
6.2.5 Gate Turn-Off Thyristor (GTO) 132
6.2.6 Insulated Gate Bipolar Transistor (IGBT) 133
6.2.7 MOS-Controlled Thyristor (MCT) 133
6.3 ac/dc Rectifiers 135
6.3.1 Circuit Topologies 135
6.3.2 Simplified Model for Three-Phase Controllable Rectifiers 138
6.4 dc to dc Converters 140
6.4.1 Boost Converters 141
6.4.1.1 Circuit Topology 141
6.4.1.2 Small-Signal State-Space Model 142
6.4.1.3 Average Model for Long-Time Simulation 144
6.4.2 Buck Converters 146
6.4.2.1 Circuit Topology 146
6.4.2.2 Small-Signal State-Space Model for Buck dc/dc Converters 148
6.4.2.3 Average Model for Long-Time Simulation 149

6.5 Three-Phase dc/ac Inverters 150
6.5.1 Circuit Topology 150
6.5.2 State-Space Model 153
6.5.3 *abc/dq* Transformation 156
6.5.4 *dq* Representation of the State-Space Model 157
6.5.5 Ideal Model for Three-Phase VSI 159
References 162

7 Control of Grid-Connected Fuel Cell Power Generation Systems 163

7.1 Introduction 163
7.2 Grid-Connected System Configuration 164
7.2.1 PEMFC Unit Configuration 166
7.2.2 SOFC Unit Configuration 166
7.3 Controller Designs for dc/dc Converters and the Inverter 168
7.3.1 Circuit and Controller Design for the Boost dc/dc Converter 168
7.3.1.1 Circuit Design 168
7.3.1.2 Controller Design 170
7.3.2 Controller Design for the Three-Phase VSI 173
7.3.2.1 Current Control Loop 174
7.3.2.2 Voltage Control Loop 176
7.3.2.3 Overall Power Control System for the Inverter 181
7.4 Simulation Results 182
7.4.1 Desired P and Q Delivered to the Grid—Heavy Loading 182
7.4.1.1 PEMFC DG 182
7.4.1.2 SOFC DG 184
7.4.2 Desired P Delivered to the Grid, Q Consumed from the Grid: Light Loading 186
7.4.2.1 PEMFC DG 187
7.4.2.2 SOFC DG 188
7.4.3 Load-Following Analysis for Fuel Cells 189
7.4.3.1 Fixed Power Supply from the Grid 189
7.4.3.2 Fixed Power Supply from the FCDG 191
7.4.4 Fault Analysis 192
7.5 Summary 195
References 195

8 Control of Stand-Alone Fuel Cell Power Generation Systems 198

8.1 Introduction 198
8.2 System Description and Control Strategy 199
8.3 Load Transient Mitigation Control 201
8.3.1 Circuit Model for Lead–Acid Batteries 202
8.3.2 Battery Charge/Discharge Controller 203
8.3.3 Filter Design 204
8.4 Simulation Results 205
8.4.1 The Load Transients 206
8.4.1.1 The dc Load Transients 206
8.4.1.2 The ac Load Transients 207
8.4.2 Load Transient Mitigation 209
8.4.2.1 PEMFC System 209
8.4.2.2 SOFC System 212
8.4.3 Battery Charge/Discharge Controller 214
8.5 Summary 216
References 216

9 Hybrid Fuel Cell Based Energy System Case Studies 219

9.1 Introduction 219
9.2 Hybrid Electronically Interfaced Systems 221
9.2.1 The dc-Coupled Systems 222
9.2.2 The ac-Coupled Systems 224
9.2.3 Stand-Alone Versus Grid-Connected Systems 225
9.3 Fuel Cells in Hybrid Combined Heat and Power Operation Mode 226
9.4 Case Study I: A Hybrid Stand-Alone Wind–PV–FC System 227
9.4.1 System Configuration 227
9.4.2 System Unit Sizing 230
9.4.3 System Component Characteristics 232
9.4.3.1 The Wind Energy Conversion System Model 233
9.4.3.2 The Photovoltaic Array Model 234
9.4.3.3 The Fuel Cell and Electrolyzer Models 235
9.4.4 System Control 236
9.4.4.1 The Overall Power Management Strategy 236

9.4.4.2 The Wind-Turbine Pitch Angle Controller 236
9.4.4.3 The PV Maximum Power Point Tracking (MPPT) Control 238
9.4.4.4 The ac Bus Voltage Regulator 240
9.4.4.5 The Electrolyzer Controller 241
9.4.5 Simulation Results 241
9.5 Case Study II: SOFC Efficiency Evaluation in Hybrid Operation Mode 247
9.5.1 Thermodynamic Laws and SOFC Efficiency 248
9.5.2 Hydrogen Fuel Heating Values 253
9.5.3 SOFC Electrical Efficiency 255
9.5.4 SOFC Efficiency in Hybrid CHP Operation Mode 256
9.6 Summary 259
References 260

10 Present Challenges and Future of Fuel Cells 265

10.1 Introduction 265
10.2 Fuel Cell System Operations 266
10.2.1 Fuel Processor 266
10.2.2 Fuel Cell Stack 267
10.2.3 Power Conditioner System 269
10.2.4 Balance of Plant (BOP) Systems 272
10.3 Present Challenges and Opportunities 272
10.3.1 Cost 272
10.3.2 Fuel and Fuel Infrastructure 273
10.3.3 Materials and Manufacturing 274
10.4 U.S. Fuel Cell R&D Programs 275
10.4.1 DOE's SOFC-Related Programs 276
10.5 Future of Fuel Cells: A Summary and Authors Opinions 278
References 279

Appendix A Instruction for Running the PEMFC and SOFC Models and Their Distributed Generation Application Models 282

Index 291

PREFACE

Fuel cells (FCs) have caught intensive attention in the last decade. Following their successful development for the manned space program by the U.S. National Aeronautical and Space Administration (NASA) in the 1960s, significant advances have been achieved in the fuel cell technology and the applications of different types of fuel cells.

Environmental concerns and world-wide social and political pressure to reduce carbon dioxide emission, and the desire to seek higher energy conversion efficiencies in electrical power generation, have been the main drivers for fuel cell technological advances and their applications, including distributed power generation and fuel cell vehicle applications. Electrical power producers are seeking ways to gain competitive position in the deregulated power market by contributing to reduced environmental emissions. In addition, as a part of the requirement to pursue technologies to reduce carbon emissions, the automobile industry has started FC vehicle development with the ultimate goal of reaching zero-emission car.

The operation of fuel cells, being based on electrochemical principles, is normally better understood by scientists (i.e., chemists, physicists, and material scientists) and chemical engineers working in this area. Since the end product of fuel cells is electricity, their operation also needs to be understood by electrical engineers working in the area to be able to design interfacing electrical circuits and controllers to achieve their improved performance. For this reason accurate and user-friendly dynamic fuel cell models are needed to evaluate their steady-state and transient performance from electrical engineering. This book bridges the gap between the two groups, namely scientists/chemical engineers and electrical engineers. It explains the principle of operation of fuel cells in a simple language understood by electrical engineers. It also explains the development of physically-based dynamic models for PEM fuel cells (PEMFCs) and tubular solid-oxide fuel cells (SOFCs), which have great potential for distributed generation (DG) and mobile applications. The main focus of the book, however, is on modeling, control and applications of the above two types of fuel cells. PEMFCs are suitable for residential and backup power as well as DG and FC vehicle applications. SOFCs are high-temperature fuel cells suitable for DG and combined heat and power (CHP) operation to achieve high system energy efficiency.

This book is the result of over 10 years of research on PEMFC and SOFC. It combines the theory, modeling, interfacing, and control of FC systems in one place. The book is intended to be a resource for all engineers, in particular, electrical, chemical, and mechanical engineers, and for all those interested in designing controllers and interfacing circuits for FC energy systems and FC vehicles of the future. An overview of the chapters covered in the book is given below:

Chapter 1 gives a brief history of U.S. electric utility formation and background about the restructured utility that leads to increased interest in DG. Then, an overview of DG and its different types, DG applications of FCs, an introduction to hydrogen economy, and a need for a FC-powered society are covered.

Chapter 2 gives a brief coverage of electrical and thermal energy, fundamentals of thermodynamics, and electrochemical processes, which have lead to the operation of FCs. The principles of operation of major types of FCs and electrolyzer are also covered.

Chapters 3–5 cover the modeling of PEMFC, tubular SOFC, and electrolyzer, respectively. The development of a physically-based dynamic model and equivalent electrical circuit model for PEMFC and PEMFC model validation are given in Chapter 3. Chapter 4 covers the development of a physically-based dynamic model for tubular SOFC under different operating conditions. The process of water electrolysis and electrolyzer modeling are introduced in Chapter 5. The emphasis in these chapters is on the electrical terminal characteristics of PEMFC, tubular SOFC, and electrolyzer.

An introduction to and modeling of power electronic switching devices and circuits for FC energy systems is given in Chapter 6. Power electronic (power conditioning) devices are important and integral part of FC systems to obtain conditioned electricity output with desired power quality. The development of the models in the rotating (*dq*) reference frame, which is useful in controller design (covered in Chapters 7 and 8) is also presented.

Fuel cells face a variety of load and/or electrical disturbances. Proper controllers must be integral parts of FC systems to achieve their reliable and durable operation in delivering electricity of desired quality. Chapters 7 and 8 cover control methodologies and controller designs for stand-alone and grid-connected operation of FCDG systems, respectively.

Hybrid alternative energy power generation systems are expected to be an important part of the power generation paradigm of the future. Chapter 9 covers two distinct examples of applying the developed FC models and controller design methodologies for designing hybrid DG systems including FCs. The chapter covers the design and performance investigation of a hybrid wind/photovoltaic (PV)/FC-electrolyzer system using PEMFC and the operation and efficiency evaluation of SOFC in CHP (combined-cycle) mode.

Three major challenges to fuel cell commercialization (cost, fuel and fuel infrastructure, and materials and manufacturing) are outlined in Chapter 10. A summary of authors' opinions on current developmental status and future potential of fuel cells is also given in the chapter.

An important feature of this book is the electronic files of the computer models provided with it, available at ftp://ftp.wiley.com/sci_tech_med/fuel_cells. They are MATLAB/SIMULINK and PSpice-based files of PEMFC and MATLAB/SIMULINK-based files of SOFC dynamic models, and their FCDG applications. The models have been developed and simulation results obtained on MATLAB/SIMULINK version 7.0.4. Instruction for running the models is given in Appendix A at the end of the book. Basic knowledge of MATLAB/SIMULINK or PSPICE is required to run the models. We hope these models and their DG applications will be useful to educators, students, and researchers worldwide in developing new methodologies for modeling and control of FC energy systems of the future.

M. Hashem Nehrir
Caisheng Wang

Bozeman, Montana
Detroit, Michigan
January 2009

ACKNOWLEDGMENTS

The authors would like to thank many people and organizations that helped them in the preparation of this book; without their help and support this book would not have been possible.

We are most grateful to Dr. Don Pierre, Professor Emeritus of Electrical & Computer Engineering (ECE) Department at Montana State University (MSU) for his invaluable suggestions and very careful editing and proof-reading of nearly the entire book. We also acknowledge the fruitful discussions we had with Dr. Steven Shaw of MSU ECE Department during the course of our research.

Dr. Paul Gannon of the MSU Chemical & Biological Engineering (CBE) Department contributed in the writing of Chapter 10 (*Present Challenges and Future of Fuel Cells*). His knowledge on fuel processing, internal operation of fuel cells and the challenges involved in bringing fuel cells into widespread use helped us bring this book to completion.

Mr. Chris Colson, currently a Ph.D. candidate in the MSU ECE Department contributed to the hybrid operation and efficiency evaluation of SOFC in combined heat and power mode. His previous work in this area co-authored by the authors of this book, as well as Professor Emeritus Max Diebert of MSU CBE and Professor Ruhul Amin of MSU Mechanical & Industrial Engineering, is used as an example in Chapter 9. Mr. Colson also helped in editing a portion of the book, for which we are thankful.

During the course of our research, we made many presentations at technical conferences on the modeling and control of PEMFC and SOFC.

We are thankful to the many people who made constructive comments and suggestions to us after attending our presentations. Their comments helped us enrich the contents of this book.

We acknowledge the financial support provided by the following agencies for work related to the contents of this book: Montana DOE-EPSCoR (1994-2001), Montana NSF-EPSCoR (1998-2000), U.S. National Science Foundation (Grant 0135229, 2002-2006), and the High Temperature Electrochemistry (HiTEC) Fuel Cell Project at MSU, funded by the U.S. Department of Energy, as a subcontract from Battle Memorial Institute and Pacific Northwest National Laboratory (Award DE-AC06-76RL01830, 2002-2008).

Hashem Nehrir is thankful to MSU for providing him with the release time (sabbatical leave) for the preparation of the first draft of the book. He is also thankful to the MSU ECE Department for its institutional support and encouragement.

We would like to thank the following fuel cell companies for providing photos of their products for this book: Ballard Power Systems, FuelCell Energy, ReliOn, Siemens, Versa Power Systems, and Hydrogenics Corp.

We are thankful to Drs. Rama Ramakumar of Oklahoma State University, Oklahoma and Kai Strunz of Technical University of Berlin, Germany for their careful review of the book; their constructive comments and suggestions made the book more comprehensive. We are also thankful to the staff of IEEE Press-Wiley and Thomson Digital for their dedication to this book project.

Last but not least, we are thankful to our families for their encouragement and support during the course of development of this book.

M. H. N.
C. W.

CHAPTER 1

INTRODUCTION

Global environmental concerns and the ever-increasing need for electrical power generation, steady progress in power deregulation, and tight constraints over the construction of new transmission lines for long distance power transmission have created increased interest in distributed generation (DG). Of particular interest are renewable DGs with free energy resources, such as wind and solar photovoltaic (PV), and alternative energy DG sources with low emission of pollutant gases, such as fuel cell (FC) and microturbine (MT) power generation devices.

In this chapter, some background about the restructured utility that lead to increased interest in DG is given first. Then, an overview of distributed generation and its different types is addressed. Distributed generation applications of fuel cells will be covered next. Finally, since all viable types of fuel cells use hydrogen (H_2) as fuel, the last part of this chapter covers the hydrogen economy, a need for a fuel-cell-powered society.

1.1 BACKGROUND: A BRIEF HISTORY OF U.S. ELECTRIC UTILITY FORMATION AND RESTRUCTURING [1–4]

Electric utilities were initially formed in the United States in late nineteenth century and established as isolated electric systems without connection to one another. In 1920s, the isolated electric systems were interconnected to help each other in load sharing and backup power. In 1934, the U.S. Congress passed the *Public Utility Holding Company Act* (*PUHCA*), where it increased the jurisdiction of the Securities Exchange Commission as well as the jurisdiction of the Federal Power Commission. This act created incentives for the isolated utilities to expand and create regional utilities, where several state utilities joined under a regional utility company. Each entity operated in its region under an investor-owned monopoly, owning generation, transmission, and distribution. However, each utility was subject to state regulation, where the utilities' rates had to be approved by the Public Utilities Commissions.

In 1977, the U.S. Department of Energy (DOE) was created to oversee the nation's energy-related activities, and under it, the Federal Energy Regulatory Commission (FERC) was formed to establish rules for generation, transport, and quality of power, among others.

The U.S. Congress passed the *Public Utilities Policy Act* (*PURPA*) in 1978. This act encouraged the construction and integration of nonutility-owned power generation technologies, including conventional and nonconventional (renewable/alternative) energy sources, to the utility grid. Under the above act, FERC sets rules for the interconnection of these power generation sources to the utility grid. Until near the end of the twentieth century, the utilities were still operating under the vertical (monopoly) structure; each utility owned generation, transmission, and distribution in a given region.

The major *Energy Policy Act*, enacted by the Congress in 1992, drove the U.S. power industry into complete restructuring; now, more than 15 years later, it is still ongoing. As a result of this act, "Exempt Wholesale Generator (EWG)" entities were created with the restriction that EWGs can only sell the power they generate on the wholesale market and not on the retail market. On the contrary, electric utilities are not required to purchase power from EWGs, but they are required to purchase power from qualified power generating facilities that include renewable/alternative energy power generation facilities. This energy policy act created a major shift in regulatory power from the regional level to the federal level with FERC continuing to be its rule making body. According to the policy act, the power generating entities had transmission access for the power they generated. In 1996, FERC issued the "Mega Rule," which spelled out how open access transmission of power is to be handled. It requires the transmission system owners to treat all transmission users on a nondiscriminatory basis and file tariffs for their transmission services.

Gradually, the vertical electric utility, where one company owned generation, transmission and distribution facilities, changed to a horizontal structure. In this new paradigm, generation, transmission and distribution companies became separate and independent, namely GENCO, TRANSCO, and DISCO. Generation being the only one of the three entities that is truly deregulated, numerous independent power producers (IPPs) were formed and found an opportunity to market their power. This change also created the opportunity for large and small power marketers to be formed to begin marketing the power produced by IPPs (GENCOs). Since the start of power deregulation in 1996, FERC has promoted the formation of regional transmission organizations (RTOs).

In 1999, FERC Order 2000 required the transmission system owners to put their transmission system under the control of RTOs. Today, several regions have established independent system operators (ISOs), or are in the planning stage to establish ISOs, to operate their transmission systems and provide transmission services. In 2005, the U.S. government passed the *Energy Policy Act of 2005*. This act authorizes the creation of an electric reliability organization (ERO), giving it the authority to enforce compliance of all market participants with the reliability standards of the National Electric Reliability Council (NERC), which was voluntary prior to 2005. In 2006, FERC certified NERC to be the U.S. ERO. Given the close interconnection of the U.S. and Canadian electric system, NERC is also seeking recognition as the ERO from the Canadian government.

Figure 1.1a shows the structure of the vertical utility of the past, where generation, transmission, and distribution systems in one region were owned by the utility in that region and sale of power within the region took place by that utility. Figure 1.1b shows the restructured horizontal utility, where different GENCOs market their power and TRANSCOs and DISCOs arrange the transport of power to customers. Figure 1.2 shows the role of ISO in the restructured utility and the deregulated power market. ISOs oversee the transport of power from generation to transmission to distribution including the marketing (buy/sell) of electric power. The structure and different entities of ISO in a region depends on the market structure in that region. At the time of writing this book, power deregulation and utility restructuring are being actively pursued worldwide.

1.2 POWER DEREGULATION AND DISTRIBUTED GENERATION [1,4]

As explained in the previous section, numerous IPPs were formed as a result of power deregulation, which also spurred the consideration of DG sources. The main reason behind this consideration was, and still is,

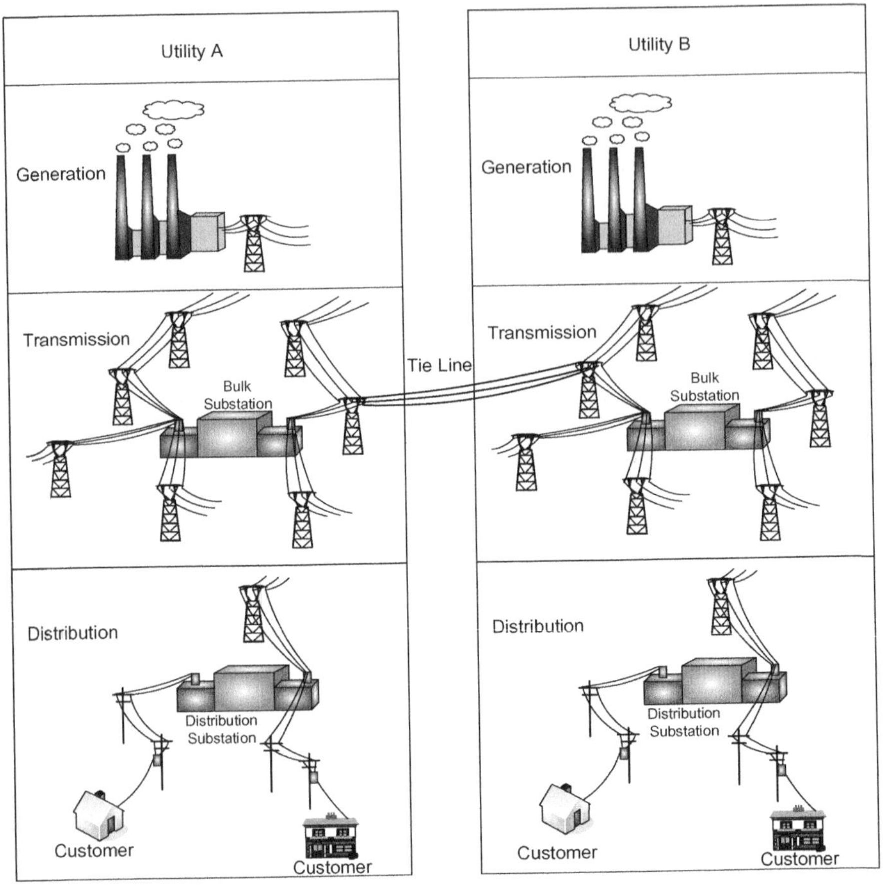

(a) Utilities of past.

FIGURE 1.1 Utility structure of past and present.

economics and the driving market forces. The fast growth in demand for electricity along with the slow growth in generation capacity in the last quarter of the twentieth century resulted in shrinking spinning reserve margins, which as a result, made power systems vulnerable and brought about the need for additional power generation. The economic constraints behind building large central power generating stations and expanding the transmission infrastructure encouraged the consideration of DGs. DGs are modular in structure and less costly to build, normally placed at the distribution level at or near load centers, and are small in size (relative to the power capacity of the system in which they are placed). When possible, DGs can be strategically (optimally) placed in distribution systems for grid

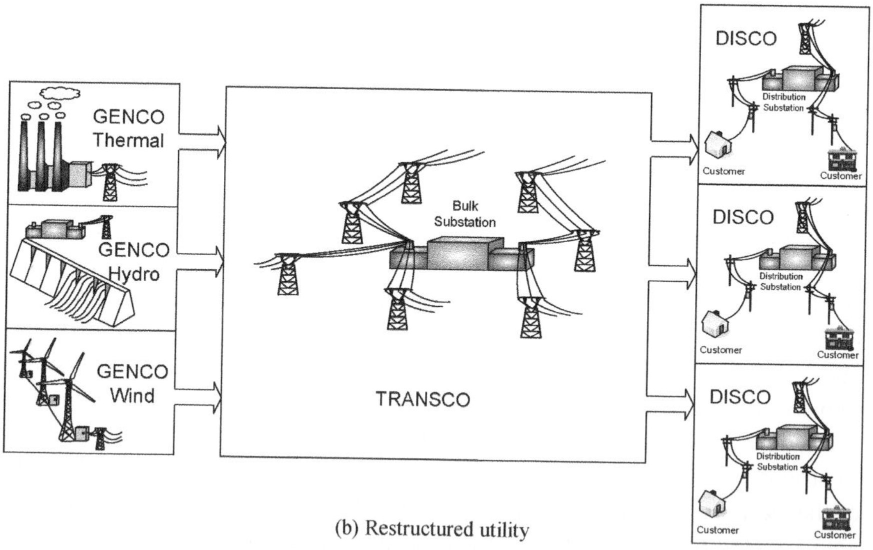

(b) Restructured utility

FIGURE 1.1 (*Continued*)

reinforcement, reducing power losses, on-peak operating costs, and improving voltage profiles and load factors. As a result, their installation can defer or eliminate the need for system upgrades, and can improve system integrity, reliability and efficiency. Because of these benefits, DG became, and still is, a priority.

DGs also have barriers and obstacles, which must be overcome before they can become a mainstream service. These barriers include technical, economic, and regulatory issues. Some of the proposed technologies have not yet entered the market; they need to meet some pricing and performance targets before entry. In addition to this, the most important

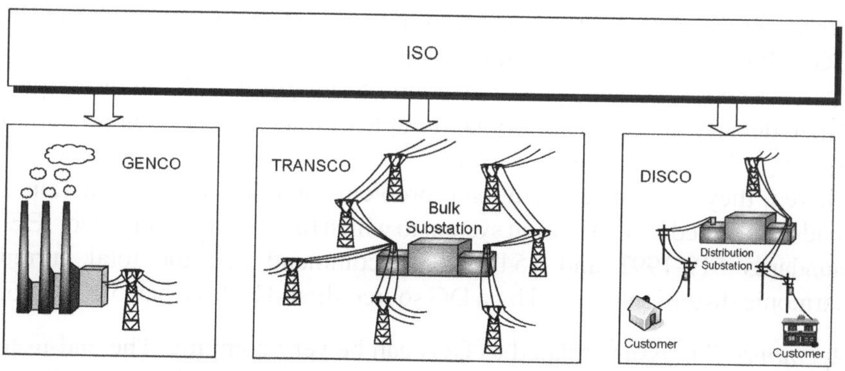

FIGURE 1.2 Role of ISO in the restructured utility.

issues facing DGs are *safety issues, operation issues, power quality issues, and accountability issues*. These issues are briefly discussed below:

Safety issues are very important. Grid-connected DGs can keep the power lines live after a grid power outage (thus making it unsafe for maintenance crews to work on the power lines) if appropriate measures are not taken to disconnect the DGs from the grid shortly after the outage. The IEEE Power Engineering Society has taken a leading role in developing standards (e.g., IEEE Standard 1547 [26]) for detection of *power islands*, that is, when the grid power goes out and DGs are still connected to the grid. According to the above standard, the islanded DGs must be disconnected from the grid within two seconds. Research is under way to develop reliable islanding detection and autonomous operation of DGs so that they can operate as isolated DGs to provide power to some critical loads, which they may be feeding. In spite of the technical and socio-economic issues, DGs, conventional and modern, are expected to become wide-spread around the globe.

Operation and reliability issues of DGs could have positive or negative impacts on the distribution systems to which they are connected. DGs are normally connected to the utility grid at the distribution level and can bring voltage support to the grid by providing reactive power support. This condition could be helpful for reliable operation of distribution systems provided that the distribution system is properly configured for inclusion of DGs. On the contrary, DGs could have negative impact on distribution systems operation. For example, DGs with variable output power (i.e., wind and solar) may not be able to provide the required power at the right time, or, DGs that use induction generators (e.g., wind), receive reactive power from the grid and could actually worsen reactive coordination and operational reliability.

Power quality issues are becoming more pronounced as the more modern DG technologies (e.g., wind, photovoltaic, fuel cells) utilize power electronic devices (i.e., dc/dc and dc/ac converters) to interface with the utility grid. These devices inject nonsinusoidal (or at least imperfect sinusoidal) current to the grid. If the harmonics generated by these devices are not properly filtered, they can cause operational problems and possible malfunction of loads connected to distribution system to which the DGs are connected. IEEE standards 519-1992 and 1547-2003 recommend that the total current harmonic distortion injected by a DG source should be less than 5% [25,26].

Accountability issues related to DGs can be very complex. The end-users may not know or care about the nature of the restructured power industry.

However, they want reliable power. In a distribution system enhanced with one or more DGs, if the DGs trip out or are not able to provide the desired amount of power to the grid, the quality of power provided to the end-users may diminish. Who is accountable to the customers for such problems, the DG owners or the end-use service provider? This is a serious problem facing the restructured power industry, which will increase in magnitude as DGs penetrate the power grid. Therefore, carefully written policies, regulations, and buy–sell agreements are needed to address such problems—thus making the role of FERC in the deregulated power market more important than ever.

1.3 DG TYPES

In this book distributed generation is referred to small generators, starting from a few kWs up to 10 MW, whether connected to the utility grid or used as stand-alone at an isolated site. Normally small DGs, in the 5–250 kW range serve households to large buildings (either in isolated or grid-connected configuration). In grid-connected configuration, DGs with larger capacities are managed by a utility or an IPP. They are located at strategic points, normally at the distribution level, near load centers, and used for such purposes as capacity support, voltage support and regulation, and line loss reduction. DG technologies can be categorized to renewable and nonrenewable DGs.

Renewable energy technologies are in general sustainable (i.e., their energy source will not run out) and cause little or no environmental damage; they include the following:

- Solar photovoltaic
- Solar thermal
- Wind
- Geothermal
- Tidal
- Low-head (small) hydro
- Biomass and biogas
- Hydrogen fuel cells (hydrogen generated from renewable resources).

Nonrenewable energy technologies are referred to those that use some type of fossil fuel such as gasoline, diesel, oil, propane, methane, natural gas, or coal as their energy source. Fossil fuel-based DGs are not considered sustainable power generation sources as their energy source will not renew. They include the following:

- Internal combustion engine (ICE)
- Combustion turbine
- Gas turbine
- Microturbine
- Fuel cells (using some type of fossil fuel, e.g., natural gas, to generate hydrogen).

Both types of DGs (renewable and nonrenewable) discussed above are popular and widely used around the world. The downside of renewable resource DGs is the intermittent nature of their renewable energy source; and the disadvantage of fossil fuel-based DGs is that they generate environmentally polluting, and in some cases poisonous exhaust gases, such as SO_2 and NO_x, which are similar to the pollutants from conventional centralized power plants. However, considering the increasing need for electricity, the benefits of the nonrenewable DG technologies with low emission of polluting gasses exceed their disadvantages and are expected to be used in the foreseeable future.

Fuel cell technology can belong to either of the above categories. If the hydrogen fuel needed to power the fuel cell is generated from a renewable source, the fuel cell power-generating unit is considered a renewable energy technology. An example of this case (i.e., wind and solar energy used to generate hydrogen to fuel a fuel cell stack) will be covered in Chapter 9. On the contrary, if hydrogen is produced from a fossil fuel source (e.g., natural gas or methane), the fuel cell is considered a nonrenewable energy technology.

Through careful design, selected fossil fuel driven DGs can be built to oxidize some of the fossil fuel (by combining with oxygen) to produce heat. Such operation modes, whether in electromechanical (rotational) or electrochemical (fuel cell) systems, are referred to as combined heat and power (CHP) operation mode.

Table 1.1 shows some existing and potential DG technologies, nonrenewable and renewable, and their capacity and efficiency ranges.

Most of the new DG technologies include power electronic devices to provide usable output power. These DGs are often referred to as *power electronically interfaced DGs*. Enormously improved power control of these generation sources has become possible by controlling their power electronic interfacing units. In a common approach the output voltage of these generation devices, whether dc or ac, is converted to a controlled

TABLE 1.1 Dispatchability, Capacity Range, and Efficiency Range of Existing and Potential DG Technologies [2,5,27,28]

DG Types	Capacity Range	Efficiency Range (%)	Dispatchable
ICE	50 kW–5 MW	25–40	Yes
Gas turbine	1–100 MW	30–40	Yes
Microturbine	10–500 kW	20–30	Yes
Wind	150 kW–5 MW	<40	No
Solar PV	200 W–10 MW	10–20	No
Biomass	20–50 MW	10–20	Yes
Fuel cells	0.5 kW–3 MW	40–65	Yes
PAFC	50 kW–1 MW	~40	
PEMFC	0.5 kW–1 MW	35–40	
SOFC[a]	5 kW–2 MW	45–65	
MCFC[a]	5 kW–3 MW	50	

[a]Efficiency of these fuel cells could reach or exceed 80% in CHP operation mode.

dc voltage and then converted to usable ac, which can be connected to a utility grid or used stand-alone.

Distributed generation devices can pose both positive and negative impacts on existing power systems. These new issues, such as islanding detection and operation (discussed earlier) and optimal size and placement of DGs in power systems, have made DG operation an important research area, which can help obtain maximum potential benefits from DGs.

Since this book's focus is on Fuel cell DGs (FCDGs), this topic will be covered in the next section.

1.4 FUEL CELL DG

Fuel cells are static energy conversion devices that convert the chemical energy of fuel directly into dc electrical energy. Fuel cells have a wide variety of potential applications including micropower, auxiliary power, transportation, stationary power for buildings, and cogeneration applications.

Since entering the twenty-first century, fuel cell technologies have experienced exponential growth; the number of installed FC units worldwide is increasing rapidly. As shown in Fig. 1.3, Government policies, public opinion, and technology advances in fuel cells all contributed to this phenomenal growth. It is expected that FC (and FCDG) technology will advance in the first half of the twenty-first century as the computer technology did in the second half of the twentieth century. However, a

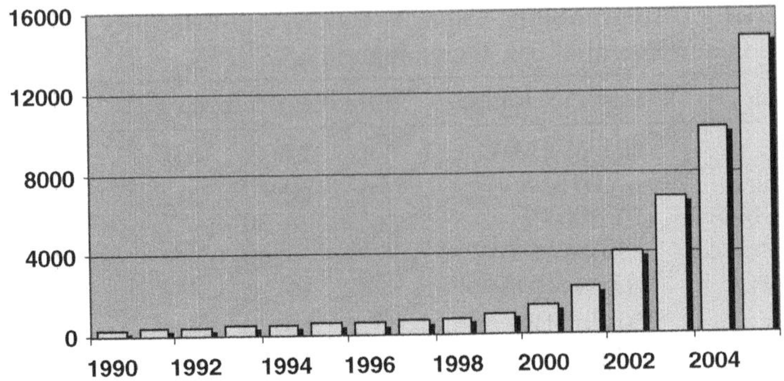

FIGURE 1.3 Cumulative installed FC units worldwide, 1990–2005. *Source*: 2005 Worldwide survey of fuel cells, www.fuelcelltoday.com [29].

number of barriers must be overcome before FCDG can be a reliable energy source. The main barriers are technical and economic issues.

In 2005, The U.S. DOE updated its "Hydrogen, Fuel Cells and Infrastructure Technologies Program's Multi-Year Research, Development and Demonstration Plan." In this plan, it lays out industry targets for key fuel cell performance indices such as cost, durability, and power density. In 1999, the Solid State Energy Conversion Alliance (SECA) was founded for bringing together government, industry, and the scientific community to promote the development of SOFC, which shows great potential for residential, auxiliary power, and DG applications. Through these years, collaborative work among the government (through the national laboratories), industry, and universities, have brought great advances in fuel cell technologies. Such supports give FCs a potential bright future, although extensive work is still needed to be done before they can play an important role in the energy market.

Among different types of fuel cells, polymer electrolyte membrane fuel cells (PEMFC), solid oxide fuel cells (SOFC), and molten carbonate fuel cells (MCFC) show great potential in DG applications. PEMFC and SOFC also show great potential in transportation applications. Compared with conventional power plants, these FCDG systems have many advantages such as high efficiency, zero, or low emission (of pollutant gases) and flexible modular structure. FCDGs can be strategically placed at any site in a power system (normally at the distribution level) for grid reinforcement, deferring or eliminating the need for system upgrades, and improving system integrity, reliability, and efficiency.

Table 1.2 gives the current status and the DOE goals for PEMFC [17,18], and Table 1.3 summarizes the current development stage of SOFC and

TABLE 1.2 Current Status and the DOE Targets for PEMFC Stack[a] in Stationary Applications[b]

Characteristics	Units	Current Status	2010 Goal
Stack power density	W/L	1330	2000
Stack specific power	W/kg	1260	2000
Stack efficiency @ 25% of rated power	%	65	65
Stack efficiency @ rated power	%	55	55
Precious metal loading	g/kW	1.3	0.3
Cost	\$/kWe	75	30
Durability with cycling	h	2200[c]	5000
Transient response (time to go from 10% to 90% of rated power)	s	1	1
Cold startup time to 90% of rated power @ −20°C ambient temperature	s	100[c]	30
Survivability (lowest ambient temperature)	°C	−40	−40

[a]Based on the technical targets reported in [17] for 80 kWe (net) transportation fuel cell stacks operating on direct hydrogen.

[b]Excludes hydrogen storage and fuel cell ancillaries: thermal, water, air management systems.

[c]Based on the performance data reported by Ballard Power Systems, 2005 [18].

SECA targets [19–21]. Considering the emphasis put on FC research, the potential for reaching the DOE and SECA goals is high.

Fuel cell energy systems include power electronic interfacing devices to provide controllable dc and ac power. Voltage and power control is normally achieved by controlling the power electronic devices. Figure 1.4 shows the major processes of a fuel cell energy system. Fuel (e.g., natural gas) containing hydrocarbons is fed to the fuel processor to be cleaned and converted into a hydrogen-rich gas. The principle of gas reforming will be briefly discussed in Section 1.5.3. Through electrochemical energy conversion (explained in Chapter 2), in each fuel cell, the energy of hydrogen is converted to dc electricity. The cells are bundled together in series and parallel combinations (called a fuel cell stack) to produce the desired power and voltage for a particular application. The power-conditioning unit converts the electric power from dc into regulated dc or ac for consumer use. An energy storage device may also be a part of the fuel cell system for energy management and/or load transient mitigation purposes, where energy flow can be bi-directional between the fuel cell and the storage system, as shown in the figure. The FC by-products include heat and clean exhaust, which can be used for water or space heating, or to produce additional electricity.

TABLE 1.3 Current Status and the SECA Targets for 3–10 kW SOFC Module in Stationary Applications

	Phase			
	I[a]	Current Status	II	III (By 2010)
Cost	800\$/kW	724\$/kW[b]	[c]	\$400/kW
Efficiency	35–55%	41%[b]	40–60%	40–60%
Steady state test hours	1,500	1,500	1,500	1,500
Availability	80%	90%[b]	85%	95%
Power degradation per 500 h	≤2%	1.3%[d]	≤1%	≤0.1%
Transient test cycles	10	n/a	50	100
Power degradation after cycle test	≤1%	n/a	≤0.5%	≤0.1%
Power density	0.3 W/cm^2	0.575 W/cm^2[e]	0.6 W/cm^2	0.6 W/cm^2
Operating temperature	800°C	700–750°C[d]	700°C	700°C

[a]The goals of Phase I have been achieved. GE is the first of six SECA industry teams to complete Phase I of the program [20].
[b]Based on the data reported by GE.
[c]Evaluate the potential to achieve \$400/kW [19].
[d]Based on the data reported by FuelCell Energy [21].
[e]Based on the data reported by Delphi [20].

Voltage and power control of a FC system can be achieved through proper control of the power-conditioning unit. Therefore, accurate models of FC power plants and the power electronic devices are necessary to evaluate the coordination between the two, and for controller design. The

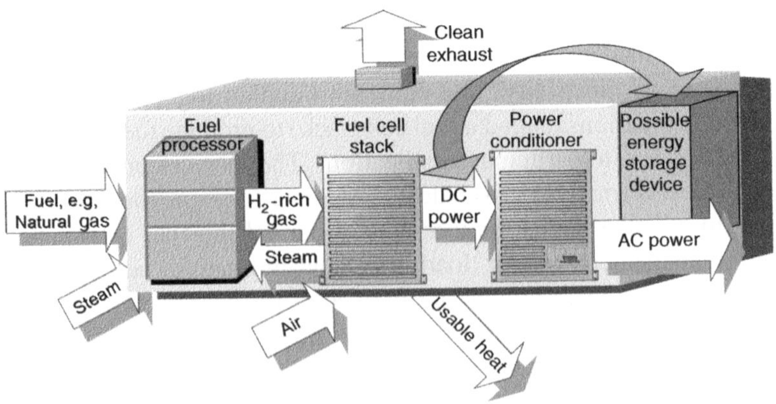

FIGURE 1.4 Major processes in a generic FC energy system.

modeling of FCDG systems and controller design for the components of the power-conditioning unit (i.e., dc/dc converter and dc/ac converter) will be discussed in detail in Chapters 3, 4, 6, 8, and 9.

1.5 THE HYDROGEN ECONOMY

1.5.1 Introduction

The concept of hydrogen energy, hydrogen production and use, and the hydrogen economy is not new. Since William Grove, a British scientist, first demonstrated the basic principle of fuel cell operation in 1839 utilizing hydrogen and oxygen (see Chapter 2), many researchers around the world have looked into different ways of hydrogen production and use. Investigations on different methods of energy production, in general, and hydrogen production, in particular, increased in the 1950s and accelerated significantly after the oil embargo, by oil producing countries, in 1973. Since then, energy production from renewable resources such as wind and solar, hydrogen production using the generated energy, and fuel cells have caught worldwide attention, for example [6–10]. Such research activities are currently ongoing and at their peak.

In 1990, the U.S. Congress passed a Hydrogen Research and Development Act, which required the Department of Energy to develop critical hardware for hydrogen technology, as an important step toward the H_2-based economy. Based on this act, a Hydrogen Technical Advisory Panel, composed of experts from industry and academia, was formed to advise the Secretary of Energy on the status and recommended direction of hydrogen energy development. In 1996, Congress passed the Hydrogen Future Act, further promoting research for the development and demonstration of hydrogen production, storage, transport, and use.

In 2001, DOE announced its National Vision for transition to a hydrogen-based economy to 2030 and beyond [22]. In response to a request from the DOE, the National Academies' National Research Council (NRC) appointed the Committee on Alternatives and Strategies for Future Hydrogen Production and Use in 2002 to address the complex subject of an H_2 economy. This committee was formed by the NRC's Board on Energy and Environmental Systems and the National Academy of Engineering (NAE) Program Office [11]. It evaluated the cost and status of technologies for H_2 production, transportation, storage, and end use. The committee also reviewed DOE's hydrogen research, development, and demonstration strategies. The visions held by this committee are still believed to be true.

The hydrogen economy is referred to an economy that relies on hydrogen as the commercial fuel that would deliver a substantial fraction of a nation's energy-based goods and services. This vision is based on two expectations: (1) that H_2 can be produced from domestic energy sources in a manner that is affordable and environmentally friendly, and (2) that applications using H_2-based FCs and FC vehicles (FCVs) can gain market share in competition with conventional power generation sources and transportation vehicles. To the extent that these expectations can be met, the entire world would benefit from reduced vulnerability to energy disruptions and improved environmental quality, especially through lower carbon emissions. However, before this vision can become a reality and the transition to such an economy can take place, many technical, social, and policy challenges must be overcome to make the use of H_2-based FCs and FCVs widespread [11–15].

H_2 is a versatile energy carrier (an energy storage medium, not a primary energy source) with the potential for use in a variety of applications, including stationary electrical power generation and transportation. H_2 is combustible and can be used as fuel in conventional internal combustion engines to produce mechanical or electrical power. It can also be combined with oxygen within FCs to produce electricity. In both cases, the overall energy efficiency of using H_2 is higher than ICEs operating with conventional fuels such as coal, diesel, or gasoline. In particular, H_2-powered FC vehicles are expected to be two to three times more efficient than gasoline-fueled ICE vehicles. Further, unlike conventional ICEs, which emit pollutant gasses as a result of combustion, H_2-powered FCs and FCVs emit only water vapor. For these reasons H_2 production has been receiving considerable attention worldwide, and FCs and FCVs are often promoted as a means to reduce dependence on oil.

1.5.2 Challenges of Transition to a Hydrogen Economy [11,24]

Expanded use of hydrogen as an energy carrier would address many of today's concerns about the use of conventional fossil fuel, including energy security and environmental quality. Despite its compelling benefits, transition to a hydrogen economy faces multiple challenges. Unlike gasoline and natural gas, hydrogen has no existing large-scale supporting infrastructure, and building one will require major investment. Although hydrogen production, storage, and delivery technologies are currently in commercial use by the chemical and refining industries, existing hydrogen storage and conversion technologies are too costly and far from widespread use in energy applications.

The transition to a hydrogen economy will likely be a lengthy period, during which FCVs and hydrogen-based FC power generation units will not be competitive (at least in cost) with conventional ICE vehicles and power generation units. Therefore, it is expected that the need for hydrogen fuel will be limited during the transition period, and hydrogen can be produced through distributed production technologies for limited use. However, many challenges related to the safe and cost-effective hydrogen production, storage and delivery must be resolved before it can be produced at small-scale, for example through steam methane reforming (SMR) of natural gas or electrolysis.

Natural gas is not renewable; greenhouse (CO_2) gas emission is also associated with its use. Therefore, it cannot be relied on as a long-term energy source for hydrogen production, especially for those countries that lack natural gas reserves. However, it is considered as a potential energy source for hydrogen production during the transition period.

Wind and solar energy are environmentally friendly options to provide some of the electricity needed for distributed hydrogen production via water electrolysis. Such production approaches will allow time for the market to develop before significant investment is set in place for centralized hydrogen production and delivery. Centralized hydrogen generation faces a multitude of challenges, including the need for hydrogen transmission pipelines, storage, and distribution infrastructure.

In the remainder of this section, key factors essential for the development of hydrogen economy, namely production, utilization, storage and distribution of hydrogen, and the U.S. DOE's hydrogen research programs to promote hydrogen economy are discussed.

1.5.3 Hydrogen Production

The various technologies that can be used to produce hydrogen include

- H_2 production by reforming natural gas.
- Conversion of coal to H_2.
- Use of nuclear energy (NE) to produce H_2.
- Electrolysis of water using electricity from the grid, or electricity produced from solar and/or wind energy.
- H_2 production from biomass.

A brief description of the above technologies and their associated technical challenges are discussed as follows.

1.5.3.1 Hydrogen Production by Reforming Natural Gas Natural gas is the most cost effective fossil fuel used for H_2 production. Currently, over 90% of the H_2 produced in the U.S. is through steam methane reforming of natural gas. The main reason for this is that the supply of natural gas is abundant. Moreover, natural gas has a high percentage of H_2 content, and the price is still affordable. Its downside is that it is a fossil fuel gas, and its increased use would eventually increase its cost. SMR can also be applied to other H_2-rich (hydrocarbn) fuels, such as methanol and gasoline. Figure 1.5 shows the block diagram of a typical SMR process. The corresponding chemical reactions are also given in the figure. The hydrocarbon fuel will first go through the purification process, where its poisonous materials, such as sulfur and chloride, are removed. The purification will not only improve the quality of product (H_2), but also increase the life of the downstream reforming and other catalysts used in the system. Then the purified hydrocarbons react with steam at high temperature (750–800°C) to produce a synthesis gas (syngas), a mixture mainly of CO and H_2. The CO produced in the previous step will be further converted into hydrogen through the water shift reaction with the aid of a catalyst. A liquid absorption system can be used to remove CO_2 from the product gas. To further remove residual traces of carbon oxides, a methanation process can be applied to produce high purity product hydrogen.

Distributed generation of H_2 from natural gas could be the lowest-cost option during the transition period. For example, when hydrogen-consuming devices, i.e., FCs and FCVs enter the market, H_2-generation

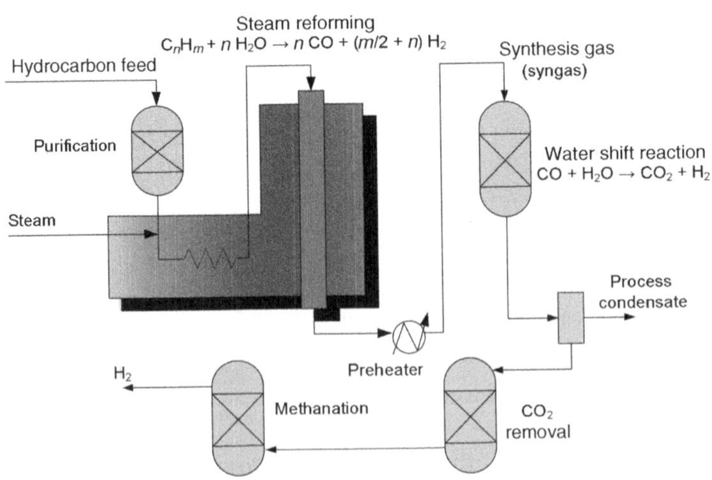

FIGURE 1.5 Steam methane reforming for H_2 generation.

appliances can be deployed in the service stations for fueling FCVs. The main challenge is to develop a reliable H_2-generation appliance that can be mass-produced and operated in the service stations. Natural gas should not be considered as a long-range fuel for centralized H_2-producing plants [11].

1.5.3.2 Hydrogen Production from Coal Coal has excellent potential for hydrogen production in large centralized plants. However, its use can be justified in the long-range, when the demand for hydrogen becomes large enough to support a large hydrogen distribution system. It has been predicted that the U.S. has enough coal to make all of the hydrogen needs of the H_2 economy for over two centuries [11], and a substantial coal infrastructure already exists to support this technology. In addition, most of the issues and technologies associated with making hydrogen from coal are similar to those associated with coal-fired electric power plants. This is particularly the case for coal gasification technology, which is the key to efficient and clean hydrogen production from coal for different hydrogen based applications. As shown in Fig. 1.6, coal gasification technology involves partial oxidation of coal with oxygen and steam in a high-temperature and elevated-pressure process as opposed to the combustion process used in conventional coal-fired power plants. Coal gasification for hydrogen production offers excellent opportunity for low-cost, high efficiency, and low emission electric power generation through a combined cycle process. Some commercial technologies for converting coal to hydrogen are already available, and the cost of hydrogen from coal is among the lowest available.

In 2003, the *FutureGen* initiative was announced by the DOE to build the world's first integrated carbon sequestration and H_2 production research power plant with zero-emissions. Research activities have been

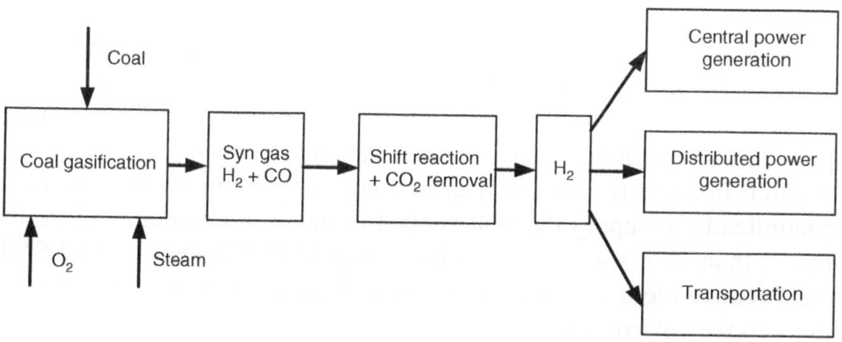

FIGURE 1.6 Schematic diagram of H_2 generation from coal.

ongoing nationwide (and in many other countries) on all aspects of carbon sequestration, and H_2 production and use.

The major drawback for making hydrogen from coal is the CO_2 emissions, which are larger than those from any other way of making hydrogen. This puts an added emphasis on the need to develop a safe and permanent storage for large amounts of CO_2 emissions (i.e., carbon sequestration techniques) before the widespread use of coal for hydrogen production [11].

In 2008, DOE announced its reconstructed approach for the FutureGen program putting more emphasis on carbon capture and storage technology compared to its 2003 plan. Coal gasification and clean coal technology stil remain the fundamental components of the future coal-based electricity production. And, high-temperature SOFCs capable of running on coal gas and developing combined cycle SPFC-gas turbine systems to generate electricity in cleaner and more efficient way is a part of the program. SECA SOFCs are one of the R&D technologies that will be tested within the FutureGen Program [30].

1.5.3.3 Hydrogen Production from Nuclear Energy Nuclear energy is a long-term energy resource that (when used safely and its waste handled safely) can serve the world for many generations for hydrogen and power production. Further, nuclear power reactors do not emit CO_2 and toxic gasses to the atmosphere as are emitted by fossil-fueled power plants. Current U.S. plants use water as the coolant and are often called light-water reactors (LWRs). They rely on the steam Rankine cycle for thermal-to-electrical power conversion. Other countries use different technologies; in the United Kingdom, CO_2-cooled reactors are used, and heavy-water-cooled reactors (HWRs) are used in Canada and India [11].

Nuclear reactors can be used for hydrogen production through water electrolysis or thermochemical processes. Efficient hydrogen production may be achieved by raising the water temperature to 700–1000°C before electrolysis or by a thermochemical process. However, LWRs and advanced LWRs operating at temperatures under 350°C cannot be used for such purposes, but advanced coolants have been proposed which can operate at temperatures above 700°C. According to the Electric Power Research Institute (EPRI), nuclear reactors can potentially be used more economically to supply the heat needed in the steam methane reforming process than their use for water electrolysis [16]. Nuclear-assisted SMR would reduce the use of natural gas as well as the emission of undesired gasses to the environment.

1.5.3.4 Hydrogen Production by Water Electrolysis The process of separating water molecules into hydrogen and oxygen is called electrolysis. This process has been in use around the world for many years, primarily in chemical plants, to meet their hydrogen needs. Electrolysis is currently more expensive than steam reforming of natural gas, but it may play an important role in the transition to a hydrogen economy. Electrolysis facilities can be placed in existing service stations to produce hydrogen for FCVs, or at residential buildings for use with FCs, provided their safe operation can be achieved. Electrolyzers can allow FCDGs to use stored hydrogen to generate additional power during utility peak-demand hours. They can use utility power to generate hydrogen and store hydrogen during off-peak hours, therefore improving utilities' load factors.

Current electrolysis technologies fall into two basic categories: solid polymer using a proton exchange membrane (PEM) and liquid electrolyte, most commonly potassium hydroxide (KOH). In both technologies, water is introduced into the reaction process and subjected to an electrical current that causes dissociation, after which the resulting hydrogen and oxygen atoms are separated.

A PEM electrolyzer is essentially a reverse-operating PEMFC. The liquid KOH electrolyte system operation is analogous to a PEM electrolyzer. In both systems, oxygen ions migrate through the electrolyte, leaving hydrogen gas dissolved in the water stream. The hydrogen is then extracted from the water and directed into a separating channel for storage. The operation principles of PEMFC and KOH electrolyzers are discussed in Chapters 2, 3, and 5.

The electricity used for electrolysis can be either from the utility grid or that generated from renewable energy sources, such as wind and solar power. Electrolysis is well matched to intermittent renewable technologies. A brief description of the technologies which use renewable energy sources to generate H_2 follows.

1.5.3.5 Solar Energy to Hydrogen Hydrogen from solar energy can be produced through conversion of solar energy to electricity using a photovoltaic cell and then to hydrogen through the electrolysis of water. In an alternate method, photoelectrochemical cells are used for direct hydrogen production. The latter method is in the early stages of development and cannot be counted on for any foreseeable future.

Currently, over 80% of commercial PV modules used for electricity production are based on single crystal or polycrystalline silicon. A second type of PV technology is based on deposition of thin films of amorphous as well as microcrystalline silicon. Thin-film technology appears to hold better promise for cost reduction. However, the cost of the thin-film technology is presently higher than the cost of silicon-based PV modules.

1.5.3.6 Wind Energy to Hydrogen Wind-generated electricity has a promising potential to be used for pollution-free water electrolysis for production of a significant amount of hydrogen, particularly during the transition, when the need for hydrogen is limited, as well as in the long term. It is the most affordable renewable technology deployed today with wind-generated electricity being as low as 4 cents/kWh. Both energy security and environmental quality, which are strong factors motivating a H_2 economy, can be addressed by the conversion of wind energy to H_2. With improved performance and efficiency of wind turbine generators (WTGs), their capacity factors, which are currently in the neighborhood of 30%, could be increased, therefore capturing the maximum amount of available wind energy. These advancements can be achieved through improved turbine design and power electronic controls.

Wind–electrolysis–H_2 systems still face many barriers to deployment and deserve continued attention by the government, including incentives both for manufacturers and end-users. Increasing the efficiency and reducing the cost of electrolyzers, and advances in hydrogen storage systems to be adapted with WTG-electrolyzer systems, are essential for a successful wind-H_2 program.

Since wind farms are generally located in rural areas, advances in H_2 distribution from the wind farms to urban areas are essential for widespread use of this technology. Further, matching (optimization) of WTGs with electrolyzers and H_2 storage systems are necessary for both stand-alone and wind farm applications. A modeling and simulation study, exploring the operation feasibility and power management of wind/PV/FC systems used for power and H_2 generation is given in Chapter 9.

1.5.3.7 Biomass Energy to Hydrogen Biomass energy can be used to produce hydrogen through biomass gasification. There are two types of biomass feedstock available for conversion to hydrogen: the bioenergy of crops, and the organic waste from agricultural farming and wood processing (referred to as biomass residues). The primary energy source for hydrogen production from the above sources is solar energy. Hydrogen production from biomass is not a thermodynamically efficient process; less than 0.5% of the total solar energy is converted to hydrogen. In addition to its low efficiency, biomass to hydrogen conversion is an expensive process; the current price of hydrogen from biomass is around

$7/kg, which is not competitive with more mature H_2 production technologies such as steam reforming of natural gas, or coal gasification, which is estimated around $1/kg. Due to the high costs of feedstock, and gathering and transporting biomass feedstock, biomass gasification plants have high operating costs. An optimistic projection estimate could bring the cost of H_2 production from biomass down to about $1.2/kg, which is still about three times the projected cost of H_2 produced in large central coal gasification plants ($0.46/kg). For the above reasons, it is unlikely that biomass gasification could play a central role in future H_2 production. On the contrary, because of its low emission of greenhouse gasses, biomass could play a significant role in meeting the goal of reducing the emission of greenhouse gasses. It is projected that biomass could be used in cofiring applications along with coal, where biomass could provide up to 15% of the total energy input of the fuel mixture [11].

1.5.4 Hydrogen Storage and Distribution

Key economic factors in any future hydrogen-based economy will be the cost and safety of the hydrogen distribution system from production sites to customers. While this is true of any fuel, it involves unique challenges in the case of hydrogen because of its high diffusivity, extremely low density (in gas or liquid form), and flammability. To overcome these unique challenges will involve special safety measures, and therefore cost. In particular, the safety measures for on-board storage of hydrogen in future FC vehicles are critical.

Hydrogen can be stored and transported as a pressurized gas or a cryogenic liquid. The common methods of storing hydrogen are as follows [3]:

- *Compressed Gas in High-Pressure Storage Tanks*: New materials have allowed the construction of pressure tanks and vessels that can store hydrogen at extremely high pressure (as high as 700 bars).
- *Hydrogen Absorbing Materials*: Hydrogen can be combined with metals and metal alloys (or with charcoal) to make a metal hydride (or charcoal) with a high hydrogen energy density. The hydrogen is released when the hydrides (or charcoal) are heated.
- *Liquid Storage*: Hydrogen will be converted to liquid when its temperature is reduced to −253°C. Storage and transportation of liquid hydrogen can be less costly compared to compressed gas, but it requires additional energy (cost) to keep the hydrogen at such low temperature. It is estimated that 25–30% of the energy content of

hydrogen is used to keep the hydrogen in liquid form. Another concern of storing liquid hydrogen is its loss through evaporation.

Hydrogen is a uniquely difficult commodity to transport on a wide scale, whether as liquid by pipeline or as pressurized gas in cylinders. On a weight basis, hydrogen has much higher energy content than gasoline (120 MJ/kg for hydrogen, 44 MJ/kg for gasoline). But on a volume basis, the energy content of gasoline is much higher than hydrogen: 3 MJ/L for hydrogen at 5000 PSI, 8 MJ/L for liquid hydrogen, and 32 MJ/L for gasoline [11].

Pipe-line transmission of hydrogen is expected to be more cost-intensive than pipeline transmission of natural gas. The diameter of pipelines for hydrogen transmission will need to be at least 150% of natural gas pipelines to achieve the equivalent energy transmission rate. Further, more costly steel and valve and metal seal connections will be required for hydrogen pipelines in order to avoid the possibilities of leakage. Major safety codes will have to be in place for hydrogen storage, transportation, and distribution as hydrogen needs grow.

1.5.5 Department of Energy Hydrogen-Related Activities

A Hydrogen Posture Plan [23], released by the DOE in December 2006, outlines the most recent coordinated plan for activities under the Hydrogen Fuel Initiative at the Department of Energy and the Department of Transportation. This document outlines the integration of ongoing and future hydrogen research, development, and demonstration activities into a focused Hydrogen Program.

The DOE Hydrogen research and development program activities are focused on advancing cost-effective, efficient production of hydrogen. It also includes a variety of related activities such as hydrogen storage, delivery, conversion (of hydrogen energy to electricity, i.e., fuel cells), applications and technology validation, safety, codes and standards, education, basic research, and systems analysis and integration. A summary of these activities and the DOE offices involved in each activity is given below.

1.5.5.1 Hydrogen Production

Hydrogen from Natural Gas The Offices of Energy Efficiency and Renewable Energy (EERE) and Fossil Energy (FE) focus on hydrogen production via steam methane reforming. The EERE Office is focused on distributed hydrogen production from natural gas and bio-derived liquid feedstocks. The FE office is focused on sub-centralized and centralized hydrogen production from natural gas. DOE does not consider natural gas as a long-term energy source for the production of hydrogen because of concerns about its long-term availability, security, and greenhouse gas emissions. However, natural gas is considered as a near-term energy source for hydrogen production by the DOE.

Hydrogen from Coal The FE Office is focused on developing the technologies needed to produce hydrogen from coal-derived synthesis gas. The FE Office also focuses on the development of zero-emission, high-efficiency power plant for coproduction of hydrogen and electricity from coal. Toward this goal, the FE Office is also investigating carbon sequestration technologies, in associated programs, as an option for managing greenhouse gas emissions from coal-fired plants.

Hydrogen from Nuclear Power The Office of Nuclear Energy (NE) is focused on developing hydrogen production technology using heat generated from nuclear energy systems. The NE Office R&D areas include high-temperature thermochemical cycles, high-temperature electrolysis, and reactor/process issues.

Hydrogen from Renewable Energy Resources The EERE Office is focused on research for developing advanced technologies for producing hydrogen from renewable energy resources. Key research areas include electrolysis, thermochemical conversion of biomass, photolytic and micro-organism systems, photoelectrochemical systems, and high-temperature chemical water splitting.

1.5.5.2 Hydrogen Basic Research The Office of Science's basic research has a major emphasis on fundamental understanding of photoinduced water splitting into hydrogen and oxygen by semiconductors and photocatalytic techniques using solar energy. This office also has emphasis on fundamental research in catalysis, membranes, and gas separation for more efficient and cost effective fossil-based hydrogen production.

1.5.5.3 Hydrogen Delivery The EERE and FE Offices, and the Office of Science are involved in infrastructure R&D for safe and cost-effective hydrogen delivery. These activities include developing improved materials for pipelines, breakthroughs in hydrogen liquefaction, light weight and strong materials for high-pressure hydrogen storage, and low-pressure solid and liquid carriers for hydrogen delivery and storage. These activities have a long-term goal of developing hydrogen delivery technologies for transportation and stationary power in a hydrogen-based economy.

1.5.5.4 Hydrogen Storage The R&D activities on hydrogen storage at high pressures and cryogenic temperatures are centered in the DOE's EERE Office. Activities in this area include on-board applications for transportation and off-board applications for refueling infrastructure and for stationary (hydrogen-based) fuel cell power generating stations.

Innovative materials development for hydrogen storage is handled by the EERE and FE Offices. These activities include development of carbon-based materials, metal-organic frameworks, and metal and chemical hydrides.

The Office of Science focuses on basic research in developing novel storage materials, including nanostructured materials.

1.5.5.5 Hydrogen Energy Conversion (Fuel Cells) The DOE activities in this area include the conversion of hydrogen to electrical or thermal power and the use of PEM fuel cells for auxiliary power units on vehicles and for stationary and backup power applications.

The Office of Science's basic research program and the EERE Office focus on improvements in cost, durability, and efficiency of PEM fuel cells. Their R&D activities in this area include improving catalysts, electrolytes (membrane), and electrode materials for PEMFCs

Although not related directly to the hydrogen initiative, R&D is also underway within DOE, on phosphoric acid, molten carbonate and solid oxide fuel cells. These technologies have a stronger tie to stationary power production and are mainly handled by the FE Office.

Other DOE hydrogen-related activities, including the development of safety codes and standards, systems analysis and integration, and education are mainly handled by the EERE Office. Basic research in these areas is handled by the Office of Basic Energy Sciences. The overview of the DOE goals for the transition plan to a hydrogen economy to year 2040 is given in Table 1.4.

Several DOE National Laboratories along with universities and industrial partners work with different DOE offices toward the goals of the above transition plan. The DOE laboratories heavily involved in hydrogen-related work are as follows:

- *National Renewable Energy Laboratory* (NREL): Hydrogen production from renewable resources, hydrogen detection and safety.

TABLE 1.4 Overview of the U.S. Transition to the Hydrogen Economy—A Projection [22]

2000		2010	2020	2030	2040
Public policy framework		• Security • Climate • H_2 safety	Outreach activities and public acceptance	Public confidence in hydrogen as an energy career	
Hydrogen industry segments	Production processes	Reforming of natural gas/biomass Electrolysis using renewable and nuclear Thermochemical water splitting using nuclear	Gasification from coal		Biophotocatalyst Photolytics to split water
	Delivery	• Pipeline • Truck, rail, barges	Onsite "distributed" facilities		Integrated central-distributed networks
	Storage	Pressurized tanks (gases and liquids)	Solid state (hydrides)	Mature technologies for mass production Solid state (carbon, glass structures)	
	Energy conversion	Combustion	• Fuel cells • Advanced combustion	Mature technologies for mass production	
	End-use energy markets	• Fuel refining • Space shuttle • Portable power	• Stationary distributed power • Bus fleets • Government fleets	• Commercial fleets • Distributed CHP • Market: introduction of personal vehicles	• Utilities systems

- *Idaho National Laboratory* (INL): Hydrogen-based internal combustion engine vehicles, hydrogen production from nuclear energy, thermal plasma and alternative fuels, hydrogen storage technologies.
- *National Energy Technology Laboratory* (NETL): Hydrogen production from coal.
- *Pacific Northwest National Laboratory* (PNNL): Advanced nanoscale materials for hydrogen production and storage.

1.5.6 The Role of This Book

As we have seen throughout the previous section, fuel cells play a critical role in the future hydrogen economy, including in FC vehicles and stationary power generation. An understanding of the FC dynamic modeling and response prediction is necessary for students, as well as practicing engineers/scientists working in related areas, to be able to evaluate FC response and to design controllers to adapt them to particular applications. This book is intended to fill the present gap for two types of fuel cells—PEMFC and SOFC—with significant promise in distributed power generation applications. PEMFC is the fuel cell of choice for transportation, residential DG and backup power applications; and SOFC has the potential for a variety of DG applications from 5 kW to MW range. The focus of this book is on dynamic modeling and controller design for the above two types of fuel cells, PEMFC and SOFC.

Chapter 2 covers the principle of operation of major types of fuel cells and gives a comparative summary of their characteristics. Dynamic modeling of PEMFC and SOFC are covered in Chapters 3 and 4, respectively. Electrolyzer modeling is introduced in Chapter 5. Introduction and modeling of power electronic interfacing circuits for fuel cell applications is covered in Chapter 6.

Fuel cells face a variety of load and/or electrical disturbances. Proper controllers need to be designed for these power generation devices so that they can handle or mitigate the disturbances and ensure their safe operation in both stand-alone and grid-connected applications. Chapters 7 and 8 cover control methodologies and controller designs for stand-alone and grid-connected operation of FCDG systems, respectively.

As an example of applying the developed models and controller design methodologies, Chapter 9 covers the design and performance investigation of a hybrid wind/PV/PEMFC-electrolyzer system and a SOFC-CHP system. Chapter 10 gives a summary of future potential of fuel cells.

Although the applications of the fuel cell models and controller design methodologies are given for stand-alone and grid-connected power generation, the models can also be used to study other FC applications, for example, transportation applications.

REFERENCES

[1] H.L. Willis and W.G. Scott, *Distributed Power Generation: Planning and Evaluation*, Marcel Decker, New York, 2000.

[2] A.-M. Borbley and J.F. Kreider (editors), *Distributed Generation: The Power Paradigm for the New Millennium*, CRC Press, Boca Raton, FL, 2001.

[3] G.M. Masters, *Renewable and Efficient Electric Power Systems*, IEEE-Wiley, Hoboken, NJ, 2004.

[4] H.B. Putgen, P.R. Macgregor, and F.C. Lambert, Distributed generation: semantic hype of the dawn of a New Era? *Power & Energy Magazine*, 1 (1), 2003.

[5] *Fuel Cell Handbook*, 7th edn, EG&G Services, Inc., Science Applications International Corporation, DOE, Office of Fossil Energy, National Energy Technology Laboratory, 2004.

[6] R. Ramakumar, H.J. Allison, and W.L. Hughes, Prospects for tapping solar energy on a large scale, *Solar Energy*, 16 (2), 107–115, 1974.

[7] R. Ramakumar, H.J. Allison, and W.L. Hughes, Solar energy conversion and storage systems for the future, *IEEE Transactions on Power Apparatus and Systems*, PAS-94 (6), 1926–1934, 1975.

[8] H.J. Allison, *A new approach to high-pressure, high-temperature hydrogen-oxygen fuel cell and electrolysis-cell design*, PhD Thesis, Oklahoma State University, Stillwater, OK, 1967.

[9] L.A. Slotin, The hydrogen economy: future policy implications, *International Journal of Hydrogen Energy*, 8 (4), 1983.

[10] J.O'M. Bockris and T.N. Veziroglu, A solar-hydrogen economy for USA, *International Journal of Hydrogen Energy*, 8 (5), 1983.

[11] *The Hydrogen Economy: Opportunities,* Costs, Barriers, and R&D Needs, Report of the committee on alternatives and strategies for future hydrogen production and use, National Research Council, National Academies Press, 2004.

[12] D. Smith, The hydrogen economy—the next great economic revolution? *REFOCUS: The International Renewable Energy Magazine*, 2003.

[13] W.W. Clark, et al. Hydrogen energy stations: along the roadside to the hydrogen economy, *Utilities Policy (Elsevier journal)*, 13, 41–50, 2005.

[14] G.W. Crabtree, et al. The Hydrogen Economy, *Physics Today*, 2004.

[15] Special Issue on the hydrogen economy, *Proceedings of the IEEE*, 94(10), 2006.

[16] L. Sandell, High Temperature gas-cooled reactors for the production of hydrogen: an assessment in support of the hydrogen economy, Electric Power Research Institute, Report No. 1007802, Palo Alto, CA, 2003.

[17] DOE Hydrogen, Fuel Cells & Infrastructure Technologies Program Multi-Year Research, Development and Demonstration Plan, 2005.

[18] Technology "Road Map," Ballard Power Systems Inc., http://www.ballard.com/be_informed/fuel_cell_technology/roadmap.

[19] SECA Program Plan, DOE Office of Fossil Energy: National Energy Technology Laboratory (NETL) and the Pacific Northwest National Laboratory, 2002.

[20] Significant Milestone Achieved in SECA Fuel Cell Development Program, http://www.netl.doe.gov/publications/TechNews/tn_ge_seca.html, 2006.

[21] *Thermally Integrated High Power Density SOFC Generator*, FuelCell Energy Inc., FY, Progress Report, 2004.

[22] *A National Vision of America's Transition to a Hydrogen Economy-to 2030 and Beyond*, U.S. Department of Energy, 2002.

[23] *Hydrogen Posture Plan: An Integrated Research, Development, and Demonstration Plan*, U.S. Department of Energy and U.S. Department of Transportation, 2006.

[24] *National Hydrogen Energy Roadmap*, U.S. Department of Energy, 2002.

[25] IEEE Std 519-1992, *IEEE Recommended Practices and Requirements for Harmonic Control in Electrical Power Systems*, 1992.

[26] IEEE Std 1547-2003, *IEEE Standard for Interconnecting Distributed Resources with Electric Power Systems*, 2003.

[27] Online http://www1.eere.energy.gov/biomass/electrical_power.html.

[28] Online http://www.fossil.energy.gov/programs/powersystems/fuelcells/.

[29] Online http://www.fuelcelltoday.com/FuelCellToday/FCTFiles/FCTArticle-Files/Article_1068_2005%20Global%20Survey.pdf.

[30] Online http://www.energy.gov/news/5912.htm.

CHAPTER 2

PRINCIPLES OF OPERATION OF FUEL CELLS

2.1 INTRODUCTION

Fuel cells (FCs) are devices that utilize an electrochemical process to convert chemical energy of a fuel into electrical energy. This electrical energy can be used to power vehicles, electronic devices, houses, or be delivered to an electrical grid. Over the last few decades increased attention has been given to fuel cell technology due to its high efficiency and clean processes. Unlike a battery that stores energy, a fuel cell converts chemical energy of its input fuel into electrical energy without the use of stored materials within its structure. Fuel cells also differ from conventional heat engines in that they produce electricity directly from chemical energy without an intermediate conversion into mechanical power. When hydrogen is used as fuel, the only by-products of the fuel cell operation are water and heat.

Sir William Grove, a British lawyer and physicist, first demonstrated the basic principle of fuel cell operation in 1839. With the use of four primitive cells, utilizing hydrogen and oxygen, Grove was able to produce electric energy. Although this was an important discovery, the technology laid dormant for over a century. The first practical application of fuel cells is credited to another British scientist, Francis Bacon. In 1950 Bacon made

Modeling and Control of Fuel Cells, By M. Hashem Nehrir and Caisheng Wang

an alkaline fuel cell prototype. Fuel cells then resurfaced and became known when the U.S. National Aeronautic and Space Administration (NASA) demonstrated the potential of fuel cell applications in the space programs in the 1950s and 1960s. After these demonstrations, commercial companies became interested in this technology. However, early research revealed technological and economic constraints on fuel cell development. In 1984, the U.S. Department of Energy began supporting research and development of fuel cell technology. As a result, some fuel cell technologies are already commercialized, although further improvements in their cost, reliability, and performance characteristics are still needed. It is expected that with further research and development, fuel cell technology will improve at a fast pace and become a reliable energy source in the first half of the twenty-first century.

Initially, in this chapter electrical and thermal energy, fundamentals of thermodynamics, and electrochemical processes are briefly explained. The principle of operation of fuel cells and a comparative summary of major types of fuel cells are presented next. Then, the principles of operation of electrolyzers, which use dc electricity to break water into hydrogen and oxygen, are presented.

2.2 CHEMICAL AND THERMAL ENERGY OF AN ELEMENT

Chemical energy is the energy held in the chemical bonds between atoms and molecules of an element. Every bond has a certain amount of energy. Chemical energy exists as stored energy, which can be converted to other forms of energy through reactions. First, energy is required to break the bonds, called *endothermic*. Then, energy will be released, called *exothermic*, when the broken bonds combine to create more stable molecules. If the total energy released is more than the energy absorbed, then the whole reaction is called *exothermic*. Otherwise, the reaction is called *endothermic*. The burning of fossil fuels (combustion) is one of the most important *exothermic* reactions for producing energy. The photosynthesis process in plants is a good example for *endothermic* reaction; the (solar) energy plants absorb to make the photosynthesis reaction happen is more than the energy released.

From a macroscopic point of view, thermal energy, or heat, can be defined as energy transferred between two systems due to a difference in temperature. From a microlevel point of view, thermal energy is associated with atomic and molecular vibration. Different forms of energy can be converted into thermal energy while the opposite conversion is limited by the second law of thermodynamics, which is briefly explained in the next section. Thermal energy can therefore be considered the most basic form of energy.

2.3 FUNDAMENTALS OF THERMODYNAMICS

Thermodynamic analysis of energy conversion processes is important when modeling electrochemical processes, for example, fuel cells. Therefore, in this section, a brief overview of the fundamental concepts and principles of thermodynamics is presented [1–6].

2.3.1 The First Law of Thermodynamics

According to the first law of thermodynamics, the energy of a system (considered in the form of heat or work) is conserved, meaning that the energy can be neither created nor destroyed, but it can be converted from one type to another. The change in the system energy ($\mathrm{d}E$) is contributed by the heat entering the system ($\mathrm{d}Q$) and the work done by the system ($\mathrm{d}W$).

$$\mathrm{d}E = \mathrm{d}Q - \mathrm{d}W \tag{2.1}$$

Note that work in the above equation is defined as work leaving the system. A system can be considered isolated from "the outside world" in the sense that changes in the "outside world" will not affect its state. The "outside world" is considered to be the system environment. Of particular interest is a *simple system* that is not affected by capillarity, distortion of solid phase, external electromagnetic or gravitational force fields, or internal adiabatic walls. For a *simple system*, the total system energy is equal to the total system internal energy U.

$$E = U \tag{2.2}$$

We only consider *simple systems* in the electrochemical and thermodynamic principles presented in this chapter.

For a *control volume* or open system, usually there is an expansion when heat is absorbed by the system at constant pressure. Part of the heat goes into internal energy and causes the system temperature to rise. The rest of the heat is used to expand the system against the pressure (P). A property of the system, called *enthalpy* (H), is used to represent the system state under a given condition. It is the sum of the system internal energy and the product of system pressure and volume (V).

$$H = U + PV \tag{2.3}$$

Enthalpy is independent of which way the system reaches its condition, and as a result, the product of the system pressure (P) and volume (V) is constant. Equation 2.3 can therefore be written as

$$\mathrm{d}H = \mathrm{d}Q - \mathrm{d}W \tag{2.4}$$

According to (2.4), any change in the enthalpy (state) of a system is a result of the difference between the change in the heat entering the system and the work leaving (done by) the system.

2.3.2 The Second Law of Thermodynamics

According to the first law of thermodynamics (2.1), work done by an ideal cyclic thermal process (heat engine), shown in Fig. 2.1, can be defined by

$$W = Q_1 - Q_2 \tag{2.5}$$

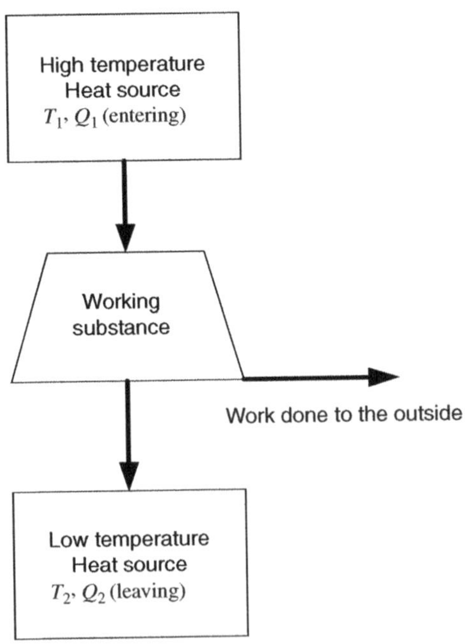

FIGURE 2.1 Block diagram of an ideal thermal process.

where W is work done by the system to the outside and Q_1 and Q_2 are the heat entering and leaving the thermal process, respectively.

In 1842, French scientist Sadi Carnot presented the relationship between the heat entering and leaving ideal thermal processes and their corresponding temperatures, T_1 and T_2, as follows.

$$\frac{Q_1}{T_1} = \frac{Q_2}{T_2} \tag{2.6}$$

The efficiency of the process then is

$$\eta_c = \frac{W}{Q_1} = \frac{Q_1 - Q_2}{Q_1} = 1 - \frac{T_2}{T_1} \tag{2.7}$$

It is clear from (2.7) that a higher efficiency can be obtained by having a larger difference between T_2 and T_1.

Another contribution from Carnot is that Q/T can be a state property of the system by showing the equivalence of Q_1/T_1 and Q_2/T_2. Based on Carnot's idea, Rudolf Clausius (a Polish scientist, 1822–1888) developed the concept of entropy and presented the second law of thermodynamics. This concept was introduced to indicate the degree of disorder of a system. If the system is undergoing an infinitesimal reversible process with $\mathrm{d}Q$ heat entering the system at temperature T, then the infinitesimal change in its *entropy* (S) is defined as

$$\mathrm{d}S = \left.\frac{\mathrm{d}Q}{T}\right|_{\mathrm{rev}} \tag{2.8}$$

A reversible process is a process that can be reversed without leaving any traces to its environment. If a cyclic process only consists of reversible processes, then there is no entropy change after a cycle (i.e., $\Delta S = 0$). On the contrary, if the cycle contains an irreversible process, then its entropy will change, which corresponds to work done (or received) by the system. Entropy is an important system property, which is very useful in describing a thermodynamic process. The temperature–entropy diagram for a Carnot Cycle is shown in Fig. 2.2, where the area abcd represents work done by the system, when the system temperature goes from T_1 to T_2. Note that $\Delta S = 0$ when there is a change in temperature, but no work is done by the system.

The second law of thermodynamics, described by Clausius for a cyclic process, can be written in an infinitesimal form as follows.

$$\mathrm{d}S \geq \frac{\mathrm{d}Q}{T} \tag{2.9}$$

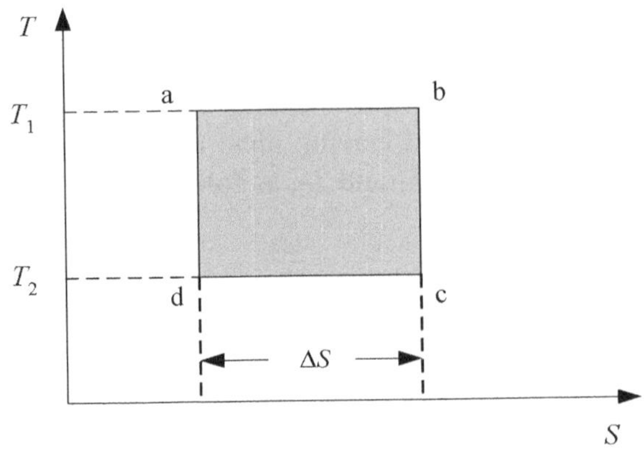

FIGURE 2.2 Temperature–entropy diagram for a Carnot cycle.

In Equation 2.9, the equality sign applies to a reversible cyclic process and the inequality sign applies to an irreversible process. That is to say, for any irreversible process, change in work is done in the direction that change in entropy is greater than dQ/T. This means that heat cannot be transferred from a low temperature to a higher temperature without a need of work from outside. The second law also reveals that no real heat engine efficiency can reach 100% due to the increase of entropy. For an isolated system, the change of entropy will always be greater than or equal to zero.

2.4 FUNDAMENTALS OF ELECTROCHEMICAL PROCESSES

2.4.1 The Gibbs Free Energy

The Gibbs free energy (G), also called the Gibbs energy or "free enthalpy," is defined as

$$G = H - TS = U + PV - TS \tag{2.10}$$

The parameters on the right side of Equation 2.10 are defined in Section 2.3. Differentiating (2.10), we obtain

$$\mathrm{d}G = \mathrm{d}H - (T\mathrm{d}S + S\mathrm{d}T) = \mathrm{d}U + P\mathrm{d}V + V\mathrm{d}P - (T\mathrm{d}S + S\mathrm{d}T) \tag{2.11}$$

Chemical reactions proceed toward the direction that minimizes the Gibbs energy. Therefore, $\mathrm{d}G$ is negative as a chemical reaction approaches its equilibrium point, and it will be zero at the equilibrium point. According to the first law of thermodynamics, Equation 2.1, for a simple system, Equation 2.11 can be rewritten as

$$dG = dH - (T\,dS + S\,dT) = dQ - dW + P\,dV + V\,dP - (T\,dS + S\,dT) \tag{2.12}$$

For systems that are restricted to performing only expansion-type of work, we can write

$$dW = P\,dV \tag{2.13}$$

Also, the following condition holds if the process is reversible:

$$dQ = T\,dS \tag{2.14}$$

Substituting (2.13) and (2.14) in Equation 2.12, four terms on the right side of Equation 2.12 will cancel out, resulting in

$$dG = V\,dP - S\,dT \tag{2.15}$$

Solving the differential equation (2.15), Gibbs energy can be obtained at a given temperature (T) and under any pressure.

$$G(T) = G^{\circ}(T) + nRT\ln\left(\frac{P}{P^{\circ}}\right) \tag{2.16}$$

In (2.16), $G^{\circ}(T)$ is the standard Gibbs energy at temperature T and at $P^{\circ} = 1$ atm.

For an electrochemical reaction, the maximum work (electricity) is determined by the change in the Gibbs energy as the reactants change to products. It can be shown that the maximum electricity production (W_e) is equal to the change in the Gibbs energy.

$$W_e = -\Delta G \tag{2.17}$$

2.5 ENERGY BALANCE IN CHEMICAL REACTIONS

In a chemical reaction, the change in enthalpy (ΔH) due to the reaction can be written as

$$\Delta H = H_P - H_R = \sum_{P_i} N_{P_i} H_{P_i} - \sum_{R_j} N_{R_j} H_{R_j} \tag{2.18}$$

where H_P is the total enthalpy of the products and H_R is the total enthalpy of the reactants. N_{P_i} is the amount of moles of the ith species in the products and N_{R_j} is the amount of moles of the jth species in the reactants.

Accordingly, H_{P_i} and H_{R_j} are the molar enthalpies of the corresponding species, which can be represented as

$$H_{P_i} = (H_f^\circ + H - H^\circ)_{P_i} \tag{2.19}$$

In Equation 2.19, H_f° is the standard molar enthalpy of formation (J/mol) and $(H - H^\circ)$ is the enthalpy due to the temperature difference.

The change of entropy due to the reaction can be written as

$$\Delta S = S_P - S_R = \sum_{P_i} N_{P_i} S_{P_i} - \sum_{R_j} N_{R_j} S_{R_j} \tag{2.20}$$

where S_P is the total entropy of the products and S_R is the total entropy of the reactants. S_{P_i} (S_{R_j}) is the molar entropy of the corresponding species.

The change in the Gibbs free energy due to the reaction (ΔG) can then be obtained from (2.10) as follows:

$$\Delta G = \Delta G_P - \Delta G_R = \Delta H - T\,\Delta S \tag{2.21}$$

Two example reactions are given below for the formation of water, in vapor form and in liquid form, from hydrogen and oxygen, to show how the change in Gibbs energy (or electric energy) in a reaction can be calculated. The enthalpy and entropy values for water, hydrogen and oxygen used in the equations are given in Table 2.1. The reactions are assumed to happen under standard conditions (1 atm and 25°C).

$$H_2 + \tfrac{1}{2}O_2 = H_2O\ (g) \tag{2.22}$$

$$H_2 + \tfrac{1}{2}O_2 = H_2O\ (l) \tag{2.23}$$

In (2.22) and (2.23), "g" and "l" indicate that the reaction product is in gas or liquid form, respectively. From (2.18) and (2.20), we have

$$\Delta H = H_P - H_R = (-241.8 - 0 - 0) = -241.8\ (\text{kJ/mol})$$

$$\Delta S = S_P - S_R = (188.7 - 130.6 - 0.5 \times 205) = -44.4\ (\text{J/mol/K})$$

TABLE 2.1 Standard Thermodynamic Properties (Enthalpy and Entropy)

	Enthalpy, H_f (kJ/mol)	Entropy, S (J/mol/K)
H_2O (g)	−241.8	188.7
H_2O (l)	−285.8	69.9
H_2	0	130.6
O_2	0	205

Therefore, the change in the Gibbs energy of the reaction given in (2.17) is

$$\Delta G = \Delta H - T\Delta S = -241.8 - 298 \times (-44.4 \times 10^{-3}) = -228.57\,\text{kJ/mol}$$

For the reaction given in (2.23), we have

$$\Delta H = H_{\text{P}} - H_{\text{R}} = (-285.8 - 0 - 0) = -285.8\,(\text{kJ/mol})$$

$$\Delta S = S_{\text{P}} - S_{\text{R}} = (69.9 - 130.6 - 0.5 \times 205) = -163.2\,(\text{J/mol/K})$$

Therefore, the change in the Gibbs energy for the formation of water in vapor form (2.17) is

$$\Delta G = \Delta H - T\Delta S = -285.8 - 298 \times (-163.2 \times 10^{-3}) = -237.16\,\text{kJ/mol}$$

From the above examples, we can see that more work is done if the product (H_2O) is in the liquid form.

2.6 THE NERNST EQUATION

Consider an electrochemical reaction under constant temperature and pressure, where reactants X and Y form products M and N as follow.

$$aX + bY \Leftrightarrow cM + dN \tag{2.24}$$

where a, b, c, and d are stoichiometric coefficients.

According to (2.21), the change in the Gibbs energy of the reaction is

$$\Delta G = G_{\text{P}} - G_{\text{R}} = cG_M + dG_N - aG_X - bG_Y \tag{2.25}$$

The change in the Gibbs energy can also be calculated at different temperatures and pressures as follows:

$$\Delta G = \Delta G^{\circ} + RT \ln\left(\frac{P_M^c P_N^d}{P_X^a P_Y^b}\right) \tag{2.26}$$

where $\Delta G^{\circ} = cG_M{}^{\circ} + dG_N{}^{\circ} - aG_X{}^{\circ} - bG_Y{}^{\circ}$.

In an electrochemical reaction, work can be considered as the electrical energy delivered by the reaction. The electrochemical work is defined by

$$W_{\text{e}} = n_{\text{e}}FE \tag{2.27}$$

where n_{e} is the number of participating electrons, F is Faraday constant (96487 C/mol) and E is the potential difference across the electrodes.

According to (2.17), the change in Gibbs energy is the negative value of the work done by the reaction.

$$\Delta G = -W_{\text{e}} = -n_{\text{e}}FE \tag{2.28}$$

Under standard condition, Equation 2.28 can be written as

$$\Delta G^{\circ} = -W_e{}^{\circ} = -n_e F E^{\circ} \tag{2.29}$$

where E° is the standard reference potential.

Using (2.28), the electrode voltage E can be calculated.

$$E = -\frac{\Delta G}{n_e F} = -\frac{\Delta G^{\circ}}{n_e F} - \frac{RT}{n_e F} \ln\left(\frac{P_M^c P_N^d}{P_X^a P_Y^b}\right) \tag{2.30}$$

Writing Equation 2.30 in terms of standard reference potential E°, the well-known electrochemical formula, the Nernst equation, for calculation of the potential difference between two electrodes can be calculated.

$$E = E^{\circ} - \frac{RT}{n_e F} \ln\left(\frac{P_M^c P_N^d}{P_X^a P_Y^b}\right) = E^{\circ} + \frac{RT}{n_e F} \ln\left(\frac{P_X^a P_Y^b}{P_M^c P_N^d}\right) \tag{2.31}$$

For a fuel cell with an overall reaction, given by (2.22), the voltage across the fuel cell electrodes (or the internal potential of the fuel cell) is given by

$$E = E^{\circ} + \frac{RT}{2F} \ln\left(\frac{P_{H_2} P_{O_2}^{0.5}}{P_{H_2O}}\right) \tag{2.32}$$

If the product (H_2O) is in liquid form, given by (2.23), then the fuel cell internal potential is

$$E = E^{\circ} + \frac{RT}{2F} \ln\left(P_{H_2} P_{O_2}^{0.5}\right) \tag{2.33}$$

2.7 FUEL CELL BASICS

Fuel cells are static energy conversion devices that convert the chemical energy of fuel directly into dc electrical energy. The physical structure of a fuel cell consists of two porous electrodes (anode and cathode) and an electrolyte layer in the middle. Figure 2.3 shows the basic workings of a fuel cell with positive ion flow through the electrolyte, which is based on electrochemical principles discussed in Sections 2.4–2.6. Hydrogen and oxygen molecules combine according to (2.22) to form water. The process is caused by the fact that charged particles migrate toward regions of lower electrochemical energy. The charged hydrogen and oxygen particles move toward each other and bond to one another because the final product of this reaction has a lower overall electrochemical energy. Electrical energy is generated as a result of the movement of the charged hydrogen and oxygen

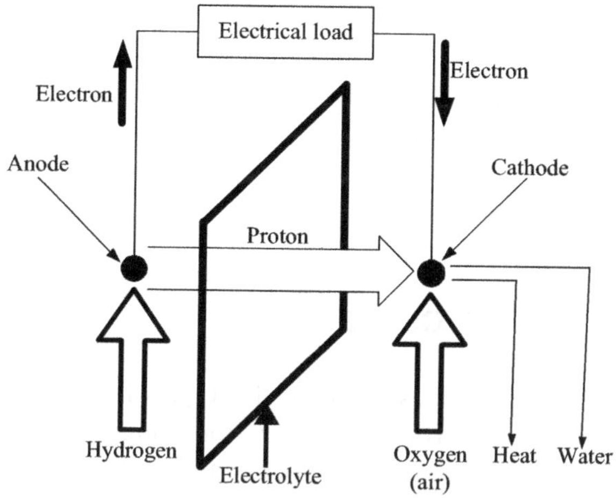

FIGURE 2.3 The basic workings of a fuel cell with proton flow through the electrolyte.

particles, which is essentially the controlled movement of electrons. By breaking the hydrogen molecules to electrons and positive ions (protons), with the help of a catalyst to facilitate faster reaction, the protons move from the cathode to anode through the membrane (electrolyte), but the electrons cannot. The electrons travel through an external electrical circuit (load) to recombine with the hydrogen protons and oxygen molecules at the cathode (again, with the help of the catalyst) to produce water. The actual chemical reaction inside a hydrogen fuel cell can be broken down into two half reactions, the oxidation half reaction and the reduction half reaction. The oxidation half reaction, represented by (2.34), shows the dissociation of hydrogen molecules to protons and electrons at the anode. After the dissociation, the protons are free and pass through the electrolyte, and recombine with the electrons (which move through the external circuit) at the cathode. In this process, which is often called the reduction half reaction, the electrons and hydrogen protons combine with the oxygen molecules from the surrounding air, according to (2.35), to form water.

$$2H_2 \Rightarrow 4H^+ + 4e^- \tag{2.34}$$

$$O_2 + 4H^+ + 4e^- \Rightarrow 2H_2O \tag{2.35}$$

The type and chemical properties of the electrolyte used in fuel cells determine their operating characteristics and internal operating temperature.

The polarity of an ion and its transport direction can differ for different fuel cells, determining the site of water production and removal. If the working ion is positive, like shown in Fig. 2.3, then water is produced at the cathode. On the contrary, if the working ion is negative, like in solid oxide fuel cell and molten carbonate fuel cell (discussed later in this chapter), water is formed at the anode. In both cases electrons pass through an external circuit and produce electric current. A summary of different types of fuel cells is given in the next section.

2.8 TYPES OF FUEL CELLS

Fuel cells are generally classified by the type of electrolyte they use, and the choice of electrolyte dictates the range of their operating temperature and the degree of fuel processing required. Low-temperature fuel cells are generally limited to temperatures below or around 200°C because high-temperature vapor causes rapid degradation of their electrolyte material. The most common type of low-temperature fuel cells are alkaline fuel cell (AFC), phosphoric acid fuel cell (PAFC), and polymer electrolyte membrane (PEMFC). In these fuel cells all the fuel must be converted to hydrogen prior to entering the fuel cell. In addition, the catalyst used in these fuel cells (mainly platinum) is strongly poisoned by carbon monoxide (CO). Therefore, the hydrogen entering these fuel cells needs to be pure. This is a downside of the low-temperature fuel cells.

In high-temperature fuel cells, CO and even hydrocarbons (e.g., CH_4) can be internally converted to hydrogen or even directly oxidized. The most common types of high-temperature fuel cells are molten carbonate fuel cell (MCFC) with operating temperature range of 600–700°C, and solid oxide fuel cell (SOFC) operating in the temperature range of 600–1000°C.

The different types of fuel cells have slightly different chemical reactions, but the same electrochemical reaction is the backbone of all of them. Because of the differences in their operating characteristics and fuel used, different types of fuel cells are suited for different applications. Table 2.2 gives a comparative summary of the anode, cathode, and overall reactions of major fuel cell types.

In addition to the above types of fuel cells, there is another category of fuel cells, which can utilize non-hydrogen fuels directly without internal or external reforming process. Two common types in this category are direct methanol fuel cell (DMFC) and direct carbon fuel cell (DCFC).

DMFC, also called direct alcohol fuel cell (DAFC), is a low-temperature polymer electrolyte fuel cell, which uses alcohol as fuel without reforming.

TABLE 2.2 Comparison of the Chemical Reaction of Major Types of Fuel Cells [9]

Type	Cathode Reaction	Anode Reaction	Overall Reaction
AFC	$1/2O_2 + H_2O + 2e \Rightarrow 2(OH)^-$	$H_2 + 2(OH)^- \Rightarrow 2H_2O + 2e^-$	$H_2 + 1/2O_2 \Rightarrow H_2O$
PEMFC	$1/2O_2 + 2H^+ + 2e^- \Rightarrow H_2O$	$H_2 \Rightarrow 2H^+ + 2e^-$	$H_2 + 1/2O_2 \Rightarrow H_2O$
PAFC	$1/2O_2 + 2H^+ + 2e^- \Rightarrow H_2O$	$H_2 \Rightarrow 2H^+ + 2e^-$	$H_2 + 1/2O_2 \Rightarrow H_2O$
MCFC	$1/2O_2 + CO_2 + 2e^- \Rightarrow CO_3^{2-}$	$H_2 + CO_3^{2-} \Rightarrow H_2O + CO_2 + 2e^-$	$H_2 + 1/2O_2 + CO_2 \Rightarrow H_2O + CO_2$
SOFC	$1/2O_2 + 2e^- \Rightarrow O_2^-$	$H_2 + 1/2O_2^- \Rightarrow H_2O + 2e^-$	$H_2 + 1/2O_2 \Rightarrow H_2O$

Its application is mainly in low power portable electronics. DCFC uses carbon (as fuel) directly in the anode, without an intermediate gasification step. The carbon can be derived from coal, biomass or pet-coke. This type of fuel cell is under development based on the AFC, MCFC, and SOFC concepts. The thermodynamics of DCFC reactions allow high efficiency conversion, and if such technology can be developed into practical fuel cell systems, it could ultimately have a significant impact on coal-based power generation.

Among the different types of fuel cells, PEMFC, MCFC, and SOFC are most likely to be used for distributed generation (DG) applications. The structure and applications of these fuel cells are covered in more detail below, and dynamic modeling of two of them, PEMFC and SOFC, is covered in Chapters 3 and 4, respectively.

PEMFCs, commonly called proton exchange membrane fuel cells, employ a solid polymer (fluorinated sulfonic acid polymer or other similar polymer) as electrolyte. This electrolyte is a Teflon-like material, which is an excellent conductor of protons and an insulator of electrons. The hydrogen molecules are broken into electrons and hydrogen protons at the anode with the help of the platinum catalyst. The hydrogen protons pass through the membrane (electrolyte) and reach the cathode surface. There, they combine with the electrons, which travel from anode to cathode through the external load, to produce water. Equations 2.36–2.38 represent the anode reaction, cathode reaction, and the overall reaction for PEMFC. The schematic diagram and chemical reactions for PEMFC are shown in Fig. 2.4.

$$H_2 \Rightarrow 2H^+ + 2e^- \quad \text{(anode reaction)} \tag{2.36}$$

$$\tfrac{1}{2}O_2 + 2H^+ + 2e^- \Rightarrow H_2O \quad \text{(cathode reaction)} \tag{2.37}$$

$$H_2 + \tfrac{1}{2}O_2 \Rightarrow H_2O \quad \text{(overall reaction)} \tag{2.38}$$

Water management of PEMFCs is critical for their efficient performance. Due to the limitation on the operating temperature of the polymer electrolyte, PEMFCs' operating temperature is normally in the range of 60–80°C. The by-product hot water is used for hydrating the membrane, however, care must be taken to ensure the membrane is not flooded. The hot water can also be used to fulfill other needs, for example, residential hot water needs. Such applications are currently being implemented in Japan [11].

PEMFCs are attractive in many applications such as transportation, stationary power, and backup generation. They have a high power density, a

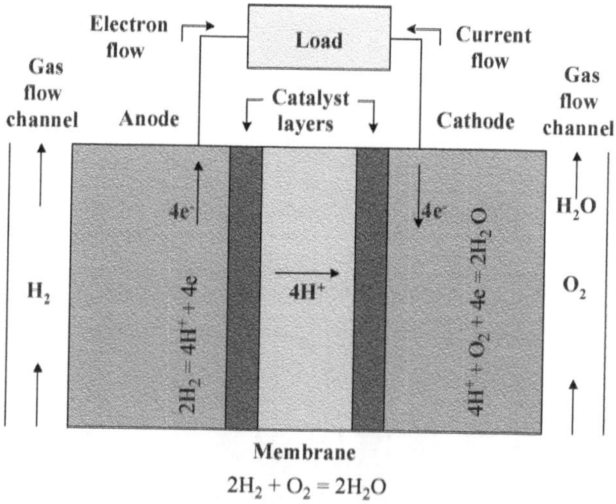

FIGURE 2.4 Schematic diagram and chemical reactions of PEMFC.

quick start, and their electrical efficiency is in the 40–50% range. Their quick start property is especially useful in transportation applications. However, their low and narrow operating temperature makes their thermal management difficult to use their rejected heat effectively. PEMFCs are quite sensitive to poisoning by contaminants, for example CO, sulfur, and ammonia. For these reasons purified hydrogen is needed for safe operation of PEMFC, hence the need for hydrogen infrastructure. If hydrocarbon fuels are used, extensive fuel processing is needed, which negatively impacts system size, complexity, and cost, and lowers the efficiency to around 35%.

In spite of the above barriers, PEMFC is the most developed fuel cell, which is already commercialized for backup generation and close to commercialization for transportation. Figure 2.5 shows a ReliOn T-1000® 1200 W PEMFC unit with six modular cartridges. The unit is designed for backup power applications within the telecommunications, utility, and government sectors. It uses a specially designed hot-swappable modular cartridge structure, which provides flexible configuration from 600 W to its full 1200 W capacity.

Figure 2.6 shows a Ballard 1-kW Mark 1030 PEMFC currently being deployed in the Japanese residential cogeneration (combined heat and power) market.

SOFCs are high temperature fuel cells operating at 600–1000°C. They use a solid (ceramic-type metal oxide) electrolyte, usually dense yttria-stabilized zirconia (Y_2O_3 stabilized with ZrO_2). This material is

FIGURE 2.5 ReliOn T-1000 1200 W PEMFC. Output: 0–1200 W at 24 or 48 VDC nominal output. Fuel: industrial grade hydrogen. *Source*: Courtesy of ReliOn, Inc.

FIGURE 2.6 Ballard's 1-kW Mark 1030 PEMFC. *Source*: Courtesy of Ballard Power Systems, Inc.

an excellent conductor of negatively charged ions (O^{2-} in the case of SOFC) at high temperatures. The anode (fuel electrode) is typically made of a cement type material, a mixture of cobalt or nickel and zirconium oxide ($Co\text{-}ZrO_2$ or $Ni\text{-}ZrO_2$). The metal (Co or Ni) provides good conductivity and the entire cement mixture provides negative ion conductivity. The cathode (air electrode) is made of a mixture of ion- and electronically conducting ceramic, typically strontium-doped lanthanum manganite (Sr-doped $LaMnO_3$) [7,12]. Both the anode and cathode surfaces are porous maintaining large surface areas for charge storage. The porous surfaces give the fuel cell a large equivalent capacitance, called the capacitance of double-layer charge effect, discussed in Section 2.10. The value of this capacitance plays an important role in short time (millisecond range) transient stability of fuel cells.

SOFCs can use hydrogen as well as hydrocarbon gases as fuel, which can be obtained from readily available fuels like natural gas or methane. The schematic diagram of a SOFC is given in Fig. 2.7. When hydrogen is used as fuel, the chemical reactions in SOFCs (also shown at the anode and cathode in Fig. 2.7) are as follows, where water is produced at the anode, as opposed to PEMFC, where water is produced at the cathode.

$$H_2 + O^{2-} \Rightarrow H_2O + 2e^- \quad \text{(anode reaction)} \tag{2.39}$$

$$\tfrac{1}{2}O_2 + 2e^- \Rightarrow O^{2-} \quad \text{(cathode reaction)} \tag{2.40}$$

$$H_2 + \tfrac{1}{2}O_2 \Rightarrow H_2O \quad \text{(overall reaction)} \tag{2.41}$$

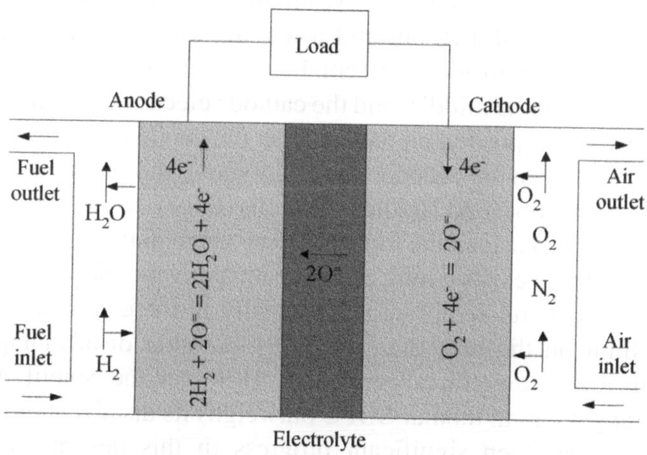

FIGURE 2.7 Schematic diagram of a SOFC.

SOFCs' high operating temperature makes them highly efficient (with efficiencies as high as 60%) and fuel flexible, and gives them the ability to use a wide variety of less expensive catalysts. This is because the chemical bonds of materials break at a much faster rate as temperature increases. The high-temperature operation also makes SOFCs attractive for combined heat and power (CHP) applications, producing electricity and/or heat, with overall efficiencies as high as 75–80%. In CHP operation mode, the SOFC exhaust heat is used for heating purposes (e.g., for space heating) or to operate a co-generation system for generating more electricity.

Figure 2.8a shows the different parts of a SOFC-CHP system, which include the fuel system, SOFC module, electrical section, exhaust, thermal management section, and recuperator. Figure 2.8b shows a 250-kW SOFC-CHP system, the largest unit of its kind, installed at the test facilities of Kinectrics, Inc. in Toronto, Canada. Figure 2.8c shows a 220-kW SOFC/gas turbine hybrid system, the world's first, operating at the National Fuel Cell Research Center at the University of California-Irvine. The hybrid system includes a 200-kW SOFC module integrated with a 20-kW gas turbine generator.

SOFCs have both a slow start up and thermal stresses due to their high operating temperature. However, they allow for internal reforming of gaseous fuel, which gives multifuel capability to SOFCs. Moreover, their solid electrolyte simplifies system design since the corrosion and management problems related to liquid electrolyte are eliminated. These merits give SOFCs a high potential to be used in stationary applications.

Two types of cell designs are being pursued for SOFC, tubular cells, and planar cells. The tubular cell design was originally pioneered by the U.S. Westinghouse Co. (later Siemens-Westinghouse, and now Siemens). The cell has a diameter of 1.27 cm and a length, which has been gradually increased from 30 cm to about 150 cm. In the recent form of this design, the air supply tube is in the middle, and the cathode electrode is deposited on it. The electrolyte is deposited on the cathode, the anode is on the electrolyte, and the fuel tube is on the outside. Figure 2.9 shows the end view of a tubular SOFC, the SOFC tube, and bundle of 24 cells (three cells in series and eight cells in parallel). Because the current flows tangentially to the electrodes (see Fig. 2.9), there exists a long current path along the tube, and the ohmic resistance (loss) is relatively large, especially in the cathode. This places an upper limit on the tube diameter [7,8]. Another disadvantage of this design is its low volumetric power density. However, the benefit of a simple sealing arrangement in tubular SOFC outweighs its disadvantages. For this reason there has been significant progress in this design, and several companies including Acumentrics (U.S.), Mitsubishi (Japan), and Rolls

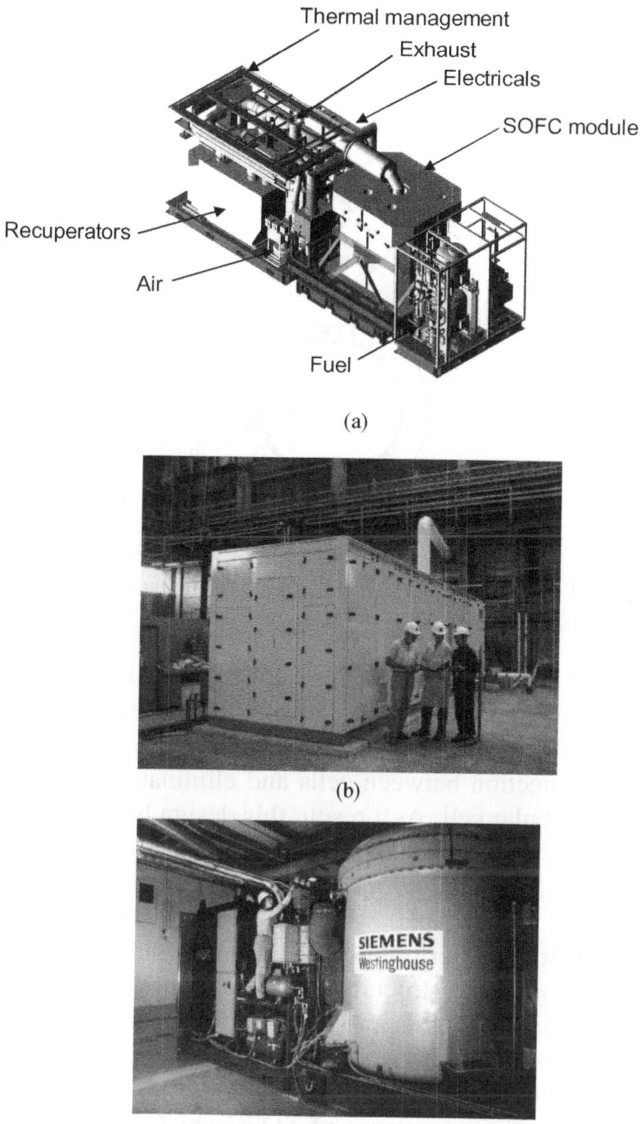

(a)

(b)

(c)

FIGURE 2.8 Siemens SOFC-CHP systems. (Photos courtesy of Siemens) (a) Major components of a Siemens 100-kW SOFC-CHP system. (b) A 250-kW SOFC-CHP system installed in Toronto, Canada. (c) The 220-kW SOFC/gas turbine hybrid system at the National Fuel Cell Research Center, University of California, Irvine.

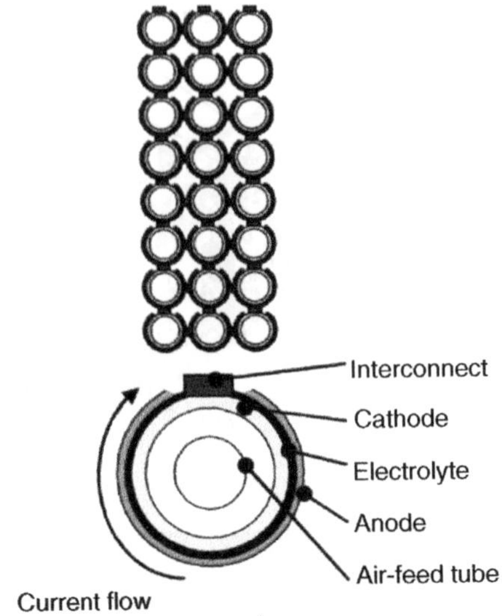

FIGURE 2.9 The end view and the tube of a Siemens tubular SOFC, and the series-parallel connection of tubes forming a SOFC stack [7].

Royce (UK) are currently manufacturing tubular SOFC stacks. The modeling of tubular SOFC is the subject of Chapter 4.

Planar SOFC design has a flat-plate structure, which enables simple electrical connection between cells and eliminates the long current path through the tubular cell. As a result, this design has lower fuel cell ohmic losses than the tubular design, and has improved power density and stack performance. There are several planar SOFC approaches currently under development, for example, with metal interconnects or ceramic interconnects, whose descriptions are beyond the scope of this book. Figure 2.10 shows the structure of a planar anode-supported SOFC with metal interconnects. This approach is currently under development in the United States with the support of the U.S. DOE SECA (Solid State Energy Conversion Alliance) Program [7].

In summary, the advantages of SOFC are fuel flexibility, inexpensive catalyst, solid electrolyte, and the availability of waste heat for CHP operation. Its disadvantages are operation at very high temperature, which leads to sealing issues. DOE-supported current research activities address these issues. Further discussion on the current status and future development of SOFC is covered in Chapter 10.

MCFCs use a molten mixture of alkaline metal carbonate as their electrolyte and an inexpensive catalyst (nickel). At high temperatures (600°C–700°C), the alkaline salt mixture is in liquid phase and is an

FIGURE 2.10 The structure of a Delphi anode-supported planar SOFC with metal interconnects [7].

excellent conductor of CO_3^{2-} ions. At the cathode, the oxygen and carbon dioxide combine with electrons to produce carbonate ions (CO_3^{2-}). At the anode, the CO_3^{2-} ions are deoxidized by hydrogen, and electrons are released at the same time. The electrons will migrate through the external circuit (generating electricity) and reach the cathode surface. In a MCFC, the anode and cathode are typically made of nickel alloy or oxide. At both electrodes, the nickel provides good catalytic activity and conductivity. The high-temperature operation range of MCFC (600–700°C), gives it fuel flexibility. While MCFCs can run on hydrogen, like other fuel cells, they can also use hydrocarbons (natural gas, methane, or alcohol) as fuel. Unlike in PEMFC, CO is not a poison and acts as a fuel in MCFC. MCFCs operate at a typical electrical efficiency of 50–55%, but their overall efficiency in CHP operation mode can reach as high as 90% [12].

Due to their high-temperature operation and slow start up, MCFCs are also suited for large-scale stationary and continuous power generation applications. Systems as large as 250 kW to 2 MW have been constructed and demonstrated. Figure 2.11 shows a 250-kW MCFC system developed by FuelCell Energy, Inc.

The schematic diagram of a MCFC is shown in Fig. 2.12. The equations representing the anode and cathode reactions, and the overall reaction of MCFCs are as follows.

$$H_2 + CO_3^{2-} \Rightarrow H_2O + CO_2 + 2e^- \quad \text{(anode reaction)} \tag{2.42}$$

$$\tfrac{1}{2}O_2 + CO_2 + 2e^- \Rightarrow CO_3^{2-} \quad \text{(cathode reaction)} \tag{2.43}$$

$$H_2 + \tfrac{1}{2}O_2 + CO_2 \Rightarrow H_2O + CO_2 \quad \text{(overall reaction)} \tag{2.44}$$

FIGURE 2.11 A MCFC-based 250-kW system made by FuelCell Energy, Inc. *Source*: Photo courtesy of FuelCell Energy, Inc.

In summary, MCFC advantages, like for SOFC, are fuel flexibility, inexpensive (metal) catalyst, and increased efficiency in CHP operation mode. Its disadvantages include corrosive molten electrolyte and relatively expensive materials and degradation issues.

Table 2.3 gives a comparison of the components and operating characteristics of the five major fuel cell types discussed earlier.

Table 2.4 gives a comparative summary of the benefits, operating characteristics, and applications of the fuel cells.

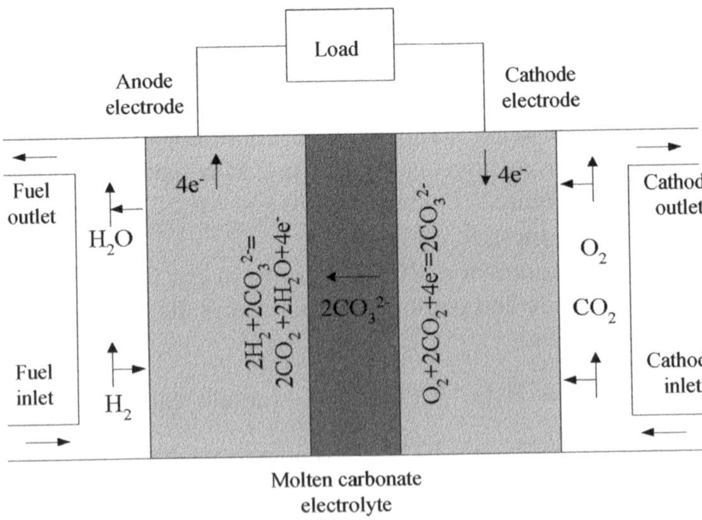

FIGURE 2.12 Schematic diagram of a MCFC.

TABLE 2.3 Comparison of Important Components and Characteristics of Different Types of Fuel Cells [7]

	AFC	PAFC	PEMFC	MCFC	SOFC
Electrodes	Transition metals	Carbon	Carbon	Nickel and nickel oxide	Metal cement
Catalyst	Platinum based	Platinum based	Platinum based	Non-noble metal	Non-noble metal
Electrolyte	Potassium Hydroxide	Liquid phosphoric acid	Solid polymer (proton exchange membranes)	Liquid molten carbonate	Dense yttria-stabilized zirconia (Ceramic)
Operating temperature range (°C)	80–260	~200	50–80	600–700	600–1000
Charge (ion) carrier	OH^-	H^+	H^+	CO_3^{2-}	O^{2-}
Product water management	Evaporative	Evaporative	Evaporative	Gaseous product	Gaseous product
Product heat management	Process gas	Process gas + liquid cooling medium or steam generation	Process gas + liquid cooling	Internal reforming + process gas medium	Internal reforming + process gas

TABLE 2.4 Comparative Summary of Major Types of Fuel Cells [12]

Fuel Cell Type	AFC	PAFC	PEMFC	MCFC	SOFC
Need for internal reforming	No	No	No	Yes	Yes
CO tolerance	No, poison (<50 ppm)	No, poison (<1%)	No, poison (<50 ppm)	Yes (fuel)	Yes (fuel)
Electrical efficiency (%)	~50	~40	40–50	45–55	50–60
Power density range (mW/cm^2)	150–400	150–300	300–1000	100–300	250–350
Power range (kW)	1–100	50–1000	10^{-3}–1000	100–10^5	5–10^5
Applications	Outer-space power supply, stationary power generation	Stationary, dispersed power plants and on-site cogeneration power plants	Portable, transportation and stationary power generation	Stationary and combined heat and power generation	Portable, transportation stationary, and combined heat and power generation

2.9 FUEL CELL EQUIVALENT CIRCUIT

In general, the voltage at the FC terminals is lower than the voltage developed inside the fuel cells, obtained from Nernst equation (2.32) and (2.33). This is due to the activation, ohmic, and concentration voltage drops inside fuel cells, as shown in Fig. 2.13. These voltage drops are functions of load current and fuel cell temperature and/or pressure. The ohmic voltage drop is a linear function of fuel cell load current, but the ohmic resistance ($R_{\text{ohm,cell}}$) is normally a function of fuel cell temperature. The activation and concentration voltage drops are nonlinear functions of load current as well as pressure and/or temperature inside the fuel cell. In this chapter these voltage drops will be discussed for two types of fuel cells, namely PEMFC and SOFC, which have promising potential for use in distributed power generation applications.

From Fig. 2.13, the output voltage of a fuel cell can be written as

$$V_{\text{cell}} = E_{\text{cell}} - V_{\text{act,cell}} - V_{\text{ohm,cell}} - V_{\text{conc,cell}} \tag{2.45}$$

where V_{cell} and E_{cell} are the fuel cell output voltage and internal voltage, respectively, and $V_{\text{act,cell}}$, $V_{\text{ohm,cell}}$, $V_{\text{conc,cell}}$ are the voltage drops discussed earlier.

In addition to fuel cell load current, pressure, and temperature, the electric charge storage capability of fuel cells also affects their dynamic response. This effect, which is common to most fuel cells, makes them behave like a large capacitor (in the order of a few Farads), called "capacitance of double-layer charge effect." The reason behind this effect and its role in fuel cell dynamic electrical performance are discussed in the next section.

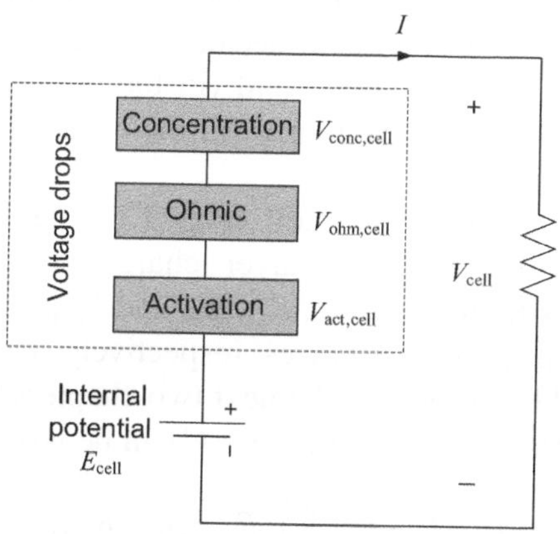

FIGURE 2.13 Fuel cell equivalent circuit.

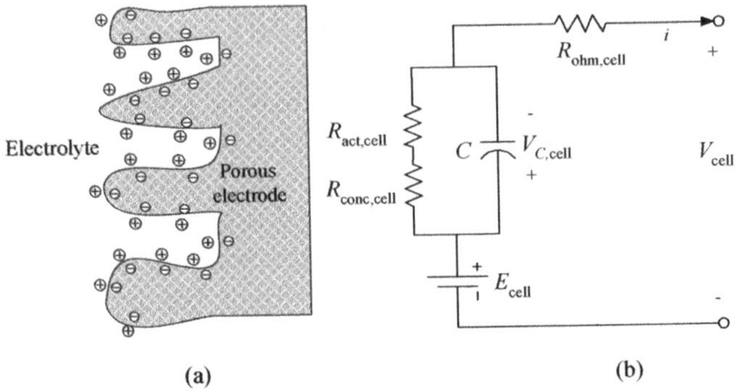

FIGURE 2.14 (a) The double-layer charge effect at the surface of a fuel cell electrode and (b) its equivalent electrical circuit.

2.10 CAPACITANCE OF DOUBLE-LAYER CHARGE EFFECT

In a fuel cell, the two electrodes are separated by the electrolyte (e.g., see Fig. 2.4) and two boundary layers are formed, that is, anode–electrolyte layer and electrolyte–cathode layers. These two charged layers of opposite polarity are formed across the boundary between the electrodes and the electrolyte. The layers, known as electrochemical double layers [8], can store electrical energy and behave like a capacitor. Figure 2.14a shows an example of how the positive and negative ions gather at the surface of a porous electrode forming the effect of a capacitor.

$$C = \varepsilon \frac{A}{l} \tag{2.46}$$

In (2.46), ε is the electrolyte's electrical permittivity, A is the effective surface area between the electrolyte and electrode, and l is the distance between the layers. In a real fuel cell, because of the porous structure of electrodes, A is large and l is very small (in the order of nanometers), resulting in a high value for capacitance (in the order of 100s of micro-Farads to a few Farads).

The equivalent circuit of a fuel cell (Fig. 2.13) can be modified to include the capacitance of double-layer charge effect as shown in Fig. 2.14b [10]. In this figure $R_{ohm,cell}$, $R_{act,cell}$, and $R_{conc,cell}$ are the ohmic, activation, and concentration resistances, respectively. The expression for these parameters will be derived in the next two chapters for PEMFC and SOFC. From Fig. 2.14b, the voltage across C can be written as follows:

$$V_{C,cell} = \left(i - C\frac{dV_{C,cell}}{dt}\right)(R_{act,cell} + R_{conc,cell}) \tag{2.47}$$

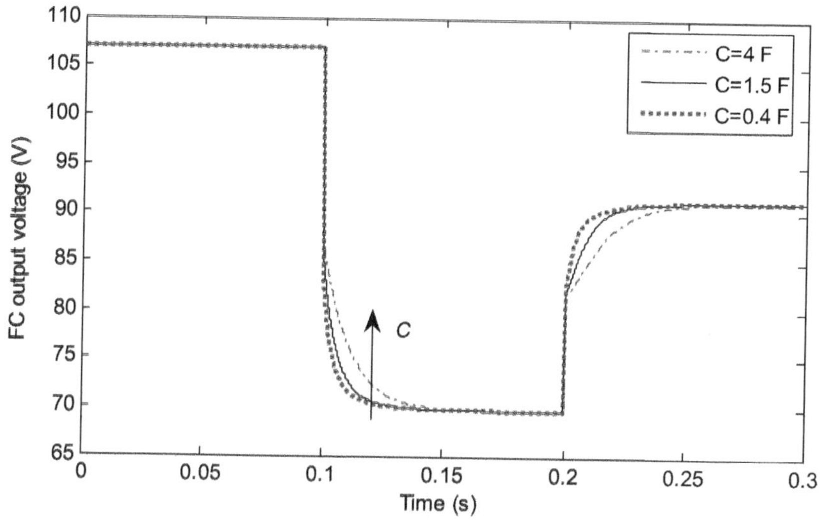

FIGURE 2.15 Fuel cell voltage response due to capacitance of double-layer charge effect.

According to Fig. 2.14b and (2.47), only the activation and concentration resistances affect the time constant $\tau = (R_{act,cell} + R_{conc,cell}) \times C$ of the circuit (the voltage across the capacitance of double-layer charge effect ($V_{C,cell}$)). This is because only changes in these two resistances are due to the electrochemical reactions inside fuel cells, and the ohmic resistance is simply the resistance due to the ohmic voltage drops.

The fuel cell output voltage can then be written as

$$V_{cell} = E_{cell} - V_{C,cell} - V_{ohm,cell} \tag{2.48}$$

From (2.48), the fuel cell output voltage is a dynamic quantity since $V_{C,cell}$ is a function of time with time constant $\tau = (R_{act,cell} + R_{conc,cell}) \times C$.

Figure 2.15 shows the voltage response of a FC module for different values of C. Based on typical fuel cell parameter values, the above time constant is in the order of milliseconds, as shown in Fig. 2.15. This time constant is much shorter than the time constants due to gas dynamics and thermal dynamics inside fuel cells, which will be discussed in Chapters 3 and 4.

2.11 SUMMARY

This chapter covered the operational principles of FCs. It gave a summary of the first and second laws of thermodynamics, fundamentals of electrochemical processes, Gibbs free energy, and the Nernst equation (based on which the internal voltage inside FCs is developed), and energy balance in chemical reactions as related to FCs. A summary of chemical reactions of five types of FCs was given, and the structure and applications of FCs with promising potential for DG applications—PEMFC, SOFC, and MCFC—were covered in more detail. Finally, FC equivalent circuit and the double-layer charge effect inside FCs, which hold for the different types of FCs, were covered.

REFERENCES

[1] A.W. Culp Jr., *Principles of Energy Conversion*, McGraw-Hill, New York, 1979.

[2] S.S. Zumdahl, *Chemical Principles*, 2nd edn, DC Heath and Company, Toronto, 1995.

[3] E.R.G. Eckert and R.M. Drake, Jr., *Analysis of Heat and Mass Transfer*, McGraw-Hill Book Company, New York, 1972.

[4] G.N. Hatsopoulos and J.H. Keenan, *Principles of General Thermodynamics*, John Wiley & Sons, Inc. , New York, 1965.

[5] G. Hoogers, *Fuel Cell Technology Handbook*, CRC Press, Boca Raton, FL, 2003.

[6] G. Kortum, *Treatise on Electrochemistry*, 2nd edn, Elsevier, Amsterdam, 1965.

[7] *Fuel Cell Handbook*, 7th edn, EG&G Services, Inc., Science Applications International Corporation, DOE, Office of Fossil Energy, National Energy Technology Laboratory, 2004.

[8] J. Larminie and A. Dicks *Fuel Cell Systems Explained*, 2nd edn, Wiley, New York, 2003.

[9] J.A. Smith, M.H. Nehrir, V. Gerez, and S.R. Shaw, A Broad Look at the Workings, Types, and Applications of Fuel Cells, Proceedings, 2002 IEEE Power Engineering Society Summer Meeting, Chicago, IL, 2002.

[10] C. Wang, M.H. Nehrir, and S.R. Shaw, Dynamic models and model validation for PEM fuel cells using electrical circuits, *IEEE Transactions on Energy Conversion*, 20 (2), 442–451, 2005.

[11] H. Aki, S. Yamamoto, Y. Ishikawa, J. Kondoh, T. Maeda, H. Yamaguchi, A. Murata, and I. Ishii, Operational strategies of networked fuel cells in residential homes, *IEEE Transactions on Power Systems*, 21 (3), 1405–1414, 2006.

[12] R. O'Hayre, S. Cha, W. Colella, and F. Prinz, *Fuel Cell Fundamentals*, Wiley, Hoboken, NJ, 2006.

CHAPTER 3

DYNAMIC MODELING AND SIMULATION OF PEM FUEL CELLS

3.1 INTRODUCTION: NEED FOR FUEL CELL DYNAMIC MODELS

Fuel cells (FCs) are good energy sources to provide reliable power at steady state, but they cannot respond to electrical load transients as fast as desired. This is mainly due to their slow internal electrochemical and thermodynamic responses and the slow responses of their mechanical accessory systems. Accurate physically based FC models are needed to predict and evaluate their steady-state and dynamic responses in various applications, such as FC vehicles and stand-alone and grid-connected FC power generation systems. Dynamic FC models are also needed for predicting and evaluating FC performance under different transient conditions such as electrical faults at the FC terminals, and FC vehicle starting, accelerating and braking. Under the above conditions, FC output voltage and power are influenced by the electrochemical and physical properties of the FC, such as the capacitance of double-layer charge effect, mass diffusion, material conservation, thermodynamic characteristics, and voltage drops inside the FCs. These characteristics will be discussed in this chapter and the next for two types of FCs that have potential for distributed power generation applications, namely PEMFC and SOFC.

Dynamic FC models are also needed for designing controllers to control FC terminal electrical quantities as desired. In addition, for ac power generation, in both stand-alone and grid-connected generation systems, the dc voltage produced by FCs is converted to ac through power electronic interfacing devices (dc/dc converters and inverters). In such applications, it is possible to control the FC performance by controlling the power electronic interfacing devices. The subject of controller design for PEMFC and SOFC power generation systems is covered in Chapters 8 and 9.

PEMFCs show great promise for use in a variety of power generation applications such as distributed generation (DG), backup generation, and automotive applications. The increased desire for power generation devices and vehicles with less emission has made PEMFCs attractive for these applications since they emit essentially no pollutants and have high power density and quick start.

In this chapter, a dynamic model is developed for PEMFC based on physical principles. First, analytical expressions are derived for the PEMFC equivalent internal voltage source, the activation, ohmic and concentration voltage drops, and their equivalent resistors. Then, the electric charge storage capability of PEMFC is modeled by the capacitance of double-layer charge effect, discussed in Chapter 2. The resulting heat from the chemical reactions inside PEMFC and heat transfer between PEMFC electrodes and membrane also affect the fuel cell response. Such effects are considered in the dynamic model development using an energy balance approach. The dynamic model developed for PEMFC is suitable for system analysis and controller design from an electrical engineering point of view, and in multi-source alternative energy power generation system design. These topics are covered in Chapters 7, 8 and 9.

3.2 NOMENCLATURE (PEMFC)

Most of the symbols used in the development of the PEMFC model are defined where they appear first. In this section, the definitions for all the symbols (including subscripts and superscripts) are tabulated alphabetically for ease of reference.

a, b	Constant terms in Tafel equation (V/K)
A_{cell}	Area of each cell (m^2)
C_i	Specific heat capacity of species i (J/[mol K])
$D_{i,j}$	Effective binary diffusivity of i–j pair of species (m^2/s)
E	Reversible potential of each cell (V)

E_0	Reference potential (V)
E_0°	Standard reference potential (V)
F	Faraday constant (96487 C/mol)
h_{cell}	Convective heat transfer coefficient (W/(m^2 K))
H_V	Vaporization heat of water (J/mol)
I, i	Current (A)
I_{den}	Current density (A/m^2)
I_{limit}	Limitation current (A)
I_0	Exchange current (A)
k_{E}	Empirical constant in calculating E_0 (V/K)
k_{RI}	Empirical constant in calculating R_{ohmic} (Ω/A)
k_{RT}	Empirical constant in calculating R_{ohmic} (Ω/K)
l_{a}	Width between anode channel to catalyst (m)
l_{c}	Width between cathode channel to catalyst (m)
M_i	Mole flow rate of species i (mol/s)
n_i	Moles of species i (mol)
N_i	Superficial gas flux of species i (mol/(m^2 s))
N_{cell}	Number of cells in the stack
p_i	Partial pressure of species i (Pa)
P	Pressure (Pa)
P_i	Pressure of compartment i (Pa)
$\bar{q}_{\text{chem}}$	Chemical, or heat energy (J)
$\bar{q}_{\text{elec}}$	Electrical energy (J)
$\bar{q}_{\text{loss}}$	Heat loss (J)
$\bar{q}_{\text{net}}$	Net heat energy (J)
$\bar{q}_{\text{sens + latent}}$	Sensible and latent heat (J)
R	Gas constant, 8.3143 (J/[mol K])
R_i	Resistance of type i (Ω)
T	Temperature (K)
T_{initial}	FC initial temperature (K)
T_{room}	Room temperature (K)
V	Volume (m^3)
V	terminal voltage (V)
V_i	Voltage drop of type i (V)
x_i	Mole fractions of species I
x	Axis x
z	Number of electrons participating
α	Electron transfer coefficient
ΔG_0	Gibbs free energy at standard condition (J/mol)
η_0	Temperature invariant part of V_{act} (V)
λ_{e}	Constant factor in calculating E_{d} (Ω)
τ_{a}	Fuel flow delay (s)
τ_{c}	Oxidant flow delay (s)
τ_{e}	Overall flow delay (s)

(Continued)

Superscripts and subscripts

a	Anode
act	Activation
c	Cathode
cell	Conditions for a single cell
channel	Conditions at the anode or cathode channel
CO_2	Carbon dioxide
conc	Concentration
consumed	Material consumed in chemical reaction
generated	Material generated in chemical reaction
H_2	Hydrogen
H_2O	Water
in	Input
l (g)	Liquid (gas) form
membrane	Conditions at the membrane
N_2	Nitrogen
O_2	Oxygen
ohm	Ohmic
out	Output
sat	Saturation conditions
*	Effective value

3.3 PEMFC DYNAMIC MODEL DEVELOPMENT

As discussed earlier, fuel cell output voltage is less than the internal voltage developed inside the fuel cell due to several voltage drops. Figure 3.1 shows a schematic diagram of cross section of a PEM fuel cell, and the voltage drops across it, from the anode side gas flow channel to the cathode side. The ohmic voltage drops associated with the anode, cathode, and membrane at certain operating point can be considered as a linear function of thickness of those segments, shown in Fig. 3.1. The activation voltage drop is due to the energy barrier that the reacting species must overcome in chemical reactions inside the FC. It can be considered as an extra voltage necessary to enable the chemical reaction to proceed at a desired rate. The concentration voltage drop is caused by the mass transfer processes from bulk channels to reaction sites in porous electrodes. Under high current densities, the concentration voltage drop will be significant since the reactant concentration could be much less than the bulk of the gas stream. For low current densities, the concentration voltage drop may be neglected.

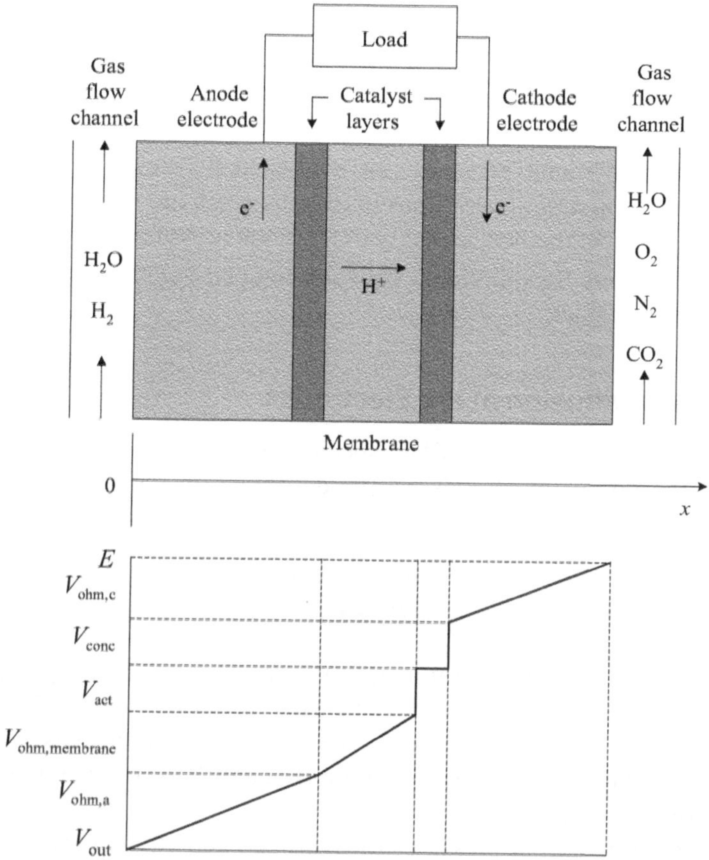

FIGURE 3.1 Schematic diagram of a PEMFC and its voltage drops.

In the following subsections, gas diffusion as a function of position (x) across PEMFC (from anode to cathode) and dynamics of flow of hydrogen and oxygen in the electrode channels are derived. It will be shown that these quantities will lead to the development of the internal voltage and voltage drops inside PEMFC.

The following assumptions, commonly made in the literature, are also used here to simplify the analysis.

(1) One-dimensional treatment of flow and distribution of gas species inside PEMFC.
(2) Ideal and uniformly distributed gases.
(3) Constant pressures in the fuel cell gas flow channels.
(4) The fuel is humidified hydrogen, and the oxidant is humidified air.

(5) On the anode side, the effective water vapor pressure is 50%, and on the cathode side, 100% of the saturated vapor pressure.

(6) PEMFC operates under 100°C and the reaction product is in liquid phase.

(7) Thermodynamic properties are evaluated at average stack temperature, temperature variations across the stack are neglected and the overall specific heat capacity of the stack is assumed to be constant.

(8) Parameters for individual cells are lumped together to represent a fuel cell stack.

3.3.1 Gas Diffusion at the Electrodes

In a gas mixture consisting of N species, the diffusion of component i through the porous electrodes can be described by the Stefan–Maxwell formulation given below [4,7].

$$\nabla x_i = \frac{RT}{P} \sum_{j=1}^{N} \frac{x_i N_j - x_j N_i}{D_{i,j}} \tag{3.1}$$

where ∇ is the gradient operator; x_i (x_j) the mole fractions of species i (j); $D_{i,j}$ the effective binary diffusivity of i–j pair of species (m^2/s); N_i (N_j) the superficial gas flux of species i (j) (mol/[m^2 s]); R the gas constant, 8.3143 J/(mol K); T the gas temperature (K); and P the overall pressure of the gas mixture (Pa).

The effective partial pressures of hydrogen and oxygen are needed in order to calculate the PEMFC output voltage. At the anode channel, the gas stream is a mixture of hydrogen and water in gas phase ($H_2O_{(g)}$). The molar flux of water (in gas phase) normal to anode surface N_{H_2O} can be set to zero according to assumptions (1)–(3) given in the previous section.

According to (3.1), assuming the one-dimensional transport process, the diffusion of water along the x axis (Fig. 3.1), can be simplified as [4]

$$\frac{dx_{H_2O}}{dx} = \frac{RT}{P_a}\left(\frac{x_{H_2O}N_{H_2} - x_{H_2}N_{H_2O}}{D_{H_2O,H_2}}\right) = \frac{RT}{P_a}\left(\frac{x_{H_2O}N_{H_2}}{D_{H_2O,H_2}}\right) \tag{3.2}$$

where P_a is the overall gas pressure at the anode (Pa).

The molar flux of H_2 can be determined by Faraday's Law [4,8].

$$N_{H_2} = \frac{I_{den}}{2F} \tag{3.3}$$

where I_{den} is the current density (A/m^2) and F is the Faraday constant (96487 C/mol).

By combining (3.2) and (3.3) and integrating the expression with respect to x from the anode channel to the catalyst surface, it gives:

$$x^*_{H_2O} = x^{ch}_{H_2O} \exp\left(\frac{RTI_{den}l_a}{2FP_a D_{H_2O,H_2}}\right) \tag{3.4}$$

where l_a is the distance from anode surface to the reaction site (m); superscript * denotes the effective value; superscript "ch" denotes the conditions at the anode or cathode channel.

According to the assumption (4) given in the previous section, the fuel consists only of hydrogen and water vapor. That is, at the anode, $x^*_{H_2O} + x^*_{H_2} = 1$. Therefore, according to the uniformity of gas distribution along the x-axis (assumption 2), the effective partial pressure of hydrogen can be written as

$$p^*_{H_2} = \frac{p^*_{H_2O}}{x^*_{H_2O}}(1 - x^*_{H_2O}) \tag{3.5}$$

According to assumption (5), the effective water vapor pressure ($p^*_{H_2O}$) at the anode is $0.5p^{sat}_{H_2O}$. Therefore, using (3.4), the effective pressure of H_2 ($p^*_{H_2}$) can then be written as [4,5]:

$$p^*_{H_2} = 0.5p^{sat}_{H_2O}\left[\frac{1}{x^{ch}_{H_2O}\exp(RTI_{den}l_a/2FP_a D_{H_2O,H_2})} - 1\right] \tag{3.6}$$

The gases flowing at the cathode channel are O_2, N_2, $H_2O_{(g)}$, and CO_2. Using (3.1), the diffusion of $H_2O_{(g)}$ at the cathode side can also be obtained as follows:

$$\frac{dx_{H_2O}}{dx} = \frac{RT}{P_c}\left(\frac{x_{O_2}N_{H_2O} - x_{H_2O}N_{O_2}}{D_{H_2O,O_2}}\right) = \frac{RT}{P_c}\left(\frac{-x_{H_2O}N_{O_2}}{D_{H_2O,O_2}}\right) \tag{3.7}$$

where P_c is the overall gas pressure at the cathode (Pa).

Similar to the analysis for anode, the effective molar fraction of H_2O, N_2, and CO_2 at the cathode catalyst interface can be derived as follows:

$$x^*_{H_2O} = x^{ch}_{H_2O} \exp\left(\frac{RTI_{den}l_c}{4FP_c D_{H_2O,O_2}}\right) \tag{3.8}$$

$$x^*_{N_2} = x^{ch}_{N_2} \exp\left(\frac{RTI_{den}l_c}{4FP_c D_{N_2,O_2}}\right) \tag{3.9}$$

$$\Delta G = \Delta H - T\,\Delta S = \Delta G_0 - RT \ln\left[\overset{*}{p}_{H_2} \cdot (\overset{*}{p}_{O_2})^{0.5}\right] \tag{3.10}$$

where l_c is the distance from cathode surface to the reaction site (m); At the cathode, the effective molar fraction of O_2 is:

$$\overset{*}{x}_{O_2} = 1 - \overset{*}{x}_{H_2O} - \overset{*}{x}_{N_2} - \overset{*}{x}_{CO_2} \tag{3.11}$$

Therefore, the corresponding effective partial pressure of O_2 can be written as

$$\overset{*}{p}_{O_2} = \frac{\overset{*}{p}_{H_2O}}{\overset{*}{x}_{H_2O}}\overset{*}{x}_{O_2} = \frac{\overset{*}{p}_{H_2O}}{\overset{*}{x}_{H_2O}}(1 - \overset{*}{x}_{H_2O} - \overset{*}{x}_{N_2} - \overset{*}{x}_{CO_2}) \tag{3.12}$$

According to assumption (5), $\overset{*}{p}_{H_2O}$ at the cathode equals $p^{sat}_{H_2O}$. Therefore, similar to (3.6), (3.12) can be rewritten as

$$\overset{*}{p}_{O_2} = p^{sat}_{H_2O}\left[\frac{1 - \overset{*}{x}_{N_2} - \overset{*}{x}_{CO_2}}{\overset{*}{x}_{H_2O}} - 1\right] \tag{3.13}$$

$\overset{*}{p}_{H_2}$ and $\overset{*}{p}_{O2}$, calculated from (3.6) and (3.13), will be used in the Nernst equation to find PEMFC output voltage. Material balance equations (also called conservation equations), also needed to determine PEMFC output voltage, will be derived in the next section.

3.3.2 Material Conservation

The dynamic equations of the effective partial pressures of hydrogen and oxygen in the anode and cathode channels can be determined through the ideal gas equations as follows [9]:

$$\frac{V_a}{RT}\frac{d\overset{*}{p}_{H_2}}{dt} = M_{H_2,in} - M_{H_2,out} - \frac{i}{2F} = M_{H_2,net} - \frac{i}{2F} \tag{3.14}$$

$$\frac{V_c}{RT}\frac{d\overset{*}{p}_{O_2}}{dt} = M_{O_2,in} - M_{O_2,out} - \frac{i}{4F} = M_{O_2,net} - \frac{i}{4F} \tag{3.15}$$

where $V_a(V_c)$ is the volume of anode (cathode) channel (m^3); $M_{H_2} = H_2$, mole flow rate (mol/s); i the fuel cell current (A); $M_{O_2} = O_2$, mole flow rate (mol/s); subscripts “in”, “out,” and “net” denote the values related to input, output, and net.

At steady-state, all partial pressures are considered to be constant, and we can write

$$\frac{dp_{H_2}^*}{dt} = \frac{dp_{O_2}^*}{dt} = 0 \tag{3.16}$$

From (3.14) and (3.15), the net mole flow rates of H_2 and O_2 at steady-state are:

$$M_{H_2,\text{net}} = 2M_{O_2,\text{net}} = \frac{I}{2F} \tag{3.17}$$

Under transient state, there are delays between the change in the load current and mol flow of fuel (H_2) and oxidant (O_2). From (3.14) and (3.15), these delays can be modeled by the following first order differential equations [10]:

$$\begin{aligned} \tau_a \frac{dM_{H_2,\text{net}}}{dt} &= \frac{i}{2F} - M_{H_2,\text{net}} \\ \tau_c \frac{dM_{O_2,\text{net}}}{dt} &= \frac{i}{4F} - M_{O_2,\text{net}} \end{aligned} \tag{3.18}$$

Time constants τ_a and τ_c represent the fuel and oxidant flow delays (in seconds) at the anode and cathode, respectively. The dynamic equation (3.18) will be used to determine the output voltage of PEMFC.

3.3.3 PEMFC Output Voltage

As discussed in Chapter 2, the overall reaction in a PEMFC is:

$$H_2 + \tfrac{1}{2}O_2 = H_2O_{(l)} \tag{3.19}$$

where subscript (l) indicates liquid form of water production (assumption 6).

The corresponding Nernst equation used to calculate the reversible potential is [8]:

$$E_{\text{cell}} = E_{0,\text{cell}} + \frac{RT}{2F}\ln\left[p_{H_2}^* \cdot (p_{O_2}^*)^{0.5}\right] \tag{3.20}$$

where E_{cell} is the reversible potential of each cell (V).

$E_{0,\text{cell}}$ is the reference potential, which is a function of temperature and can be expressed as follows [1]:

$$E_{0,\text{cell}} = E_{0,\text{cell}}^{\circ} - k_E(T - 298) \tag{3.21}$$

where $E_{0,\text{cell}}^{\circ}$ is the standard reference potential at standard state (298 K and 1 atm pressure).

The overall effect of the fuel and oxidant delays is represented by a voltage $E_{d,cell}$, which is subtracted from the right side of (3.20). This voltage, which is a function of time, shows the influence of the fuel and oxidant delays on the fuel cell output voltage during load transients. In Laplace domain, $E_{d,cell}$ can be written as:

$$E_{d,cell}(s) = \lambda_e I(s)\left(1 - \frac{1}{\tau_e s + 1}\right) = \lambda_e I(s)\frac{\tau_e s}{\tau_e s + 1} \tag{3.22}$$

where λ_e is a constant (Ω), and τ_e is the overall flow delay (s).

Note from Equation 3.22 that the steady-state value of $E_{d,cell}$, is zero as s approaches zero. From Laplace transform theory, $s \to 0$ in the s-domain corresponds to $t \to \infty$ in time domain, and multiplication in the s-domain corresponds to convolution in the time domain. Therefore, converting (3.22) from s-domain into time domain, we get:

$$E_{d,cell} = \lambda_e[i(t) - i(t) \otimes \exp(-t/\tau_e)] \tag{3.23}$$

where "$\otimes$" is the convolution operator, and $\exp(-t/\tau_e)$ is the time lag function corresponding to $1/(\tau_e s + 1)$ in s-domain.

$E_{d,cell}$ is now subtracted from the right side of (3.20) to take into account the effect of oxidant delays. The internal potential (or the PEMFC open-circuit voltage) E_{cell}, given by (3.20) can now be modified as follows:

$$E_{cell} = E_{0,cell} + \frac{RT}{2F}\ln\left[p^*_{H_2} \cdot (p^*_{O_2})^{0.5}\right] - E_{d,cell} \tag{3.24}$$

As shown in Fig. 3.1 under normal operating conditions, the fuel cell output voltage is less than E_{cell} due to the fuel cell activation loss, ohmic voltage drop, and concentration overpotential (voltage drop) across the fuel cell. Therefore,

$$V_{cell} = E_{cell} - V_{act,cell} - V_{ohm,cell} - V_{conc,cell} \tag{3.25}$$

where V_{cell}, $V_{act,cell}$, $V_{ohm,cell}$, and $V_{conc,cell}$ are the cell output voltage, activation voltage drop, ohmic voltage drop, and concentration voltage drop, respectively.

Applying assumption (8), the output voltage of a fuel cell stack can then be obtained as

$$V_{out} = N_{cell}V_{cell} = E - V_{act} - V_{ohm} - V_{conc} \tag{3.26}$$

where V_{out} is the output voltage of a fuel cell stack (V), N_{cell} the number of cells in the stack (V), E the fuel cell stack internal potential (V), V_{act} the overall activation voltage drop (V), V_{ohm} the overall ohmic voltage drop (V), and V_{conc} the overall concentration voltage drop (V).

Equations 3.25 and 3.26 give the output voltage of a single cell, and a fuel cell stack in dynamic form; they will be used in PEMFC computer model development.

3.3.4 PEMFC Voltage Drops

Fuel cell voltage drops shown in (3.26) are calculated as follows:

Activation Voltage Drop: The activation voltage drop is a function of the PEMFC current and temperature, as described empirically by Tafel equation given below [1].

$$V_{act} = \frac{RT}{\alpha z F} \ln\left(\frac{I}{I_0}\right) = T \cdot [a + b \ln(I)] \tag{3.27}$$

V_{act} can be described as a sum of V_{act1} and V_{act2} as follows:

$$V_{act} = \eta_0 + (T - 298) \cdot a + T \cdot b \ln(I) = V_{act1} + V_{act2} \tag{3.28}$$

where η_0, a, b are empirical constants, $V_{act1} = (\eta_0 + (T - 298)\ a)$ is the voltage drop affected only by the fuel cell internal temperature (is not current dependent), and $V_{act2} = (T \cdot b \cdot \ln(I))$ is both current and temperature dependent.

The equivalent resistance of activation is defined by the ratio of V_{act2} and the fuel cell current. From the equation given below, it is noted that the resistance is both temperature and current dependent.

$$R_{act} = \frac{V_{act2}}{I} = \frac{T \cdot b \ln(I)}{I} \tag{3.29}$$

Ohmic Voltage Drop: The ohmic resistance of a PEMFC consists of the resistance of the polymer membrane, the conducting resistance between the membrane and electrodes, and the resistances of electrodes. The overall ohmic voltage drop can be expressed as:

$$V_{ohm} = V_{ohm,a} + V_{ohm,membrane} + V_{ohm,c} = IR_{ohm} \tag{3.30}$$

R_{ohm} is a function of current and temperature and can be expressed by [1].

$$R_{ohm} = R_{ohm0} + k_{RI} I - k_{RT} T \tag{3.31}$$

where R_{ohm0} is the constant part of R_{ohm}, k_{RI} the empirical constant for calculating R_{ohm} (Ω/A), and k_{RT} the empirical constant for calculating R_{ohm} (Ω/K).

Concentration Voltage Drop: During the reaction process, concentration gradients can be formed due to mass diffusions from the gas flow channels to the reaction sites (at the catalyst surfaces). At high current densities, slow transportation of reactants (products) to (from) the reaction sites is the main reason for the concentration voltage drop [1]. Any water film covering the catalyst surfaces at the anode and cathode can be another contributor to this voltage drop [4]. The concentration overpotential in the fuel cell is defined as follows [1]:

$$V_{\text{conc}} = -\frac{RT}{zF}\ln\frac{C_S}{C_B} \tag{3.32}$$

where C_S is the surface concentration at the reaction sites, C_B the bulk concentration in gas channels, and z the number of participating electrons.

According to Fick's first law and Faraday's law [8], Equation 3.32 can be written as a function of current as follows:

$$V_{\text{conc}} = -\frac{RT}{zF}\ln\left(1-\frac{I}{I_{\text{limit}}}\right) \tag{3.33}$$

I_{limit} is the fuel cell current limit (A).

The equivalent resistance for the concentration loss can therefore be defined as:

$$R_{\text{conc}} = \frac{V_{\text{conc}}}{I} = -\frac{RT}{zFI}\ln\left(1-\frac{I}{I_{\text{limit}}}\right) \tag{3.34}$$

Double-Layer Charge Effect: In PEMFC, the electrodes are separated by a solid membrane (Fig. 3.1), which only allows the hydrogen protons to pass through it, but blocks the electron flow. The electrons will flow from the anode through the external load and gather at the surface of the cathode, to which the hydrogen protons are attracted. Thus, two charged layers of opposite polarity are formed across the boundary between the porous cathode and the membrane [2]. The layers, known as electrochemical double-layer, can store electrical energy and behave like a super-capacitor. The equivalent circuit of fuel cell considering this effect is given in Fig. 3.2. This figure is similar to the circuit given in Fig. 2.14b in Chapter 2, but the values of R_{act} (see Eq. 3.29) and the voltage source are different from what are given in Fig. 2.14b and in [2]. In Fig. 3.2, R_{act} is defined according to (3.29), as V_{act2}/I, and the internal voltage is defined as $E–V_{\text{act1}}$. V_{act1} is the temperature-dependent part of V_{act}; and R_{conc} is the equivalent resistance of concentration voltage drop, which can be calculated according to (3.34). The

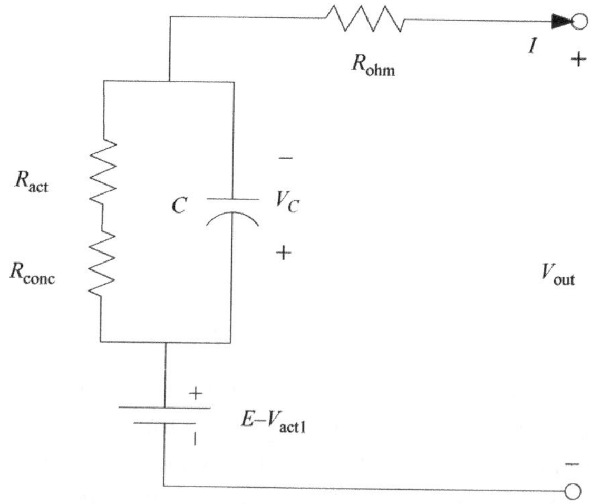

FIGURE 3.2 Equivalent circuit of the double-layer charging effect inside a PEMFC.

voltage across C in Fig. 3.2 can be written as:

$$V_C = \left(I - C\frac{dV_C}{dt}\right)(R_{act} + R_{conc}) \tag{3.35}$$

The double-layer charge effect is integrated into the modeling, by using V_C for calculating V_{out}. From Fig. 3.2, the fuel cell output voltage can be written as follows:

$$V_{out} = E - V_{act1} - V_C - V_{ohm} \tag{3.36}$$

3.3.5 Thermodynamic Energy Balance for PEMFC

The net heat generation rate due to the chemical reaction inside the fuel cell, which causes its temperature to rise or fall, can be written as [6]:

$$\bar{q}_{net} = \bar{q}_{chem} - \bar{q}_{elec} - \bar{q}_{sens+latent} - \bar{q}_{loss} \tag{3.37}$$

where $\bar{q}_{net}$ is the net heat energy (J), $\bar{q}_{chem}$ the chemical (or heat) energy (J), $\bar{q}_{elec}$ the electrical energy (J), $\bar{q}_{sens+latent}$ the sensible and latent heat (J), and $\bar{q}_{loss}$ the heat loss (J).

The power released due to chemical reaction due to the change in the enthalpy of the chemical reaction inside the fuel cell (ΔH) can be written as:

$$\bar{q}_{chem} = \bar{n}_{H_2,consumed} \cdot \Delta H \tag{3.38}$$

where $\bar{n}_{H_2,consumed} = H_2$ consumption rate (mol/s).

The maximum available electrical energy can be calculated from the Gibbs free energy as follows [1,8]:

$$\Delta G = \Delta H - T\,\Delta S = \Delta G_0 - RT\ln\left[p^*_{H_2}\cdot\left(p^*_{O_2}\right)^{0.5}\right] \tag{3.39}$$

where ΔG is the Gibbs free energy (J/mol); ΔG_0 the Gibbs free energy at standard condition[1] (J/mol). ΔS the entropy change (J/(mol/K)).

The electrical output power can be written as:

$$\dot{q}_{\text{elec}} = V_{\text{out}}\cdot I \tag{3.40}$$

Sensible heat is the heat energy that is transferred by a body that has a temperature higher than its surroundings. Sensible heat transportation rate is the product of the species mol flow rate, its specific heat capacity, and its temperature and the room temperature. Latent heat is the amount of energy in the form of heat released or absorbed by a substance during a change of state or phase. Heat of vaporization is used to indicate the amount of energy required when a substance changes its state into gas. Assuming the inlet temperate is the same as the room temperature, the sensible and latent heat absorbed during the process can be estimated by the following equation [6,9]:

$$\begin{aligned}\dot{q}_{\text{sens + latent}} &= \dot{n}_{H_2,\text{out}}(T - T_{\text{room}})\cdot C_{H_2} + \dot{n}_{O_2,\text{out}}(T - T_{\text{room}})\cdot C_{O_2}\\ &\quad + \dot{n}_{H_2O,\text{generated}}\cdot(T - T_{\text{room}})\cdot C_{H_2O,l} + \dot{n}_{H_2O,\text{generated}}\cdot H_V\end{aligned} \tag{3.41}$$

where $\bar{q}_{\text{sens + latent}}$ is the sensible and latent heat (J), $\bar{n}_i$ the flow rate of species i (mol/s), C_i the specific heat capacity of species i (J/(mol K)), H_V the vaporization heat of water (J/mol), and T_{room} the room temperature (K).

The heat loss, which is mainly transferred by air convection, can be estimated as follows [9]:

$$\dot{q}_{\text{loss}} = h_{\text{cell}}(T - T_{\text{room}})N_{\text{cell}}A_{\text{cell}} \tag{3.42}$$

where h_{cell} is the convective heat transfer coefficient (W/(m^2 K)); it can be obtained through experiment [6].

At steady state, the fuel cell operates at constant temperature and $\bar{q}_{\text{net}} = 0$. During transitions, fuel cell temperature will rise or drop according to the fuel cell specific heat capacity and its net heat rate as follows [9]:

$$M_{FC}C_{FC}\frac{dT}{dt} = \dot{q}_{\text{net}} \tag{3.43}$$

[1] 1 atm, 298 K.

where M_{FC} is the total mass of the fuel cell stack and C_{FC} is the overall specific heat capacity of the stack.

3.4 PEMFC MODEL STRUCTURE

A computer model can be developed for PEMFC based on its electrochemical and thermodynamic characteristics, discussed in Section 3.3, to predict the fuel cell dynamic response. The fuel cell output voltage is a function of temperature and load current. Since the fuel cell temperature and the voltage across the equivalent capacitance of double-layer charge effect are a function of time during a transient state, the resultant fuel cell output voltage is a dynamic quantity.

Figure 3.3 shows a block diagram, based on which a computer model can be developed for PEMFC. In this figure, the input quantities are anode and cathode pressures, the initial fuel cell temperature, and room temperature. At any given load current and time, the internal temperature T is determined and both the load current and temperature are fed back to different blocks, which take part in the calculation of the fuel cell output voltage.

In the block diagram of Fig. 3.3, mass diffusion equations (3.1)–(3.13) are used to calculate the effective partial pressure of hydrogen and oxygen. Then, the Nernst equation (3.20), (3.24) and the overall fuel and oxidant delay effect, (3.22) and (3.23), are employed to determine the internal potential (E) of the fuel cell. The activation voltage drop equations (3.27)–(3.29), ohmic voltage drop equations (3.30), (3.31), and concentration

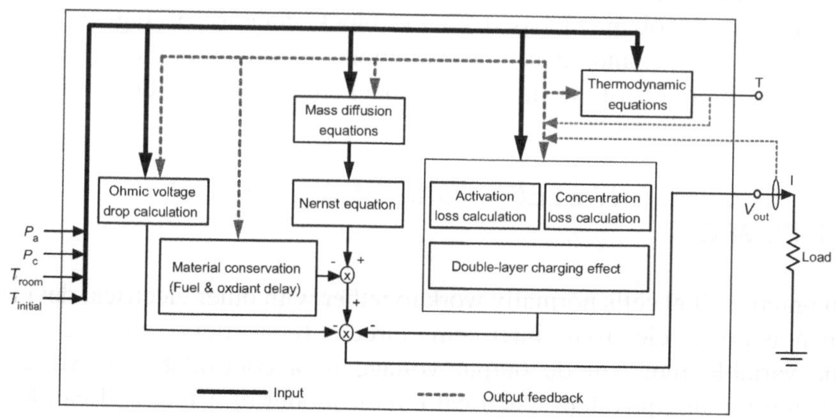

FIGURE 3.3 Block diagram for building a dynamic model of PEMFC.

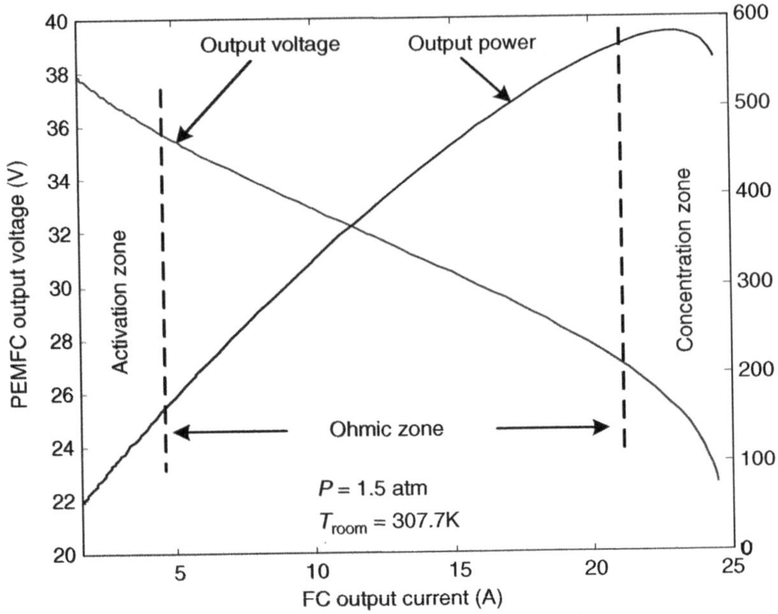

FIGURE 3.4 *V–I* and *P–I* characteristic curves for a SIMULINK-based simulation of a 500 W PEM fuel cell stack.

voltage drop equations (3.33), (3.34) together with the voltage of the equivalent capacitance of double-layer charge effect (3.35) are applied to determine the terminal (output) voltage of the PEMFC stack. Thermodynamic effects are also considered using energy balance equations (3.37)–(3.43). Figure 3.4 shows the MATLAB/SIMULUNK-based simulation of steady-state (*V–I* and *P–I*) characteristics of a 500 W PEMFC stack, obtained based on the block diagram of Fig. 3.3. The activation, ohmic, and concentration regions are shown in the figure. The dynamic responses of the simulation model are given in Section 3.6.

3.5 EQUIVALENT ELECTRICAL CIRCUIT MODEL OF PEMFC

In practice, fuel cells normally work together with other electrical devices such as power electronic interfacing circuits (converters), which convert the variable fuel cell dc output voltage to a controllable ac voltage. Therefore, an equivalent electrical circuit model of a fuel cell can be a valuable tool used with the circuit models of the interfacing components to

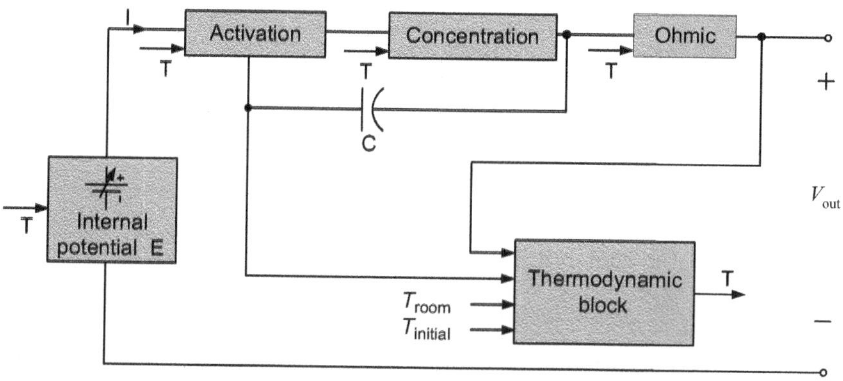

FIGURE 3.5 Block diagram for building an electrical circuit model for PEMFC.

study the performance of the fuel cell power generation unit. A block diagram to build such a model for PEMFC (which is also based on the PEMFC physical characteristics discussed in Section 3.3) is given in Fig. 3.5. The block diagram consists of the following blocks: internal potential, activation, concentration, ohmic, and thermodynamic blocks, each discussed below. C is the capacitance of the double layer charge effect discussed earlier.

Equivalent Circuit for the Internal Potential Block: According to Equations 3.20–3.26, the fuel cell internal potential E is a function of load current and temperature. An equivalent electrical circuit for the internal potential block is given in Fig. 3.6, where its input $E_0{}^\circ$ is the standard reference potential at standard state, $T = 298$ K and $P = 1$ atm, and its output is the PEMFC stack internal potential E. The current- and temperature-controlled voltage source $f_1(I,T)$ represents the current and temperature dependent part of (3.20). Note that the effective partial pressures ($p^*_{H_2}$ and $p^*_{O_2}$) given in Equation 3.20 are functions of PEMFC current and temperature; see (3.14) and (3.15). The expression for the current- and temperature-controlled voltage source $f_1(I,T)$ can be obtained by substituting (3.21) in (3.24). The current-controlled voltage source $f_2(I)$

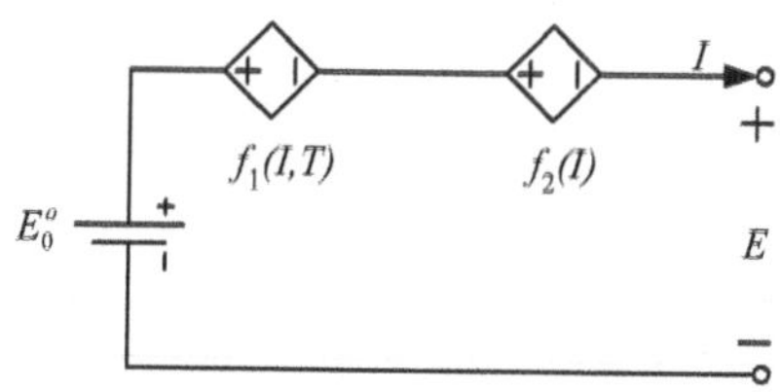

FIGURE 3.6 Electrical circuit for internal potential E.

represents $E_{d,cell}$ in (3.22) for the total number of cells (N_{cell}) in series. These expressions are given below.

$$f_1(I,T) = -\frac{N_{cell}RT}{2F} \ln\left[p^*_{H_2} \cdot (p^*_{O_2})^{0.5}\right] + N_{cell}k_E(T-298) \quad (3.44)$$

$$f_2(I) = N_{cell}E_{d,cell} \quad (3.45)$$

From Fig. 3.6, the fuel cell internal potential can be written as

$$E = E_0{}^\circ - f_1(I,T) - f_2(I) \quad (3.46)$$

Equivalent Circuit for the Activation Loss: From (3.28), the activation voltage drop can be divided into two parts, V_{act1} and V_{act2}. V_{act1} can be modeled by a constant voltage source in series with a temperature-controlled voltage source. V_{act2} can be represented by the voltage drop across a temperature- and current-dependent resistor R_{act}. This resistor is assumed to be composed of a fixed resistor (R_{act0}), a current-dependent resistor ($R_{act1}(I)$) and a current- and temperature-dependent resistor ($R_{act2}(I,T)$). A current-dependent resistor can be built, by using a polynomial current-controlled voltage source model. For example, in PSPICE, it can be built by using the polynomial current-controlled voltage source "*HPOLY*". A temperature-dependent resistor can be built in PSPICE by Analog Behavioral Models (ABMs). ABMs use mathematical and logic expressions that may consist of circuit voltages, currents, time, temperature, and other analog parameters to set their outputs.

An electrical equivalent circuit for activation voltage drop is shown in Fig. 3.7, where with reference to (3.28), V_{act1} and V_{act2} are:

$$V_{act1} = \eta_0 + f_3(T) = \eta_0 + (T-298)a \quad (3.47a)$$

$$V_{act2} = (R_{act})(I) = (R_{act0} + R_{act1} + R_{act2})(I) \quad (3.47b)$$

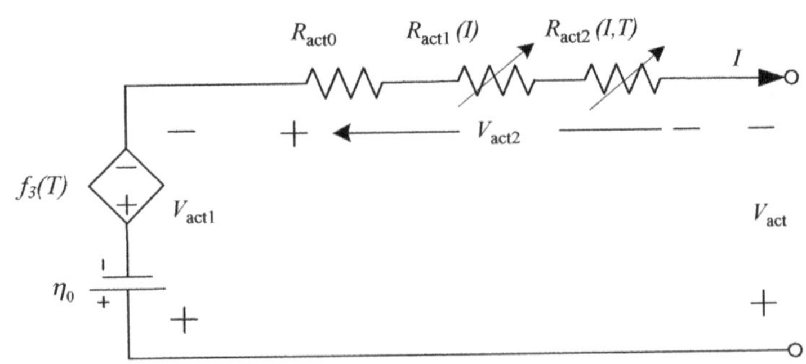

FIGURE 3.7 Electrical circuit model for activation loss in PEMFC.

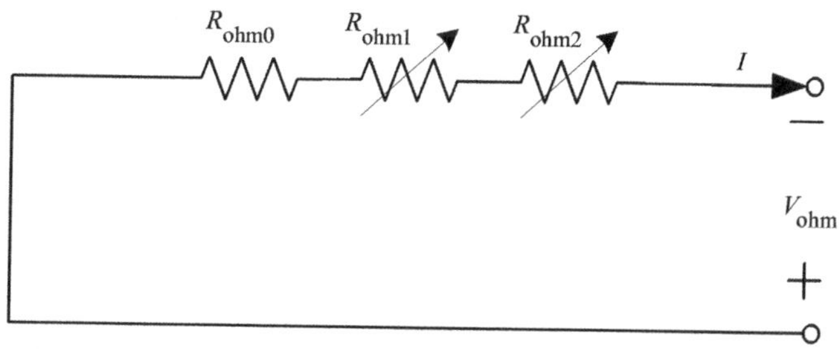

FIGURE 3.8 Electrical circuit model for ohmic voltage drop in PEMFC.

Equivalent Circuit for Ohmic Voltage Drop: Fig. 3.8 shows the equivalent electrical circuit for the fuel cell ohmic voltage drop. According to (3.31), R_{ohm} is a current- and temperature-dependent resistor, which can be modeled the same way R_{act} is modeled. From Fig. 3.8,

$$R_{ohm} = R_{ohm0} + R_{ohm1} + R_{ohm2} \tag{3.48}$$

where R_{ohm0} is the constant part of R_{ohm} with reference to (3.31); $R_{ohm1} = k_{RI}\, I$ is the current-dependent part of R_{ohm}; and $R_{ohm2} = -k_{RT}T$ is its temperature-dependent part.

Equivalent Circuit for Concentration Voltage Drop: The electrical circuit model for concentration voltage drop is shown in Fig. 3.9. With reference to (3.34), and similar to (3.48), the concentration equivalent resistance can be written as:

$$R_{conc} = \frac{V_{conc}}{I} = R_{conc0} + R_{conc1} + R_{conc2} \tag{3.49}$$

where R_{conc0} is the constant part of R_{conc}, and R_{conc1} and R_{conc2} are its current-dependent and temperature-dependent parts. There is no voltage source in the equivalent circuits for the ohmic and concentration voltage drops since there is no generation associated with these circuits; they strictly represent voltage drops.

Equivalent Circuit model for Capacitance of Double-Layer Charge Effect: In Fig. 3.5, C is the equivalent capacitance of the double-layer charge

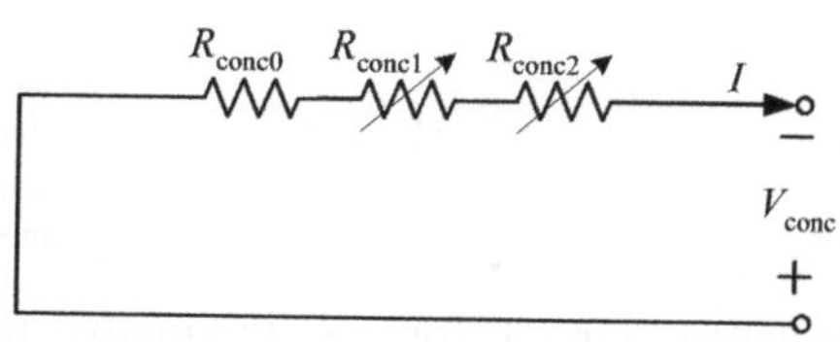

FIGURE 3.9 Electrical circuit for concentration voltage drop in PEMFC.

TABLE 3.1 Analogies Between Thermodynamic and Electrical Quantities

Electrical potential: U (V)	Temperature: T (K)
Electrical current: I (A)	Heat flow rate: P_h (W)
Electrical resistance: R (Ohm)	Thermal resistance: θ (K/W)
Electrical capacitance: C (F)	Heat capacity: C_h (J/K)
$RI = U$	$\theta P_h = T$
$I = C\frac{dU}{dt}$	$P_h = C_h\frac{dT}{dt}$

effect. This capacitance can be estimated through measurement of the dynamic response of a real fuel cell stack in a very short time range (millisecond) [2]. Figure 3.2 shows how this equivalent capacitance effect is combined with the activation, concentration, and ohmic resistances to make the fuel cell response dynamic due to the circuit time constant.

Circuit Model for the Thermodynamic Block: The analogies between the thermodynamic quantities and electrical quantities used are given in Table 3.1. These analogies can be used to develop an electrical circuit model for the thermodynamic block shown in Fig. 3.5. The thermodynamic property inside the fuel cell is described by (3.37) through (3.43). These equations can be simulated by the R–C circuit shown in Fig. 3.10. The power consumed by the activation, ohmic and concentration losses, that is $(E - V_{out})*I$, is considered as a heat source which causes the fuel cell temperature to rise. Therefore, the rate of change of the heat source ($\bar{q}_{in}$) is:

$$\dot{q}_{in} = (E - V_{out}) \cdot I \tag{3.50}$$

The thermal resistance due to air convection in the fuel cell can be written as:

$$R_T = \frac{1}{h_{cell} \cdot N_{cell} \cdot A_{cell}} \tag{3.51}$$

where h_{cell} is the convective heat transfer coefficient (W/(m^2 K)).

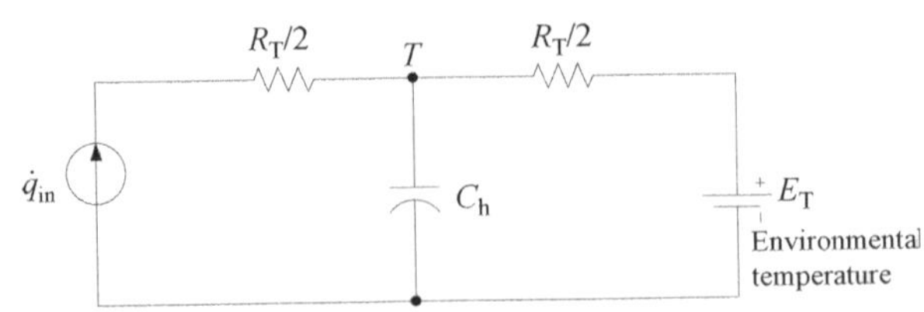

FIGURE 3.10 Equivalent circuit for the thermodynamic property inside PEMFC.

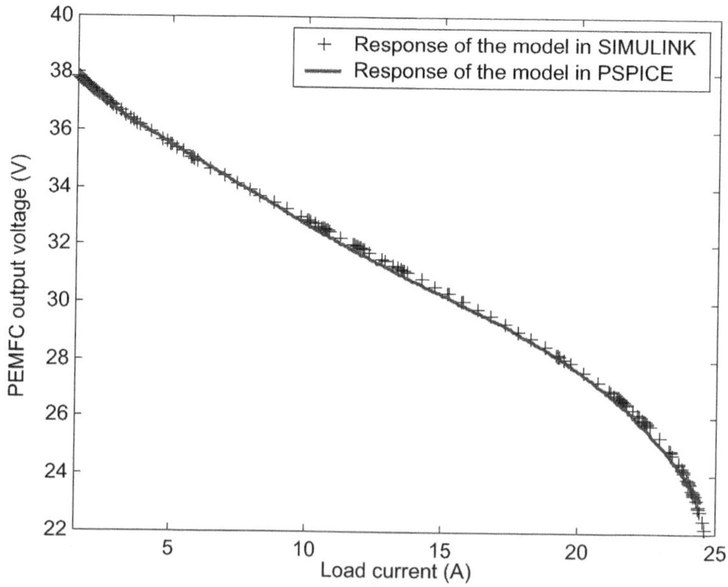

FIGURE 3.11 Comparison of *V–I* characteristic of PEMFC stack from SIMULINK and PSPICE models.

In Fig. 3.10, the constant voltage source E_T represents the environmental temperature, R_T is the fuel cell equivalent thermal resistance defined by (3.51), and C_h corresponds to the fuel cell heat capacity. The voltage across C_h corresponds to the overall temperature of the fuel cell stack, T.

The equivalent circuits given in Figs. 3.2, 3.6–3.10 can be used in a circuit-based computer simulation environment (e.g., PSPICE) to simulate the performance characteristics of PEMFC. Figures 3.11 and 3.12 show a comparison of the steady-state V–I and P–I characteristic curves of a 500 W PEMFC stack obtained from PSPICE and SIMULINK simulation.

3.6 PEMFC MODEL VALIDATION

The data reported in this section were obtained from a 500 W SR-12 PEMFC stack manufactured by Avista Labs (currently ReliOn, Inc.). Figure 3.13 shows a photo of such stack, on which the experimental work was performed. The dimensions and other specifications of this fuel cell stack are given in Table 3.2 [11]. The environmental air was driven by the SR-12's internal fan to pass through the cathode to provide oxygen, and pure hydrogen (99.9% pure) flew at manifold pressure of about 7 psi through the anode.

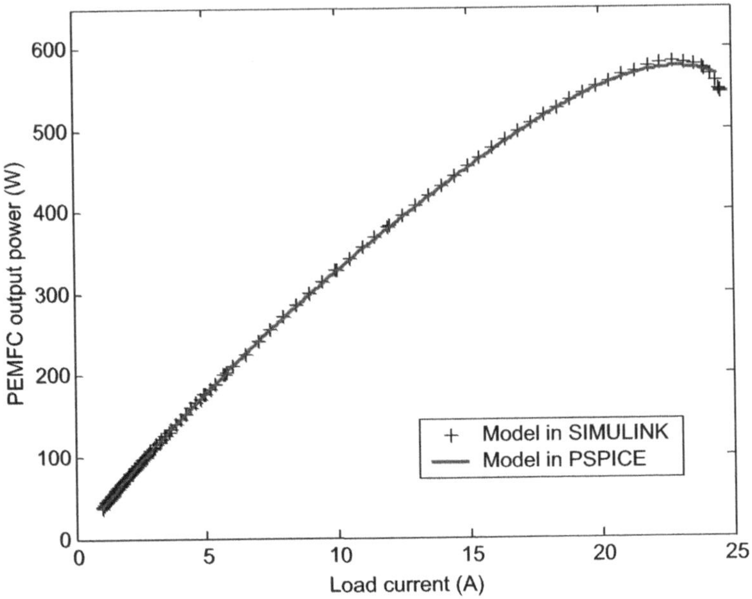

FIGURE 3.12 Comparison of *P–I* characteristic of PEMFC from the SIMULINK and PSPICE models.

FIGURE 3.13 Avista Labs (now ReliON) SR-12 500 W PEMFC stack. *Source*: Courtesy of ReliOn, Inc., Spokane, Washington.

TABLE 3.2 Specifications of SR-12 500 W PEMFC Stack [11]

Description	Value
Capacity	500 W
Number of cells	48
Operating environmental temperature	5–35°C
Operating pressures*	$P_{H_2} \approx 1.5$ atm $P_{cathode} \approx 1.0$ atm
Unit dimensions ($W \times D \times H$)	56.5 cm × 61.5 cm × 34.5 cm
Weight	44 kg

Figures 3.14 and 3.15 show comparisons of the average V-I and P-I characteristic curves for the SR-12 PEMFC stack, obtained experimentally, and the SIMULINK and PSPICE simulation results obtained for the models developed. The lower and upper curves shown in the above figures are the lower and upper ranges of the experimental data, which had ripples. The raw data with ripples have been filtered out and the average characteristic curves are shown for easier comparison. In Fig. 3.14, the

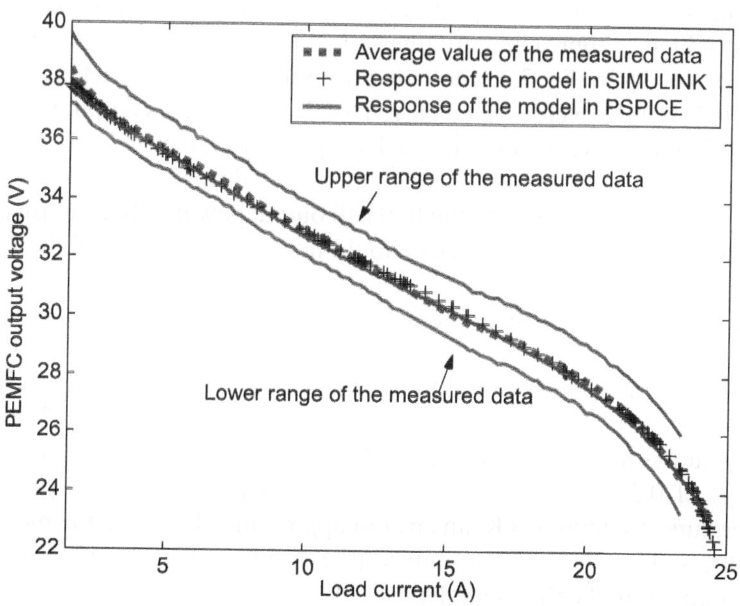

FIGURE 3.14 *V–I* characteristics of SR-12 PEMFC and the SIMULINK and PSPICE models.

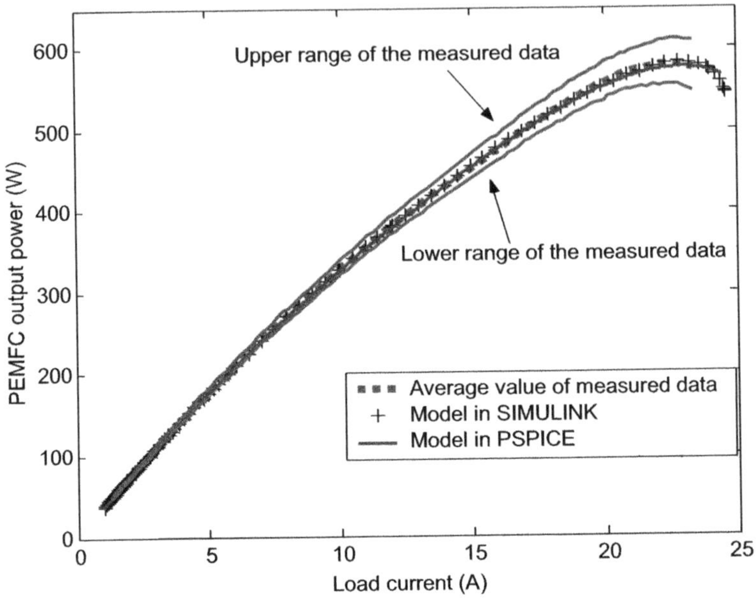

FIGURE 3.15 *P–I* characteristics of SR-12 PEMFC and the SIMULINK and PSPICE models.

voltage drops at the left end and right end of the curves are due to activation and concentration losses, respectively; the voltage drop in the middle of the curves (which is approximately linear) is due to the ohmic loss in the fuel cell stack [1–3]. In Fig. 3.15, note that the 500 W fuel cell stack model can be loaded slightly over its rated current of 20 A, where its output voltage is approximately 25 V. Beyond the peak output power, the fuel cell goes into the concentration zone, and its output voltage drops sharply as the load current increases. As a result, the fuel cell output power will decrease as the load current increases. Fuel cells should not be loaded beyond their peak power points.

The electrical model parameter values used in the simulation studies reported in this section are given in Table 3.3.

The temperature response of the PEMFC stack, obtained experimentally and from the SIMULINK and PSPICE models, are shown in Fig. 3.16. The temperature is measured as the fuel cell current increases from 1.1 to 20.5 A in steps of 0.2 A every 40 s. The models can predict the fuel cell stack temperature to within 3–4 K, an error of approximately 1%. In the measured data, the sudden decrease in temperature is due to the operation of a fan internal to the fuel cell assembly.

The steady-state characteristics of SR-12 were obtained by increasing the load current. Both the stack current and terminal current were

TABLE 3.3 Electrical Model Parameter Values for SR-12 PEMFC Stack

$E_0°$(V)	58.9
k_E (V/K)	0.00085
τ_e (s)	80.0
λ_e (Ω)	0.00333
η_0 (V)	20.145
a (V/K)	−0.1373
R_{act0} (Ω)	1.2581
R_{act2} (Ω)	$0.00112 \times (T-298)$
R_{act1} (Ω)	$-1.6777 \times 10^{-6}I^5 + 1.2232 \times 10^{-4}I^4 - 3.4 \times 10^{-3}I^3 + 0.04545I^2 - 0.3116I$
R_{conc1} (Ω)	$5.2211 \times 10^{-8}I^6 - 3.4578 \times 10^{-6}I^5 + 8.6437 \times 10^{-5}I^4 - 0.010089I^3 + 0.005554I^2 - 0.010542I$
C_h (F)	22000
R_T (Ω)	0.0347
C (F)	0.1F (4.8 F for each cell)
R_{ohm0} (Ω)	0.2793
R_{ohm1} (Ω)	$0.001872 \times I$
R_{ohm2} (Ω)	$-0.0023712 \times (T-298)$
R_{conc0} (Ω)	0.080312
R_{conc2} (Ω)	$0.0002747 \times (T-298)$

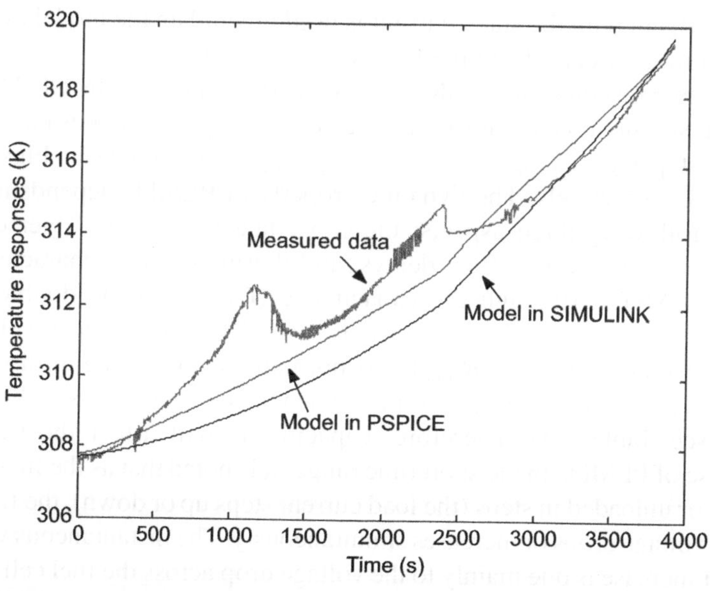

FIGURE 3.16 Temperature responses of the SR-12 PEMFC stack and the SIMULINK and SPICE models.

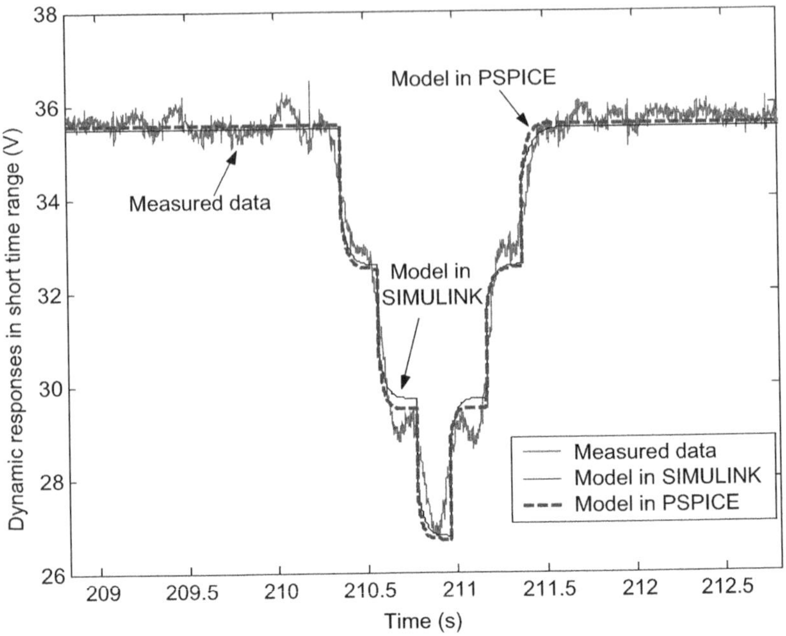

FIGURE 3.17 Transient responses of the SR-12 PEMFC and the SIMULINK and PSPICE models in short time range.

measured, but only the stack current is used to validate the models since it is the current directly from the fuel cell stack [10].

Dynamic response of the SR-12 PEMFC stack and the SIMULINK and PSPICE simulation models to a fast step load change (4 seconds range) and a slow step load change (1500 seconds range) are shown in Figs. 3.17 and 3.18, respectively. The dynamic property of PEMFC depends mainly on the following three aspects: the capacitance of double-layer charge effect, fuel and oxidant flow delays, and thermodynamic characteristics inside PEMFC. Although the capacitance (*C*) due to the double-layer change effect is large (in the order of several Farads for each cell), the time constant $\tau = (R_{\rm act} + R_{\rm conc})\,C$ is normally small (in the milliseconds range) because $(R_{\rm act} + R_{\rm conc})$ is small when the fuel cell works in the linear zone (see Table 3.3). Therefore, capacitor *C* will affect the transient response of PEMFC in the short time range. It is noted that as the fuel cell is loaded or unloaded in steps (the load current steps up or down), the fuel cell output voltage drops or increases simultaneously. The instantaneous voltage drop or increase is due mainly to the voltage drop across the fuel cell ohmic resistance, which increases as a result of loading the fuel cell and is decreased when the load on the fuel cell decreases. After the instantaneous

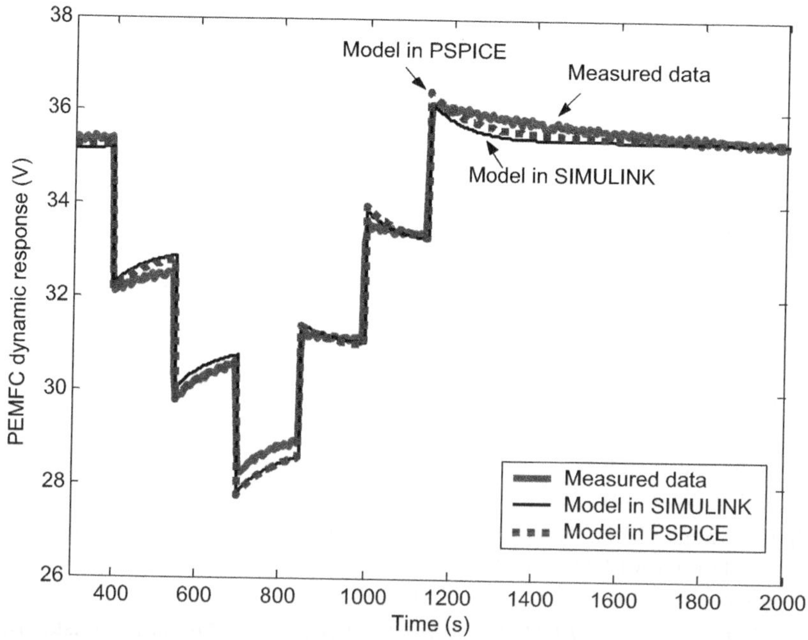

FIGURE 3.18 Transient responses of the SR-12 PEMFC and the SIMULINK and PSPICE models in long time range.

change, the fuel cell output voltage increases exponentially to its steady-state value when the fuel cell is loaded (and decreases exponentially when the fuel cell is unloaded). In the short time (millisecond) range, these exponential variations are mainly due to the time constant associated with the capacitance of double-layer charge effect. In the long time range, the PEMFC transient is mainly affected by fuel and oxidant flow delays, which can range from several tens of seconds to several minutes, and by the fuel cell thermodynamic time constant, which can be in the order of minutes.

REFERENCES

[1] *Fuel Cell Handbook,* 7th edn, EG&G Services, Inc., Science Applications International Corporation, DOE, Office of Fossil Energy, National Energy Technology Laboratory, 2004.

[2] J. Larminie and A. *Dicks Fuel Cell Systems Explained*, 2nd edn, Wiley, Chichester, UK, 2003.

[3] R. O'Hayare, S. Cha, W. Colella, and B. Fritz, *Fuel Cell Fundamentals*, Wiley, Hoboken, NJ, 2006.

[4] J.C. Amphlett, R.M. Baumert, R.F. Mann, B.A. Peppley, and P.R. Roberge, Performance modeling of the Ballard Mark IV solid polymer electrolyte fuel

cell I. Mechanistic model development, *Journal of the Electrochemical Society*, 142 (1), 1–8, 1995.

[5] J.C. Amphlett, R.M. Baumert, R.F. Mann, B.A. Peppley, and P.R. Roberge, Performance modeling of the Ballard Mark IV solid polymer electrolyte fuel cell. II. Empirical model development, *Journal of the Electrochemical Society*, 142 (1), 9–15, 1995.

[6] J.C. Amphlett, R.F. Mann, B.A. Peppley, P.R. Roberge, and A. Rodrigues, A model predicting transient responses of proton exchange membrane fuel cells, *Journal of Power Sources*, 61 (1–2), 183–188, 1996.

[7] A. Rowe and X. Li, Mathematical modeling of proton exchange membrane fuel cells, *Journal of Power Sources*, 102 (1–2), 82–96, 2001.

[8] G. Kortum, *Treatise on Electrochemistry*, 2nd edn, Elsevier, Amsterdam, 1965.

[9] G. N. Hatsopoulos and J.H. Keenan, *Principles of General Thermodynamics*, Wiley, New York, 1965.

[10] C. Wang, M.H. Nehrir, and S.R. Shaw, Dynamic models and model validation for PEM fuel cells using electrical circuits, *IEEE Transactions on Energy Conversion*, 20 (2), 442–451, 2005.

[11] *SR-12 Modular PEM Generator™ Operator's Manual*, Avista Labs (ReliOn), 2000.

CHAPTER 4

DYNAMIC MODELING AND SIMULATION OF SOLID OXIDE FUEL CELLS

4.1 INTRODUCTION

As explained in Chapter 2, Solid Oxide Fuel Cells (SOFCs) are high-temperature (600–1000°C) direct energy conversion devices, which transport the chemical energy of a fuel to electrical energy. Due to their high-temperature operation, they allow internal reforming of gaseous fuel inside the fuel cell, which gives them multifuel capability. Their energy conversion efficiency is higher than that of the traditional combustion methods and can reach up to 65%. In addition, SOFCs can be used in combined heat and power (CHP) applications. They take advantage of their high-temperature exhaust gas streams for uses in such applications as residential or commercial heating or for further electricity production, for example, in integrated SOFC-combustion turbine systems. In such cases, the overall efficiency of SOFCs can reach 80%. Their disadvantage is their relatively slow start up and thermal stresses due to the high operating temperature. Two SOFC designs were introduced in Chapter 2, namely tubular and planar designs; the tubular design is more advanced technologically while the planar design is still in the research and development state. In this chapter, the development of a physically-based model for tubular SOFC is presented. This approach can be modified

Modeling and Control of Fuel Cells, By M. Hashem Nehrir and Caisheng Wang

and adapted to planar SOFC design when more information and data are available for that design.

SOFC modeling is of interest for its performance prediction and controller design. Increased interest in SOFC power plant design and control has led to a need for appropriate application oriented SOFC models [6–10]. In this chapter, the development of a dynamic model of a tubular SOFC stack is presented based on its electrochemical and thermodynamic properties, and on the mass and energy conservation laws with emphasis on the fuel cell terminal electrical characteristics. The model consists of an electrochemical sub-model and a thermodynamic sub-model. The double-layer charge effect, which was discussed in Chapter 2, is also taken into account in the model.

The underlying fuel cell electrochemical principles discussed in Chapter 2 and some of the details discussed in the previous chapter for PEMFC, also hold true for SOFC. Some of the equations and figures are repeated for SOFC for continuity purposes and for ease of reference. Important parameters used in equations are defined the first time they are used. However, for quick reference, the symbols used in the model development, including subscripts and superscripts and their definitions are tabulated in the next section.

4.2 NOMENCLATURE (SOFC)

The symbols used in the SOFC model development are given below for ease of reference.

a	Constant in calculating material resistance (Ω m)
A	Area (m^2)
b	Constant in calculating material resistance (K)
C_i	Specific heat capacity of species i (J/(mol K))
$D_{i,j}$	Effective binary diffusivity of i–j pair (m^2/s)
E	Reversible potential of each cell (V)
E_0	Reference potential (V)
$E_0{}^\circ$	Standard reference potential (V)
F	Faraday constant (96487 C/mol)
h	Heat transfer coefficient (W/(m^2 K))
i_0	Exchange current (A)
i_{den}	Current density (A/m^2)
i, I	Current (A)
k_{a1}, k_{a2}	Empirical constants in calculating i_0
k_E	Empirical constant in calculating E_0 (V/K)

l_a	Distance from anode surface to the reaction site (m)
l_c	Distance from cathode surface to the reaction site (m)
m	Mass (kg)
M_i	Mole flow rate of species i (mol/s)
N_i	Superficial gas flux of species i (mol/(m^2 s))
N_{cell}	Number of cells in the stack
p, P	Pressure (Pa)
q_{chem}	Energy/heat (J)
R	Gas constant, 8.3143 J/(mol K), or resistance (Ω)
T	Temperature (K)
V	Volume (m^3), or voltage (V)
x	Mole fraction or axis x
z	Number of participating electrons
β	Electron transfer coefficient
δ	Length/thickness (m)
ΔH	Enthalpy change (J/mol)
ε	Emissivity (dimensionless)
η_0	Temperature invariant part of V_{act} (V)
σ	Stefan–Boltzmann constant, 5.6696×10^{-8} W/(m^2 K^4)
τ	Time constant (s)
ξ_0	Constant term of activation voltage drop (V)
ξ_1	Temperature coefficient of second term in activation voltage drop (V/K)

Superscripts and subscripts

a	Anode
act	Activation
air	Conditions for air
ann	Annulus of cell
AST	Air supply tube
c	Cathode
cell	Conditions for a single cell
ch	Conditions at the anode or cathode channel
chem	Chemical
CO_2	Carbon dioxide
conc	Concentration
consumed	Material consumed in chemical reaction
conv	Convective
elec	Electricity
elecyt	Electrolyte
gas (g)	Gas, or in gas phase
gen	Material (energy) generated in chemical reaction
flow	Flow heat exchange
fuel	Conditions for fuel

(*Continued*)

H_2	Hydrogen
H_2O	Water
in	Conditions of input/inlet
inner/outer	Inner/outer conditions
interc	Interconnection between cells
mw	Molecular weight (kg/mol)
N_2	Nitrogen
net	Net values
O_2	Oxygen
ohm	Ohmic
out	Output
rad	Radiation
*	Effective value

4.3 SOFC DYNAMIC MODEL DEVELOPMENT

In this section a physically based approach is presented for building a dynamic model for a tubular SOFC fuel cell stack. The following assumptions are made, as commonly seen in the literature [6–9,14,15], to simplify the analysis.

(1) One-dimensional treatment of flow of gas species.
(2) Oxygen ion ($O^{=}$) conducting electrolytes.
(3) Ideal gases.
(4) Hydrogen is used as fuel.
(5) Large stoichiometric quantity of oxygen (air) available at cathode.
(6) Under normal operation, partial pressures of hydrogen and oxygen are uniformly decreased along the anode channel while partial pressure of water vapor is uniformly increased.
(7) Lumped thermal capacitance is assumed in the thermodynamic analysis.
(8) The effective temperature of gases flowing in the anode and cathode channels is represented by their arithmetic mean value, that is $T_{\text{gas}}^{\text{ch}} = \left(T_{\text{gas}}^{\text{in}} + T_{\text{gas}}^{\text{out}}\right)/2$.
(9) Combustion is not modeled in the thermal model, and the fuel and air are assumed to be preheated.
(10) Individual cell parameters are lumped together to represent a fuel cell stack.

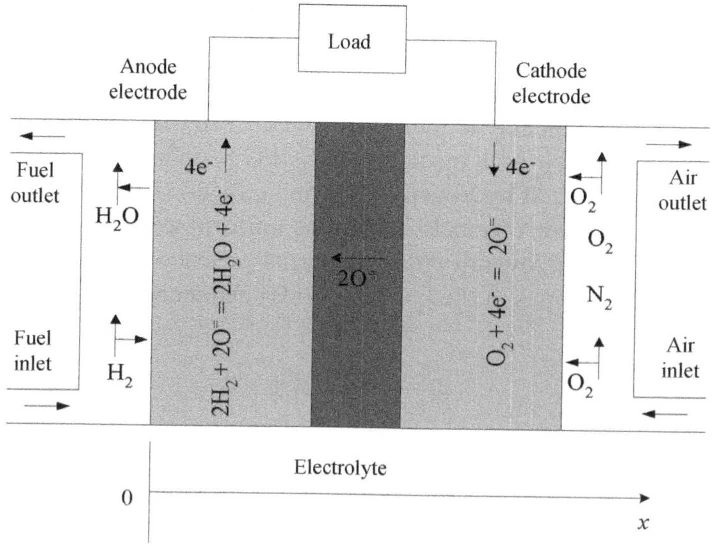

FIGURE 4.1 Schematic diagram of a solid oxide fuel cell.

The schematic diagram of a SOFC, similar to that given in Chapter 2, is repeated in Fig. 4.1, where the horizontal (x) axis from anode to cathode is also shown. Later, the variation of partial gas pressures along the x-axis will be derived and analyzed. The porous electrodes (anode and cathode) are separated by a solid ceramic electrolyte. The electrolyte material (normally dense yttria-stabilized zirconia, YSZ) is an excellent conductor for negatively charged ions ($O^=$ for SOFC) at high temperatures [1]. At the cathode, oxygen molecules accept electrons from the external circuit (electric current flows through the load) and change to oxygen ($O^=$) ions. These negative ions pass across the electrolyte and combine with hydrogen, at the anode, to produce water. Further details on the principle of operation of SOFC, can be found in [1,2], and other literature, for example [11–13].

4.3.1 Effective Partial Pressures

In this section, expressions for partial pressures of H_2, O_2 and H_2O at the anode and cathode channels of SOFC, and their effective values at actual reaction sites are derived in terms of SOFC operating parameters such as fuel and oxidant flow rates, anode and cathode inlet pressures, SOFC temperature, and SOFC physical and electrochemical parameters. The expressions for partial pressures will be used to calculate the fuel cell output voltage.

As shown in Fig. 4.1, when load current is being drawn from the fuel cell, H_2 and O_2 will diffuse through the porous anode and cathode, respectively, to reach the reaction sites. The reactant H_2O, which is the product of the chemical reactions, will then diffuse from the reaction sites to the anode channel. As a result, when SOFC is under load, gradients of the partial pressures of hydrogen, water and oxygen will be formed along the anode and cathode channels. Assuming uniform variations of these gas partial pressures, the overall effective partial pressures are given by their arithmetic mean values in the gas (electrode) channels. Hence,

$$p_{\mathrm{H_2}}^{\mathrm{ch}} = \frac{p_{\mathrm{H_2}}^{\mathrm{in}} + p_{\mathrm{H_2}}^{\mathrm{out}}}{2} \tag{4.1a}$$

$$p_{\mathrm{H_2O}}^{\mathrm{ch}} = \frac{p_{\mathrm{H_2O}}^{\mathrm{in}} + p_{\mathrm{H_2O}}^{\mathrm{out}}}{2} \tag{4.1b}$$

$$p_{\mathrm{O_2}}^{\mathrm{ch}} = \frac{p_{\mathrm{O_2}}^{\mathrm{in}} + p_{\mathrm{O_2}}^{\mathrm{out}}}{2} \tag{4.1c}$$

The effective partial pressures of hydrogen and oxygen at the actual reaction sites will be less than those in the gas flow channels due to mass diffusion. In contrast, partial pressure of the vapor at the reaction sites is higher than that in the anode flow channel. In order to calculate the fuel cell output voltage, the effective partial pressures of hydrogen, water and oxygen at the reaction sites need to be determined. As in PEM fuel cells, in a gas mixture consisting of N species, the gradient of diffusion of gas species i through the porous electrodes can be described by Stefan–Maxwell formulation as follows [3]:

$$\nabla x_i = \frac{RT}{P} \sum_{j=1}^{N} \frac{x_i N_j - x_j N_i}{D_{i,j}} \tag{4.2}$$

where ∇ is the gradient operator; x_i (x_j) the mole fractions of species i (j); $D_{i,j}$ the effective binary diffusivity of i–j pair of gas species (m^2/s); N_i(N_j) the superficial gas flux of species i (j) (mol/(m^2 s)); R the gas constant, 8.3143 J/(mol K); T the gas temperature (K); and P the overall pressure of the gas mixture (Pa).

In the anode channel, the gas stream is a mixture of hydrogen and water. In the one-dimensional transport process assumed along the x-axis as shown in Fig. 4.1, using (4.2), the diffusion of hydrogen can be written as follows:

$$\frac{\mathrm{d}x_{\mathrm{H_2}}}{\mathrm{d}x} = \frac{RT}{P_{\mathrm{a}}^{\mathrm{ch}}} \left(\frac{x_{\mathrm{H_2}} N_{\mathrm{H_2O}} - x_{\mathrm{H_2O}} N_{\mathrm{H_2}}}{D_{\mathrm{H_2,H_2O}}} \right) \tag{4.3}$$

The molar flux of hydrogen and water vapor at the anode channel can be determined by Faraday's Law as follows [4]:

$$N_{\mathrm{H_2}} = -N_{\mathrm{H_2O}} = \frac{i_{\mathrm{den}}}{2F} \tag{4.4}$$

where, i_{den} is the current density (A/m^2) and F is the Faraday's constant (96487 C/mol).

The gas at the anode channel is assumed to be a mixture of H_2 and H_2O, hence, $x_{\mathrm{H_2}} + x_{\mathrm{H_2O}} = 1$. Remembering that $\mathrm{d}p_{\mathrm{H_2}} = \mathrm{d}x_{\mathrm{H_2}} \times P_{\mathrm{a}}^{\mathrm{ch}}$ and $\mathrm{d}p_{\mathrm{H_2O}} = \mathrm{d}x_{\mathrm{H_2O}} \times P_{\mathrm{a}}^{\mathrm{ch}}$, then (4.3) and (4.4) can be combined to give the partial pressures of H_2 and H_2O with respect to x as follows:

$$\frac{\mathrm{d}p_{\mathrm{H_2}}}{\mathrm{d}x} = -\frac{RT}{D_{\mathrm{H_2,H_2O}}} \frac{i_{\mathrm{den}}}{2F} \tag{4.5}$$

$$\frac{\mathrm{d}p_{\mathrm{H_2O}}}{\mathrm{d}x} = \frac{RT}{D_{\mathrm{H_2,H_2O}}} \frac{i_{\mathrm{den}}}{2F} \tag{4.6}$$

Integrating Equations 4.5 and 4.6 with respect to x from the anode channel surface to the actual reaction site will yield

$$p_{\mathrm{H_2}}^{*} = \int_{0}^{l_{\mathrm{a}}} \frac{\mathrm{d}p_{\mathrm{H_2}}}{\mathrm{d}x} + p_{\mathrm{H_2}}^{\mathrm{ch}} = p_{\mathrm{H_2}}^{\mathrm{ch}} - \frac{RTl_{\mathrm{a}}}{2FD_{\mathrm{H_2,H_2O}}} i_{\mathrm{den}} \tag{4.7}$$

$$p_{\mathrm{H_2O}}^{*} = \int_{0}^{l_{\mathrm{a}}} \frac{\mathrm{d}p_{\mathrm{H_2O}}}{\mathrm{d}x} + p_{\mathrm{H_2O}}^{\mathrm{ch}} = p_{\mathrm{H_2O}}^{\mathrm{ch}} + \frac{RTl_{\mathrm{a}}}{2FD_{\mathrm{H_2,H_2O}}} i_{\mathrm{den}} \tag{4.8}$$

where $p_{\mathrm{H_2O}}^{\mathrm{ch}}$, $p_{\mathrm{H_2O}}^{\mathrm{ch}}$ are the partial pressures of H_2 and H_2O at the anode channel.

In the cathode channel, the oxidant is air that mainly consists of O_2, and N_2, that is, $x_{\mathrm{O_2}} + x_{\mathrm{N_2}} \approx 1$. Applying a similar procedure for the cathode, the one-dimensional Stefan–Maxwell diffusion equation for oxygen can be expressed as follows:

$$\frac{\mathrm{d}x_{\mathrm{O_2}}}{\mathrm{d}x} = \frac{RT}{P_{\mathrm{c}}^{\mathrm{ch}}} \left(\frac{x_{\mathrm{O_2}} N_{\mathrm{N_2}} - x_{\mathrm{N_2}} N_{\mathrm{O_2}}}{D_{\mathrm{O_2,N_2}}} \right) \tag{4.9}$$

Since nitrogen does not take part in the chemical reaction, the net nitrogen molar flux normal to the cathode surface is assumed to be zero ($N_{N_2} = 0$). The molar flux of O_2 is then determined by Faraday's Law as follows:

$$N_{O_2} = \frac{i_{den}}{4F} \tag{4.10}$$

The number of participating electrons is 4 for the reaction at the cathode, which is also shown in Fig. 4.1. That is, one mole of oxygen needs four moles of electrons to complete the reaction.

Equation 4.9 can then be rewritten as

$$\frac{dx_{O_2}}{dx} = \frac{RTi_{den}}{4FP_c^{ch}D_{O_2,N_2}}(x_{O_2} - 1) \tag{4.11}$$

The effective partial pressure of oxygen at the reaction site can be found similar to the analysis for the anode.[1]

$$p_{O_2}^* = P_c^{ch} - \left(P_c^{ch} - p_{O_2}^{ch}\right)\exp\left(\frac{RTi_{den}l_c}{4FP_c^{ch}D_{O_2,N_2}}\right) \tag{4.12}$$

The effective partial pressures ($p_{H_2}^*$, $p_{H_2O}^*$, and $p_{O_2}^*$), calculated from (4.7), (4.8) and (4.12), will be used in the Nernst equation to find the SOFC output voltage.

4.3.2 Material Conservation

Similar to PEMFC, the instantaneous change in the effective partial pressures of hydrogen and water vapor in the anode gas flow channel and oxygen in the cathode can be determined through the ideal gas equations as follows [5]:

$$\frac{V_a}{RT}\frac{dp_{H_2}^{ch}}{dt} = M_{H_2}^{in} - M_{H_2}^{out} - \frac{i}{2F} \tag{4.13}$$

$$\frac{V_a}{RT}\frac{dp_{H_2O}^{ch}}{dt} = M_{H_2O}^{in} - M_{H_2O}^{out} + \frac{i}{2F} \tag{4.14}$$

[1] $\int_{channel}^{reactionsite} \frac{d(x_{O_2}-1)}{x_{O_2}-1} = \int_0^{l_c} \frac{RTi_{den}}{4FP_c^{ch}D_{O_2,N_2}}dx \Rightarrow \ln\left(\frac{p_{O_2}^* - P_c^{ch}}{p_{O_2}^{ch} - -P_c^{ch}}\right) = \frac{RTi_{den}l_c}{4FP_c^{ch}D_{O_2,N2}}$

$$\frac{V_c}{RT}\frac{\mathrm{d}p_{O_2}^{ch}}{\mathrm{d}t} = M_{O_2}^{in} - M_{O_2}^{out} - \frac{i}{4F} \tag{4.15}$$

Assuming full availability of hydrogen fuel and large stoichiometric quantities of water vapor and oxygen at the cathode (assumption 5), mass flow rate for H_2, H_2O and O_2 at the inlet and outlet of the anode and cathode flow channels can be expressed as follows:

$$\begin{cases} M_{H_2}^{in} = M_a \cdot x_{H_2}^{in} = M_a \cdot \dfrac{p_{H_2}^{in}}{P_a^{ch}} \\ M_{H_2}^{out} = M_a \cdot x_{H_2}^{out} = M_a \cdot \dfrac{p_{H_2}^{out}}{P_a^{ch}} \end{cases} \tag{4.16}$$

$$\begin{cases} M_{H_2O}^{in} = M_a \cdot x_{H_2O}^{in} = M_a \cdot \dfrac{p_{H_2O}^{in}}{P_a^{ch}} \\ M_{H_2O}^{out} = M_a \cdot x_{H_2O}^{out} = M_a \cdot \dfrac{p_{H_2O}^{out}}{P_a^{ch}} \end{cases} \tag{4.17}$$

$$\begin{cases} M_{O_2}^{in} = M_c \cdot x_{O_2}^{in} = M_c \cdot \dfrac{p_{O_2}^{in}}{P_c^{ch}} \\ M_{O_2}^{out} = M_c \cdot x_{O_2}^{out} = M_c \cdot \dfrac{p_{O_2}^{out}}{P_c^{ch}} \end{cases} \tag{4.18}$$

Substituting the mole flow rate terms using (4.16)–(4.18), Equations 4.13–4.15 can be written as follows:

$$\frac{\mathrm{d}p_{H_2}^{ch}}{\mathrm{d}t} = \frac{2M_aRT}{V_aP_a^{ch}}p_{H_2}^{in} - \frac{2M_aRT}{V_aP_a^{ch}}p_{H_2}^{ch} - \frac{RT}{2FV_a}i \tag{4.19}$$

$$\frac{\mathrm{d}p_{H_2O}^{ch}}{\mathrm{d}t} = \frac{2M_aRT}{V_aP_a^{ch}}p_{H_2O}^{in} - \frac{2M_aRT}{V_aP_a^{ch}}p_{H_2O}^{ch} + \frac{RT}{2FV_a}i \tag{4.20}$$

$$\frac{\mathrm{d}p_{O_2}^{ch}}{\mathrm{d}t} = \frac{2M_cRT}{V_cP_c^{ch}}p_{O_2}^{in} - \frac{2M_cRT}{V_cP_c^{ch}}p_{O_2}^{ch} - \frac{RT}{4FV_c}i \tag{4.21}$$

The above differential equations (4.19)–(4.21), can be written in the Laplace transform domain as follows:

$$P_{H_2}^{ch}(s) = \frac{1}{(1+\tau_a s)}\left[P_{H_2}^{in}(s) + \tau_a P_{H_2}^{ch}(0) - \frac{P_a^{ch}}{4FM_a}I(s)\right] \tag{4.22}$$

$$P_{H_2O}^{ch}(s) = \frac{1}{(1+\tau_a s)}\left[P_{H_2O}^{in}(s) + \tau_a P_{H_2O}^{ch}(0) + \frac{P_a^{ch}}{4FM_a}I(s)\right] \tag{4.23}$$

$$P_{O_2}^{ch}(s) = \frac{1}{(1+\tau_c s)}\left[P_{O_2}^{in}(s) + \tau_c P_{O_2}^{ch}(0) - \frac{P_c^{ch}}{8FM}I(s)\right] \tag{4.24}$$

The time constants $\tau_a = V_a P_a^{ch}/2M_a RT$ and $\tau_c = V_c P_c^{ch}/2M_c RT$ (with unit of seconds) are associated with the pressure at the anode and cathode, respectively, giving the rate at which the pressure of hydrogen and water vapor (in the anode channel), and oxygen (in the cathode channel) vary as a result of load variations. The physical meaning of the time constant, τ_a, is that it will take τ_a seconds to fill a tank of volume $V_a/2$ at pressure P_a^{ch} if the mass flow rate is M_a. Similar physical meaning holds for τ_c.

4.3.3 SOFC Output Voltage

As explained in Chapter 2, the overall reaction in a solid oxide fuel cell is

$$2H_2 + O_2 = 2H_2O_{(g)} \tag{4.25}$$

where subscript "g" indicates the product H_2O is in gas form. The Nernst equation used to calculate the reversible potential for SOFC, as derived in Chapter 2, is

$$E_{cell} = E_{0,cell} + \frac{RT}{4F}\ln\left[\frac{\left(p_{H_2}^{ch}\right)^2 \cdot p_{O_2}^{ch}}{\left(p_{H_2O}^{ch}\right)^2}\right] \tag{4.26}$$

$E_{0,cell}$ consists of a constant term and a temperature-dependent term as follows [1]:

$$E_{0,cell} = E_{0,cell}{}^{\circ} - k_E(T-298) \tag{4.27}$$

where, $E_{0,cell}{}^{\circ}$ is the standard reference potential at standard state, 298 K and 1 atm pressure.

E_{cell}, calculated from (4.26), is the SOFC open-circuit voltage, which is both temperature- and pressure-dependent. When the fuel cell is under load, its output voltage is less than E_{cell} due to activation loss, ohmic resistance voltage drop and concentration overpotential. As was for PEMFC, the SOFC output voltage can therefore be written as:

$$V_{cell} = E_{cell} - V_{act,cell} - V_{ohm,cell} - V_{conc,cell} \tag{4.28}$$

The output voltage of the individual cells connected in series can be lumped together (assumption 10) to obtain the output voltage of a SOFC stack as follows:

$$V_{\rm out} = N_{\rm cell} V_{\rm cell} = E - V_{\rm act} - V_{\rm ohm} - V_{\rm conc} \qquad (4.29)$$

The SOFC stack output voltage can therefore be calculated, once the voltage drops are calculated. The expressions for these voltage drops are derived next.

4.3.3.1 Activation Voltage Drop At low currents, when chemical reaction starts to take place inside SOFC, a voltage loss (drop) results due to an activation energy barrier that must be overcome. This drop is called the *activation voltage drop*. The Butler–Volmer equation (4.30) is used to calculate the activation voltage drop [6]:

$$i = i_0 \left\{ \exp\left(\beta \frac{zFV_{\rm act}}{RT} \right) - \exp\left(-(1-\beta) \frac{zFV_{\rm act}}{RT} \right) \right\} \qquad (4.30)$$

where i_0 is the apparent exchange current (A); β the electron transfer coefficient ($\beta \approx 0.5$ for SOFC); z the number of participating electrons.

The apparent exchange current is a function of temperature, and can be expressed as an exponential function as follows [14,15]:

$$i_0 = k_{\rm a1} T \exp\left(-\frac{k_{\rm a2}}{T} \right) \qquad (4.31)$$

where $k_{\rm a1}$, $k_{\rm a2}$ are empirical constants.

Under high activation condition, the first term in (4.30) is much larger than the second term and the activation voltage drop for each cell can be obtained by neglecting the second term in (4.30) as follows [1]:

$$V_{\rm act,cell} = \frac{RT}{z\beta F} \ln\left(\frac{i}{i_0} \right) \qquad (4.32)$$

Equation 4.32, which is the same as (3.27), is the well-known Tafel equation, is only good for high activation conditions. Otherwise, according to (4.32), the value of $V_{\rm act}$ approaches ∞ at $i = 0$, which is unreasonable.

The value of β is about 0.5 for fuel cell applications[14], and simplifying (4.30), the cell activation voltage drop (4.30) can be written as follows:[2]

[2] $i = i_0 \{ \exp(0.5 \frac{zFV_{\rm act}}{RT}) - \exp(-0.5 \frac{zFV_{\rm act}}{RT}) \} \Rightarrow i = 2i_0 \, sinh(\frac{zFV_{\rm act}}{2RT})$

$$V_{\text{act,cell}} = \frac{2RT}{zF} \sinh^{-1}\left(\frac{i}{2i_0}\right) \tag{4.33}$$

The equivalent activation resistance can then be defined as

$$R_{\text{act,cell}} = \frac{V_{\text{act,cell}}}{i} = \frac{2RT}{zFi} \sinh^{-1}\left(\frac{i}{2i_0}\right) \tag{4.34}$$

According to (4.33), the activation voltage drop is zero when load current is zero. However, even at open-circuit, SOFC output voltage is less than the theoretical value given by (4.26) [2]. Therefore, similar to the calculation of the activation voltage developed for PEMFC [3], a constant and a temperature-dependent term can also be added to (4.34) for calculating the SOFC activation voltage drop.

$$V_{\text{act,cell}} = \xi_0 + \xi_1 T + iR_{\text{act,cell}} = V_{\text{act0,cell}} + V_{\text{act1,cell}} \tag{4.35}$$

In (4.35), ξ_0 is the constant term of activation voltage drop (V) and ξ_1 the temperature coefficient (V/K); $V_{\text{act0,cell}} = \xi_0 + \xi_1 T$ is the part of activation drop affected only by the fuel cell internal temperature, while $V_{\text{act1,cell}} = iR_{\text{act,cell}}$ is both current- and temperature-dependent.

Ohmic Voltage Drop The ohmic resistance of SOFC consists mainly of the resistance of the electrodes, electrolyte, and the interconnection between cells. The overall ohmic voltage drop inside SOFC can be expressed as

$$V_{\text{ohm,cell}} = V_{\text{electrodes}} + V_{\text{ohm,elecyt}} + V_{\text{ohm,interc}} = iR_{\text{ohm,cell}} \tag{4.36}$$

In general, the resistance of electrodes is negligible compared to that of electrolyte and interconnects; hence, only ohmic losses of electrolyte and interconnects are included in calculating the ohmic voltage drop. The ohmic resistance of electrolyte and interconnects decreases exponentially with temperature, and $R_{\text{ohm,cell}}$ can be written as follows:

$$R_{\text{ohm,cell}} = \frac{a_{\text{elecyt}} \exp(b_{\text{elecyt}}/T)}{A_{\text{cell}}} \delta_{\text{elecyt}} + \frac{a_{\text{interc}} \exp(b_{\text{interc}}/T)}{A_{\text{interc}}} \delta_{\text{interc}} \tag{4.37}$$

where a (Ω m) and b (K) are constants; δ is the thickness (m); A the effective cross-sectional area (m^2).

Concentration Voltage Drop During the reaction process, concentration gradients can be formed due to mass diffusion from the flow channels to the reaction sites. As a result, the effective partial pressures of hydrogen and oxygen at the reaction site are less than those in the electrode channels. On the contrary, the effective partial pressure of water vapor at the reaction site is higher than that in the anode channel since water is produced at the reaction sites. Therefore, the actual internal voltage of the fuel cell is smaller than that obtained from Equation 4.26, still using the channel partial pressures. The difference is called concentration voltage drop or concentration overpotential. At high current densities, due to the slow diffusion process, reactants cannot be transported to the reaction site (and products cannot be transported from the reaction site) as fast as needed. As a result, the concentration voltage drop becomes more significant [1,2]. The concentration voltage drop in the fuel cell can be obtained as

$$V_{\mathrm{conc,cell}} = \frac{RT}{4F}\left\{\ln\left[\frac{\left(p_{\mathrm{H_2}}^{\mathrm{ch}}\right)^2 \cdot p_{\mathrm{O_2}}^{\mathrm{ch}}}{\left(p_{\mathrm{H_2O}}^{\mathrm{ch}}\right)^2}\right] - \ln\left[\frac{\left(p_{\mathrm{H_2}}^{*}\right)^2 \cdot p_{\mathrm{O_2}}^{*}}{\left(p_{\mathrm{H_2O}}^{*}\right)^2}\right]\right\} = V_{\mathrm{conc,a}} + V_{\mathrm{conc,c}} \tag{4.38}$$

where,

$$V_{\mathrm{conc,a}} = \frac{RT}{2F}\ln\left[\frac{1+\left(RTl_{\mathrm{a}}i_{\mathrm{den}}\right)/\left(2FD_{\mathrm{H_2O,H_2}}p_{\mathrm{H_2O}}^{\mathrm{ch}}\right)}{1-\left(RTl_{\mathrm{a}}i_{\mathrm{den}}\right)/\left(2FD_{\mathrm{H_2O,H_2}}p_{\mathrm{H_2}}^{\mathrm{ch}}\right)}\right] \tag{4.39}$$

$$V_{\mathrm{conc,c}} = -\frac{RT}{4F}\ln\left\{\frac{1}{p_{\mathrm{O_2}}^{\mathrm{ch}}}\left[p_{\mathrm{c}}^{\mathrm{ch}} - \left(p_{\mathrm{c}}^{\mathrm{ch}} - p_{\mathrm{O_2}}^{\mathrm{ch}}\right)\exp\left(\frac{RTi_{\mathrm{den}}l_{\mathrm{c}}}{4Fp_{\mathrm{c}}^{\mathrm{ch}}D_{\mathrm{O_2,N_2}}}\right)\right]\right\} \tag{4.40}$$

Note that the concentration voltage drop, obtained from Equation 4.38, is the difference between two reversible potentials calculated using the Nernst equation (4.26) with two sets of parameters. One set uses the channel partial pressures, and the other uses the actual effective partial pressures obtained through the mass diffusion equations (4.7), (4.8) and (4.12). $V_{\mathrm{conc,a}}$ is the part of the concentration voltage drop due to the differences in the partial pressures of H_2 and H_2O while $V_{\mathrm{conc,c}}$ is the part due to the partial pressure of oxygen.

The equivalent resistance for the concentration voltage drop can be determined by dividing the concentration voltage drop (4.38) by the SOFC current.

$$R_{\mathrm{conc,cell}} = \frac{V_{\mathrm{conc,cell}}}{i} \tag{4.41}$$

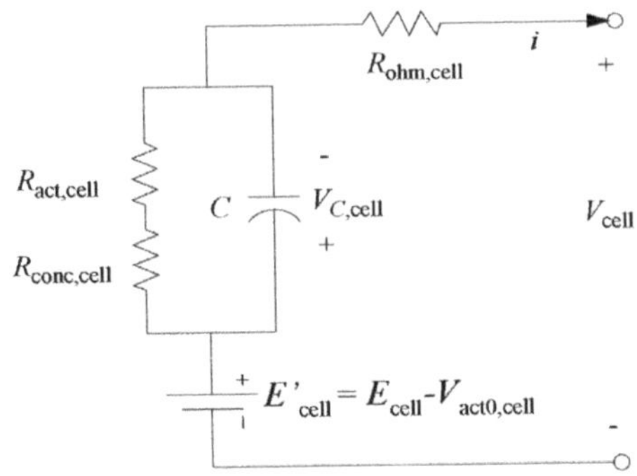

FIGURE 4.2 Equivalent electrical circuit of the double-layer charge effect inside a SOFC.

Double-Layer Charge Effect As explained for PEMFC, in a SOFC also, the two electrodes are separated by the electrolyte, and two boundary layers are formed, that is, anode-electrolyte layer and electrolyte-cathode layer (Fig. 4.1). Due to the polarization effect, these layers, known as electrochemical double-layer charge effect, can store electrical energy and behave like a super-capacitor. The model for SOFC considering this effect is similar to that for PEMFC and is given by the equivalent circuit of Fig. 4.2 [2].

In the above circuit, C is the equivalent capacitance of double-layer charge effect, and $R_{act,cell}$, $R_{ohm,cell}$, and $R_{conc,cell}$ are equivalent resistances of activation, ohmic and concentration voltage drops, which can be calculated according to (4.34), (4.37) and (4.41), respectively. Since the SOFC electrodes are porous, the value of C is large and can be in the order of several Farads [2]. In Fig. 4.2, the voltage across C can be written as:

$$V_{C,cell} = (i - i_C)(R_{act,cell} + R_{conc,cell}) = \left(i - C\frac{dV_{C,cell}}{dt}\right)(R_{act,cell} + R_{conc,cell}) \tag{4.42}$$

Therefore, the SOFC output voltage (V_{cell}) turns out to be

$$V_{cell} = E_{cell} - V_{C,cell} - V_{act0,cell} - V_{ohm,cell} \tag{4.43}$$

The output voltage of SOFC stack (V_{out}) is then obtained according to (4.29).

4.3.4 Thermodynamic Energy Balance for Tubular SOFC

The structure of tubular SOFC was presented in Chapter 2. The cross section and heat transfer (flow) inside a tubular SOFC are shown in

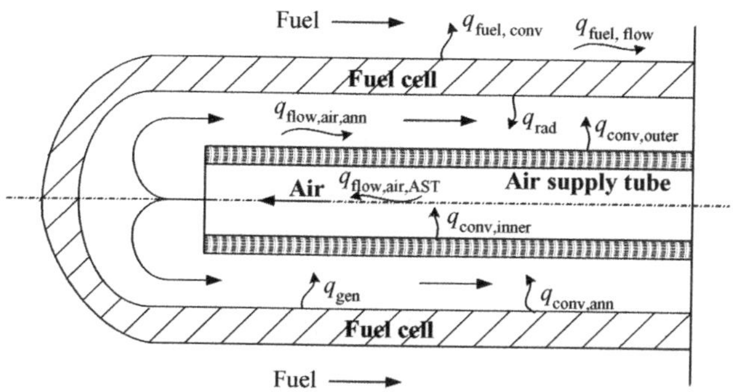

FIGURE 4.3 Heat flow and transfer inside a tubular solid oxide fuel cell.

Fig. 4.3. One advantage of this tubular structure is that it eliminates the seal problems between cells since the support tube of each cell is closed at one end [1]. The air is fed through a central air supply tube (AST) and forced to flow back past the interior of the cell (cathode surface) to the open end. The fuel (hydrogen and water vapor mixture) flows past the exterior of the cell (anode surface) and in parallel with air [1, 10].

The temperature difference between adjacent fuel cells can be neglected, and heat transport inside each fuel cell occurs mainly by means of radiation, convection and mass flow. The thermal analysis presented below is for the fuel cell only. Other components, such as fuel reformer, combustor, and heat exchanges between cells are not included in the model.

The heat generation rate by the chemical reaction inside SOFC can be written as

$$\dot{q}_{gen} = \bar{q}_{chem} - \dot{q}_{elec} \tag{4.44}$$

The available power released due to the chemical reaction is calculated by

$$\dot{q}_{chem} = \dot{n}_{H_2,consumed} \cdot \Delta H \tag{4.45}$$

where ΔH is the change in enthalpy of the chemical reaction inside SOFC. The fuel cell electrical output power is defined as

$$\dot{q}_{elec} = V_{out} \cdot i \tag{4.46}$$

The thermodynamic and energy balance analysis for the different parts inside the tubular SOFC, as shown in Fig. 4.3, are discussed below.

4.3.4.1 The Fuel Cell Tube

$$\dot{q}_{in,cell} = \dot{q}_{gen} = \dot{q}_{chem} - \dot{q}_{elec} \tag{4.47a}$$

$$\dot{q}_{\text{out,cell}} = \dot{q}_{\text{rad}} + \dot{q}_{\text{conv,ann}} + \dot{q}_{\text{flow,air,ann}} + \dot{q}_{\text{conv,fuel}} + \dot{q}_{\text{flow,fuel}} \tag{4.47b}$$

$$\dot{q}_{\text{net,cell}} = \dot{q}_{\text{in,cell}} - \dot{q}_{\text{out,cell}} = m_{\text{cell}} C_{\text{cell}} \frac{dT_{\text{cell}}}{dt} \tag{4.47c}$$

$$\dot{q}_{\text{rad}} = \varepsilon^{*}_{\text{AST}} \sigma A_{\text{AST,outer}} (T^4_{\text{cell}} - T^4_{\text{AST}}) \tag{4.47d}$$

$$\dot{q}_{\text{conv,ann}} = h_{\text{cell}} A_{\text{cell,inner}} (T_{\text{cell}} - T_{\text{air,ann}}) \tag{4.47e}$$

$$\dot{q}_{\text{flow,air,ann}} = M_{\text{air}} M_{\text{mw,air}} C_{\text{air}} (T_{\text{in,air,ann}} - T_{\text{out,air,ann}}) \tag{4.47f}$$

$$\dot{q}_{\text{conv,fuel}} = h_{\text{cell}} A_{\text{cell,outer}} (T_{\text{cell}} - T_{\text{fuel}}) \tag{4.47g}$$

$$\dot{q}_{\text{flow,fuel}} = \dot{q}_{\text{flow,H}_2} + \dot{q}_{\text{flow,H}_2\text{O}} \tag{4.47h}$$

$$\dot{q}_{\text{flow,H}_2} = (M^{\text{in}}_{\text{H}_2} + M^{\text{out}}_{\text{H}_2})(T^{\text{out}}_{\text{fuel}} - T^{\text{in}}_{\text{fuel}}) C_{\text{H}_2} M_{\text{mw,H}_2} \tag{4.47i}$$

$$\dot{q}_{\text{flow,H}_2\text{O}} = (M^{\text{in}}_{\text{H}_2\text{O}} + M^{\text{out}}_{\text{H}_2\text{O}})(T^{\text{out}}_{\text{fuel}} - T^{\text{in}}_{\text{fuel}}) C_{\text{H}_2\text{O}} M_{\text{mw,H}_2\text{O}} \tag{4.47j}$$

4.3.4.2 Fuel

$$\dot{q}_{\text{in,fuel}} = \dot{q}_{\text{conv,fuel}} + \dot{q}_{\text{flow,fuel}} \tag{4.48a}$$

$$\dot{q}_{\text{out,fuel}} = \dot{q}_{\text{flow,fuel}} \tag{4.48b}$$

$$\dot{q}_{\text{net,fuel}} = \dot{q}_{\text{in,fuel}} - \dot{q}_{\text{out,fuel}} = m_{\text{fuel}} C_{\text{fuel}} \frac{dT_{\text{fuel}}}{dt} \tag{4.48c}$$

4.3.4.3 Air Between Cell and Air Supply Tube (AST)

$$\dot{q}_{\text{in,air,ann}} = \dot{q}_{\text{conv,cell,ann}} + \dot{q}_{\text{flow,air,ann}} \tag{4.49a}$$

$$\dot{q}_{\text{out,air,ann}} = \dot{q}_{\text{flow,air,ann}} \tag{4.49b}$$

$$\dot{q}_{\text{net,air,ann}} = \dot{q}_{\text{in,air,ann}} - \dot{q}_{\text{out,air,ann}} = m_{\text{air,ann}} C_{\text{air}} \frac{dT_{\text{air,ann}}}{dt} \tag{4.49c}$$

4.3.4.4 Air Supply Tube

$$\dot{q}_{\text{in,AST}} = \dot{q}_{\text{rad}} + \dot{q}_{\text{conv,AST,outer}} \tag{4.50a}$$

$$\dot{q}_{\text{out,AST}} = \dot{q}_{\text{conv,AST,inner}} + \dot{q}_{\text{flow,air,AST}} \tag{4.50b}$$

$$\dot{q}_{\text{net,AST}} = \dot{q}_{\text{in,AST}} - \dot{q}_{\text{out,AST}} = C_{\text{AST}} m_{\text{AST}} \frac{\mathrm{d}T_{\text{AST}}}{\mathrm{d}t} \tag{4.50c}$$

$$\dot{q}_{\text{conv,AST,outer}} = h_{\text{AST,outer}} A_{\text{AST,outer}} (T_{\text{air,cell}} - T_{\text{AST}}) \tag{4.50d}$$

$$\dot{q}_{\text{conv,AST,inner}} = h_{\text{AST,inner}} A_{\text{AST,inner}} (T_{\text{AST}} - T_{\text{air,AST}}) \tag{4.50e}$$

$$\dot{q}_{\text{air,flow,AST}} = M_{\text{air}} C_{\text{air}} (T_{\text{in,air,AST}} - T_{\text{out,air,AST}}) \tag{4.50f}$$

4.3.4.5 Air in AST

$$\dot{q}_{\text{in,air,AST}} = \dot{q}_{\text{out,AST}} = \dot{q}_{\text{conv,AST,inner}} + \dot{q}_{\text{flow,air,AST}} \tag{4.51a}$$

$$\dot{q}_{\text{out,air,AST}} = \dot{q}_{\text{flow,air,AST}} \tag{4.51b}$$

$$\dot{q}_{\text{net,air,AST}} = \dot{q}_{\text{in,air,AST}} - \dot{q}_{\text{out,air,AST}} = C_{\text{air}} m_{\text{air,AST}} \frac{\mathrm{d}T_{\text{air,AST}}}{\mathrm{d}t} \tag{4.51c}$$

The symbols in the above equations (4.47)–(4.51) are defined below. m is the mass (kg); C_i the specific heat capacity of species i (J/(mol K), or J/(kg K)); h the heat transfer coefficient (W/(m^2 K)); A the area (m^2); M_i the mole flow rate of species i (mol/s); $M_{\text{mw},i}$ the molecular weight of species i (kg/mol); ε the emissivity; σ the Stefan–Boltzmann constant, 5.6696×10^{-8} W/(m^2 K^4) The superscripts and subscripts in the above equations are defined as:

air	Conditions for air
ann	Annulus of cell
AST	Air supply tube
cell	Conditions for a single cell
ch	Conditions at the anode or cathode channel
chem	Chemical
conv	Convective
conc	Concentration
consumed	Material consumed in chemical reaction

(*Continued*)

gen	Material (energy) generated in chemical reaction
flow	Heat exchange due to mass flow
fuel	Conditions for fuel
H_2	Hydrogen
H_2O	Water
in	Conditions of input/inlet
inner/outer	Inner/outer conditions
mw	Molecular weight (kg/mol)
net	Net values
O_2	Oxygen
out	Output
rad	Radiation
*	Effective value

The thermal model for the tubular SOFC can be developed according to (4.44)–(4.51).

4.4 SOFC DYNAMIC MODEL STRUCTURE

A dynamic model for a tubular SOFC stack can be developed based on the electrochemical and thermodynamic characteristics of SOFC discussed in the previous section. The cell/stack output voltage depends on conditions including fuel composition, fuel and oxidant flows, anode and cathode pressures, electrical and thermal properties of the cell materials, cell temperature, and load current.

Figure 4.4 shows a block diagram (similar to that for PEMFC), based on which a SOFC computer model can be developed. There are seven

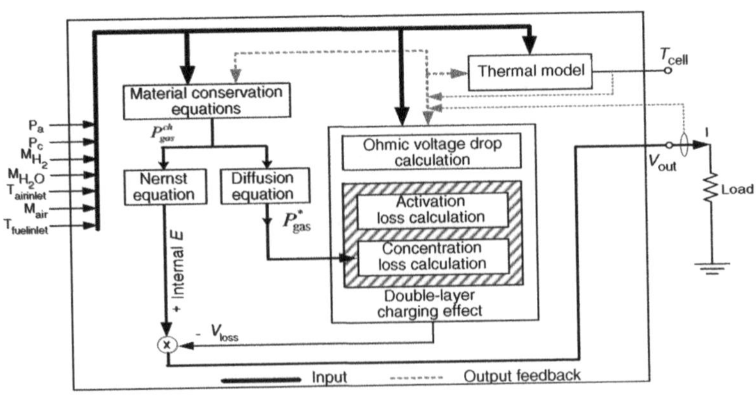

FIGURE 4.4 Block diagram for building a dynamic model of SOFC.

input quantities: anode and cathode pressures (P_a and P_c), fuel (hydrogen) flow rate (M_{H_2}), water vapor flow rate (M_{H_2O}), air flow rate (M_{air}) and the initial fuel cell and air temperature ($T_{fuelinlet}$ and $T_{airinlet}$). These input quantities are fed to different blocks shown in Fig. 4.4, namely the material conservation block, the voltage drop block, and the thermodynamic block. The cell temperature (T_{cell}) is determined at any given load current and time, and both the load current and temperature are fed back to the different blocks, which take part in the calculation of the fuel cell output voltage.

The material conservation block is based on Equations 4.13–4.24, which are used to calculate the partial pressures of hydrogen, water vapor, and oxygen in the gas flow channels. Then, the Nernst equation (4.26) is used to calculate SOFC internal potential (E). The material conservation equations along with diffusion equations (4.2)–(4.12) will give the SOFC concentration loss. Equations 4.35, 4.36, and (4.38) are used to calculate the activation, ohmic and concentration voltage drops. The SOFC terminal voltage is determined based on (4.29) and (4.43), and the voltage due to the capacitance of double-layer charge effect, (4.42). The thermal model is developed via energy balance equations (4.44)–(4.51).

The steady-state and dynamic responses of a 5-kW SOFC stack model, implemented in MATLAB/SIMULINK, are presented in the next section.

4.5 SOFC MODEL RESPONSE—CONSTANT FUEL FLOW OPERATION

In this section the steady-state and dynamic responses of a 5-kW SOFC stack model obtained from the SOFC model (whose development was presented in the previous section) are given. These responses are under constant fuel flow (constant hydrogen flow) operation, that is, constant hydrogen flow regardless of the amount of load on the SOFC stack. The effects of SOFC operating pressure and temperature on its response are also given.

4.5.1 Steady-State Characteristics

The steady-state terminal voltage versus current (V–I) characteristic curves of the SOFC stack model at different temperatures are shown in Fig. 4.5 [16]. As discussed in Section 4.3.3, in the low-current region, a voltage drop results due to an activation energy barrier that must be overcome before the chemical reaction starts inside SOFC. This voltage

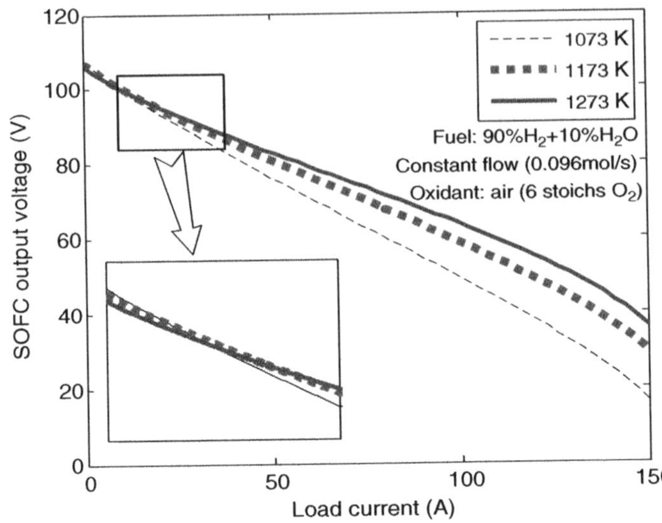

FIGURE 4.5 *V–I* characteristics of the SOFC model at different temperatures.

drop, called the SOFC activation voltage drop, dominates the total SOFC voltage drop at low currents. As load current increases, the ohmic voltage drop (IR_{ohm}) increases proportional to the current and dominates the total SOFC voltage drop, as shown in Fig. 4.6. When the load current exceeds a certain (theoretical maximum) value, the SOFC output voltage and power

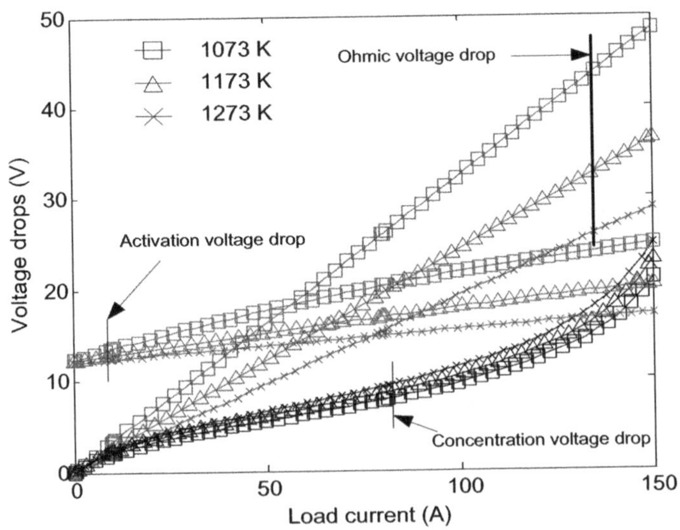

FIGURE 4.6 SOFC stack activation, ohmic, and concentration voltage drops as a function of load current at different temperatures.

delivering capability will drop sharply due to large ohmic and concentration voltage drops.

Figure 4.5 shows the effect of temperature on the SOFC *V–I* characteristic. In the low-current region, the SOFC output voltage is slightly higher at lower temperatures, while in the high current region, the voltage is higher at higher temperatures. The main reason for this is the negative term in the expression for the open-circuit internal potential $E_{0,\mathrm{cell}}$, Equation 4.27, and also the temperature-dependent activation and ohmic voltage drops. As shown in Fig. 4.6, both the activation and ohmic voltage drops increase with load current while they decrease as the SOFC temperature increases.

The output power versus current (*P–I*) curves of the model at different temperatures are shown in Fig. 4.7. At higher load currents, higher output power can be achieved at higher operating temperatures. Under each operating temperature, there is a critical current, where the model output power reaches its maximum value. This critical current is 95 A at 1073 K, 110 A at 1173 K, and 120 A at 1273 K. Beyond these points, an increase in the load current will reduce the output power due to large ohmic and concentration voltage drops.

The steady-state *V–I* and *P–I* characteristics of the 5-kW (96-cell) SOFC stack model were compared with the steady-state performance data from a General Electric (GE) 40-cell SOFC stack reported at the 2006 Solid State

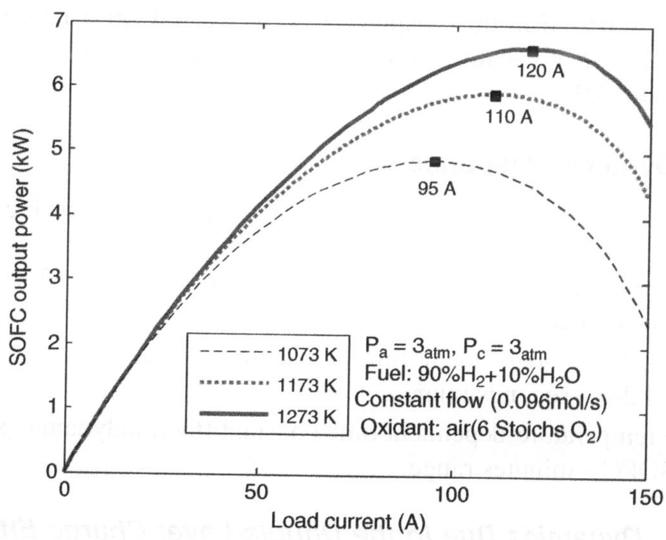

FIGURE 4.7 *P–I* characteristics of the SOFC stack model at different temperatures.

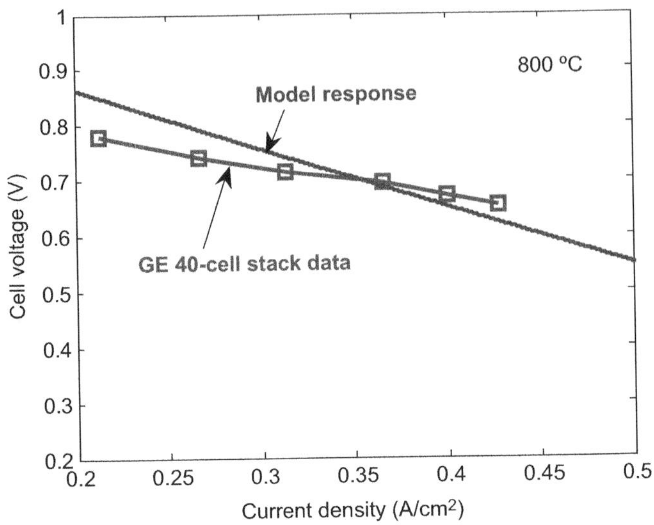

FIGURE 4.8 Steady-state SOFC cell voltage versus current density—comparison of the model response and GE data.

Energy Conversion Alliance (SECA) Annual Review Meeting [18]. The actual GE's SOFC stack data were not available to use in the simulation model for model validation purposes. Never the less, the steady-state cell voltage versus current density and power density versus current density curves obtained from the simulation model (using the SOFC parameters previously reported in this chapter) were in general agreement with the actual SOFC steady-state characteristics reported by GE, as shown in Figs. 4.8 and 4.9.

4.5.2 Dynamic Response

The SOFC dynamic response is affected mainly by the following:

- The time constant due to the double-layer charge effect—milli-seconds range.
- The pressure-dependent time constants τ_a and τ_c (see Eqs. 4.22 and 4.24)—seconds range.
- The temperature-dependent time constant (thermodynamic property) of SOFC—minutes range.

4.5.2.1 Dynamics Due to the Double-Layer Charge Effect In practice, the capacitance of the double-layer charge effect (C) is large (in the order of several Farads), and the sum of the activation and ohmic resistances ($R_{act,cell} + R_{conc,cell}$) is small (a few m$\Omega$ for a single cell) when

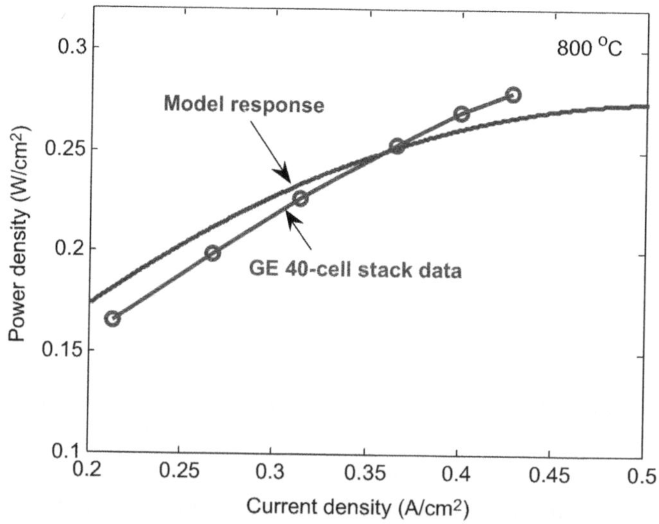

FIGURE 4.9 Steady-state SOFC power density versus current density—comparison of the model response and GE data.

SOFC operates in the linear zone (Fig. 4.5). The time constant of the double-layer charge effect, $\tau_{dlc} = (R_{act,cell} + R_{conc,cell})C$, is normally in the order of 10^{-2}–10^{-3} s. Therefore, this time constant will only affect the SOFC dynamic response in the range of a few to tens of milliseconds. Figure 4.10

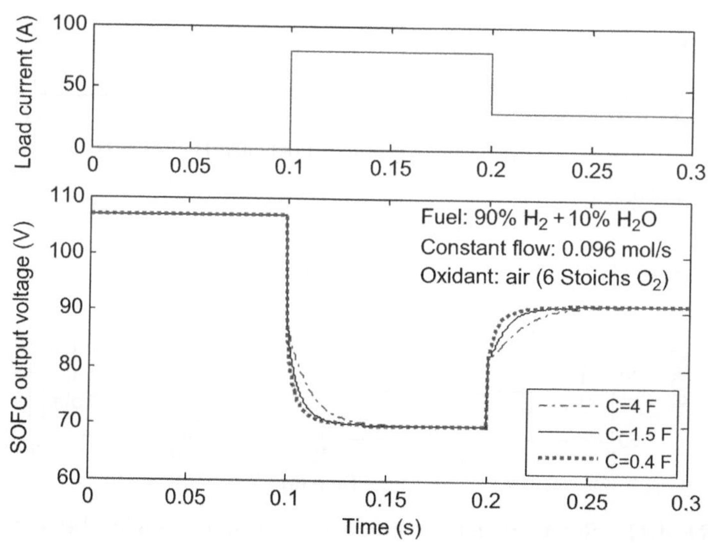

FIGURE 4.10 SOFC model dynamic voltage response due to the double-layer charge effect.

shows the model dynamic response in the small (ms) time scale. The load current steps up from 0 to 80 A at 0.1 s and then steps down to 30 A at 0.2 s. The lower part of the figure shows the corresponding SOFC output voltage responses for different values of the capacitance of double-layer charge effect. When the load current steps up (or down), the fuel cell output voltage drops (or rises) immediately due to the ohmic voltage drop. Then the voltage drops (or rises) "smoothly" to its final value. It is noted that the larger is the capacitance *C*, the slower the output voltage reaches its final value, which is because the effective time constant (τ_{dlc}) becomes larger.

4.5.2.2 Dynamics Due to the Effect of Pressure The time constants $\tau_{\text{a}} = V_{\text{a}}P_{\text{a}}^{\text{ch}}/2M_{\text{a}}RT$ and $\tau_{\text{c}} = V_{\text{c}}P_{\text{c}}^{\text{ch}}/2M_{\text{c}}RT$ are in the seconds range; therefore, they affect the model dynamic response in the seconds time scale, which we will call medium time scale. Figure 4.11 shows the SOFC voltage response in the medium time scale under different operating pressures; the anode and cathode channel pressure are assumed equal ($P_{\text{a}}=P_{\text{c}}=P$). The load current steps up to 80 A at 1 s and then steps down to 30 A at 11 s. It is noted that the SOFC output voltage is higher under a higher operating pressure, which is consistent with Equation 4.26. Also, it is noted that the time constant associated with the output voltage

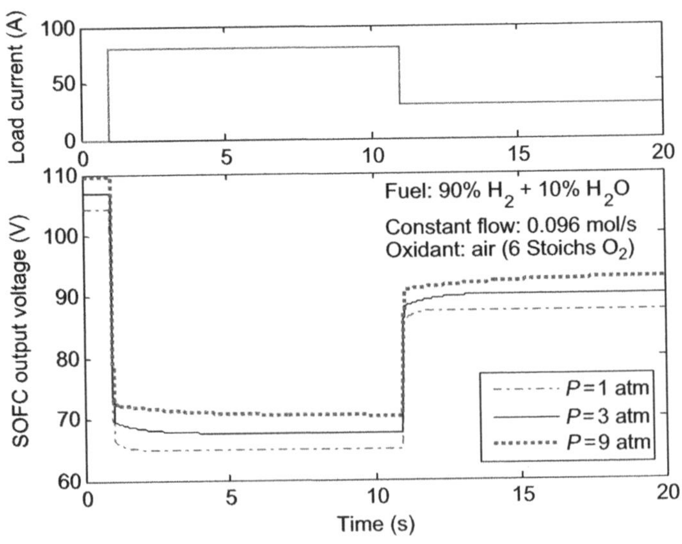

FIGURE 4.11 SOFC model dynamic voltage response under different operating pressures.

response increases with an increase in the operating pressure, which is consistent with the expressions for the time constants τ_a and τ_c given above.

SOFC operating temperature also appears in the expressions for the time constants τ_a and τ_c. However, as we will see shortly, the SOFC temperature response time is in the order of minutes or longer, which is much longer than its response time due to gas dynamics. Therefore, the operating temperature can be assumed to be constant when dealing with dynamic responses in short and medium time scales (i.e., the dynamic responses due to double-layer charge effect and the gas dynamics).

4.5.2.3 Dynamics Due to the Effect of Temperature The equivalent thermodynamic time constant of a SOFC can be in the order of tens of minutes [17]. In large time scale (10^2–10^3 s), the dynamics of double-layer charge effect and operating pressure are diminished, and the temperature effect (due to SOFC thermodynamic characteristic) will dominate SOFC dynamic response. Figure 4.12 shows the dynamic voltage response of the SOFC model when subjected to a step increase in the load current from zero to 100 A at $t = 10$ min and then back to 30 A at $t = 120$ min. The fuel and air inlet temperatures are assumed to be equal and are given in the figure as T_{inlet}. When the load current steps up, the SOFC output voltage

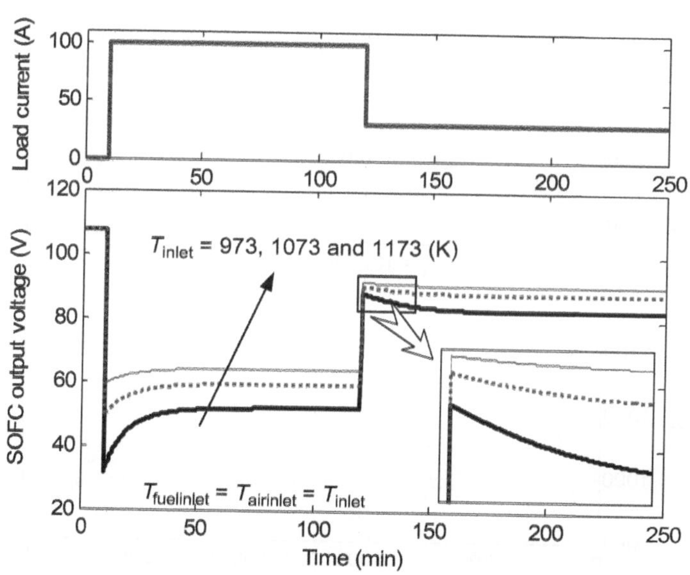

FIGURE 4.12 Model dynamic responses under different inlet temperatures in large time scale.

drops sharply, and then rises to its final value. The sharp voltage drop is due to the relatively slow chemical reactions inside SOFC. As a result, sufficient fuel will not reach the reaction site inside the SOFC; this phenomenon is sometimes called *fuel starvation*. The SOFC output voltage will then increase slowly due to an increase in the SOFC operating temperature (due to a large thermodynamic time constant in the minutes range), which is a result of an increase in the SOFC current. Similar argument holds when there is a sudden decrease in the SOFC load current, which causes the fuel cell operating temperature, and as a result the SOFC output voltage to decrease slowly. The SOFC model temperature responses for different inlet temperatures under step increase/decrease in the load current are shown in Fig. 4.13. From this figure, the equivalent thermodynamic time constant is around 15 min.

Other parameters equally important in affecting SOFC temperature response are the voltage drops inside SOFC, that is, activation, ohmic, and concentration voltage drops. In the normal operating range of SOFC, the activation and ohmic voltage drops (Eqs. 4.33 and 4.36) are dominant. From the expression for the activation resistance (4.34 and using 4.31) and ohmic resistance (4.37), it is noted that these resistances increase with load current. On the contrary, the resistances will decrease as SOFC internal temperature increases. Therfore, both activation and ohmic

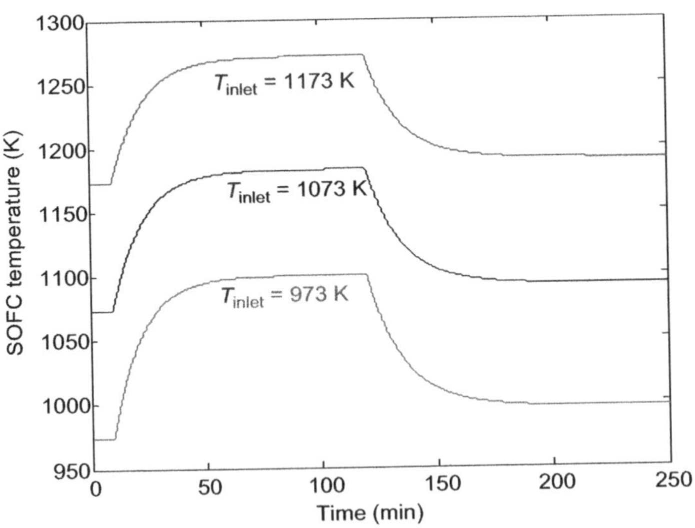

FIGURE 4.13 SOFC model temperature response.

voltage drops increase with SOFC current but decrease with its temperature as shown in Fig. 4.6

4.6 SOFC MODEL RESPONSE—CONSTANT FUEL UTILIZATION OPERATION

SOFC can also be operated in constant fuel utilization mode, where the utilization factor (u), defined as the ratio of mole flow rate of hydrogen consumed to the mole flow rate of hydrogen fuel input to the SOFC, is kept constant [1]. Using Faraday's law (4.4) for the mole flow rate of hydrogen consumed inside SOFC, the fuel utilization factor can be written as a function of SOFC current:

$$u = \frac{M_{\mathrm{H_2,consumed}}}{M_{\mathrm{H_2,in}}} = \frac{2F}{M_{\mathrm{H_2,in}}} i \tag{4.52}$$

Based on (4.52), constant fuel utilization operation can be achieved by feeding back the fuel cell current (i) through a proportional gain $1/(2F \times u)$ to control fuel flow to the fuel processor through a valve, as shown in Fig. 4.14. As a result, the amount of fuel input to the fuel processor will vary proportionally as the fuel cell current varies to keep the fuel utilization factor constant.

The fuel processor shown in Fig. 4.14 can be modeled as a simple delay transfer function $1/(\tau_f s + 1)$ with τ_f is in the order of seconds.

In the remainder of this section, the steady-state and dynamic responses of the SOFC model under constant fuel utilization operation are presented, and the responses are compared with those obtained under constant fuel flow operation.

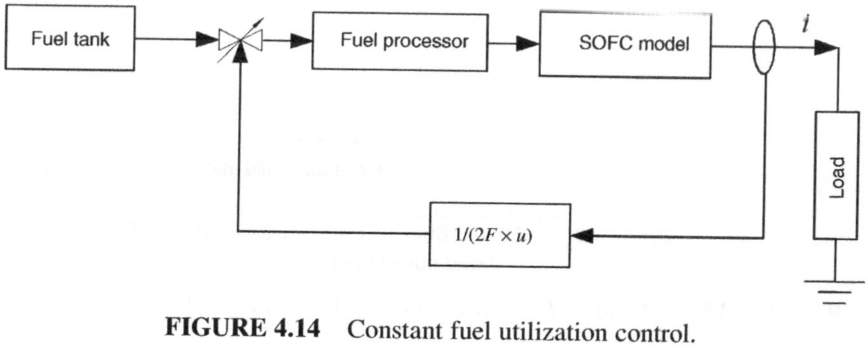

FIGURE 4.14 Constant fuel utilization control.

4.6.1 Steady-State Characteristics

The *V–I* and *P–I* characteristic curves of the SOFC model under constant fuel utilization of 85% and constant fuel flow operation are shown in Fig. 4.15. At low current, the output voltage under constant fuel flow operation is higher than under constant fuel utilization operation for the same load current. This is due to the fact that under constant fuel flow operation, and at low currents, more hydrogen fuel is fed to the SOFC than needed. As a result, the output voltage is higher at low currents. Because of the higher voltage, the output power under constant fuel flow operation is also slightly higher than under constant fuel utilization. The difference between the two voltages gets smaller as load current increases. The reason for this is that the utilization factor in the constant fuel flow operation increases with load current and gets closer to the utilization factor set for the constant fuel utilization operation near the nominal operating condition, around the maximum power point on the *P–I* characteristic curve.

The curves of hydrogen (fuel) versus load current under constant fuel flow and constant fuel utilization operation are shown in Fig. 4.16. Constant fuel flow operation requires more fuel input at light loading. The input fuel is about the same in both cases around nominal operating condition. Under constant fuel flow operation, the unused hydrogen (at light loading) can be recycled to be used again.

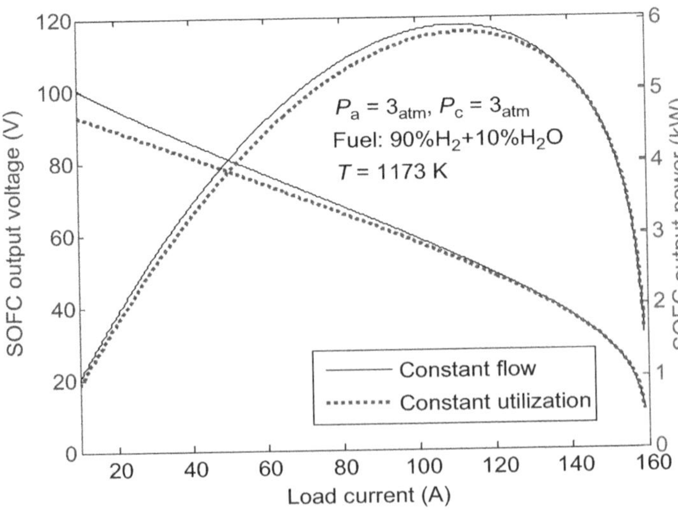

FIGURE 4.15 *V–I* and *P–I* characteristics of the SOFC model under constant fuel utilization and constant fuel flow operations.

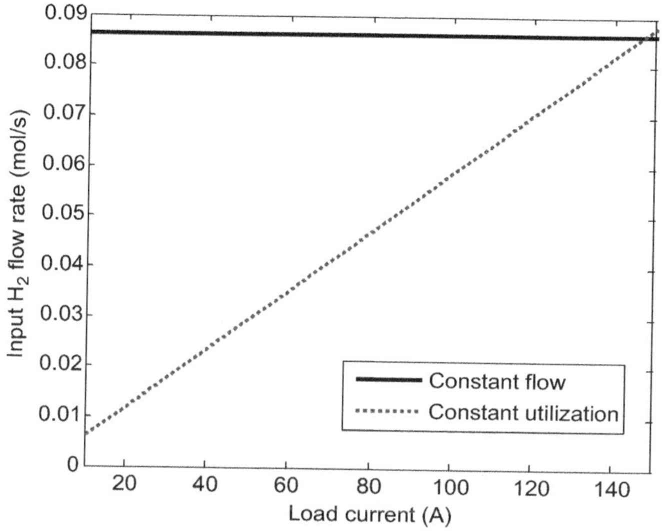

FIGURE 4.16 Hydrogen fuel input for constant utilization and constant flow operation of SOFC.

4.6.2 Dynamic Response

Dynamic response of the SOFC model in small time scale is dominated by the double-layer charge effect, and in large time scale is mainly influenced by the thermodynamic properties of the SOFC. Both the double-layer charge effect and the thermal characteristics of SOFC are determined by the fuel cell's physical and electrochemical properties. These properties are not heavily affected by whether SOFC is operating under constant fuel flow or constant fuel utilization. On the contrary, SOFC dynamic response in medium time scale (seconds range) is affected by whether the fuel cell is under constant fuel flow or constant fuel utilization. Under constant fuel flow operation, the time constant $\tau_a = V_a P_a^{ch} / 2M_a RT$ is not affected by SOFC current because fuel flow is constant and independent of SOFC current. However, under constant fuel utilization operation, fuel flow is adjusted as SOFC load current varies to result in a desired fuel utilization factor. As a result, τ_a becomes a load-dependent quantity.

Figure 4.17 shows the SOFC model dynamic voltage response under constant fuel utilization operation in the medium time scale under similar step current changes used in the constant fuel flow operation. The lower part of the figure shows the model output voltage responses under different operating pressures. Similar to what was shown for constant fuel flow operation (Fig. 4.11), a higher operating pressure results in a higher output voltage and larger time constants τ_a and τ_c. In this case, the dynamic

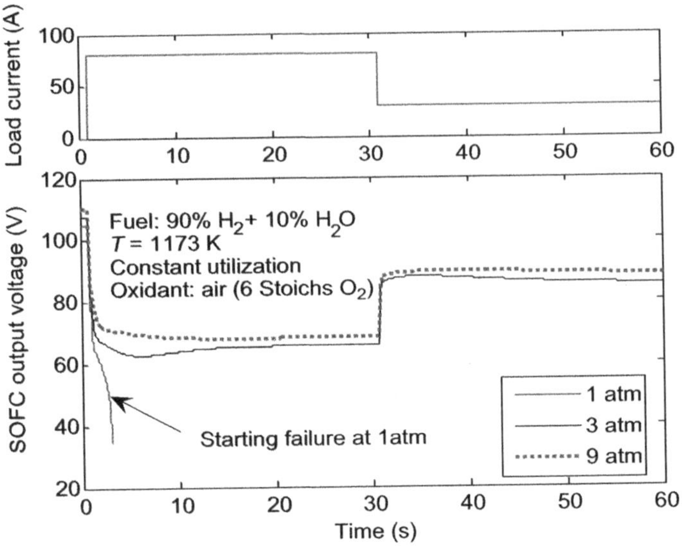

FIGURE 4.17 SOFC dynamic voltage response under constant fuel utilization operation at different operating pressures in medium time (s) scale.

responses are not determined only by τ_a and τ_c, but also by the dynamics of the fuel processor, which in this case is modeled by a first order delay. Hence, the output voltage curves (Fig. 4.17) show the typical characteristic of a second order system. On the contrary, the dynamic responses in medium time scale for constant fuel flow operation (Fig. 4.11) show the characteristic of a first order system since the dynamics of the fuel processor was not included in that case. It is noted from Fig. 4.17 that the SOFC fails to start up at 1 atm operating pressure due to the delay in the fuel processor in providing sufficient fuel supply for the fuel cell continued operation. The selection of a SOFC operating mode depends on desired SOFC performance requirements.

REFERENCES

[1] *Fuel Cell Handbook*, 7th edn, EG&G Services, Inc., Science Applications International Corporation, DOE, Office of Fossil Energy, National Energy Technology Laboratory, 2004.

[2] J. Larminie and A. Dicks, *Fuel Cell Systems Explained*, 2nd edn, Wiley Chichester, UK, 2003.

[3] J.C. Amphlett, R.M. Baumert, R.F. Mann, B.A. Peppley, and P.R. Roberge, Performance modeling of the Ballard Mark IV solid polymer electrolyte fuel

cell I. Mechanistic model development *Journal of the Electrochemical Society*, 142 (1), 1–8, 1995.

[4] G. Kortum, *Treatise on Electrochemistry*, 2nd edn, Elsevier, Amsterdam, 1965.

[5] G.N. Hatsopoulos and J.H. Keenan, *Principles of General Thermodynamics*, Wiley, NewYork, 1965.

[6] S.H. Chan, K.A. Khor, and Z.T. Xia, A complete polarization model of a solid oxide fuel cell and its sensitivity to the change of cell component thickness, *Journal of Power Sources*, 93 (1–2), 130–140, 2001.

[7] P. Aguiar, D. Chadwick, and L. Kershenbaum, Modeling of an indirect internal reforming solid oxide fuel cell, *Chemical Engineering Science*, 57, 1665–1677, 2002.

[8] P. Costamagna and K. Honegger, Modeling of solid oxide heat exchanger integrated stacks and simulation at high fuel utilization, *Journal of the Electrochemical Society*, 145 (11), 3995–4007, 1998.

[9] J. Padullés, G.W. Ault, and J.R. McDonald, An integrated SOFC plant dynamic model for power system simulation, *Journal of Power Sources*, 495–500, 2000.

[10] D.J. Hall and R.G. Colclaser, Transient modeling and simulation of a tubular solid oxide fuel cell, *IEEE Transactions on Energy Conversion*, 14 (3), 749–753, 1999.

[11] S.C. Singhal, Solid oxide fuel cells for stationary, mobile and military applications, *Solid State Ionics*, 152–153, 405–410, 2002.

[12] S.C. Singhal, Advances in tubular solid oxide fuel cell technology, *Solid State Ionics*, 135, 305–313, 2000.

[13] O. Yamamoto, Solid oxide fuel cells: fundamental aspects and prospects, *Electrochimica Acta*, 45 (15–16), 2423–2435, 2000.

[14] S.H. Chan, C.F. Low, and O.L. Ding, Energy and exergy analysis of simple solid-oxide fuel-cell power systems, *Journal of Power Sources*, 103 (2), 188–200, 2002.

[15] S. Nagata, A. Momma, T. Kato, and Y. Kasuga, Numerical analysis of output characteristics of tubular SOFC with internal reformer, *Journal of Power Sources*, 101, 60–71, 2001.

[16] C. Wang and M.H. Nehrir, A Physically-Based Dynamic Model for Solid Oxide Fuel Cells, *IEEE Transactions on Energy Conversion*, 22 (4), 2007.

[17] E. Achenbach, Response of a solid oxide fuel cell to load change, *Journal of Power Sources*, 57 (1–2), 105–109, 1995.

[18] *Proceedings of the 7th Annual SECA Workshop and Peer Review Meeting*, Philadelphia, PA, September 12–14, 2006.

CHAPTER 5

PRINCIPLES OF OPERATION AND MODELING OF ELECTROLYZERS

5.1 PRINCIPLE OF OPERATION OF ELECTROLYZERS

An electrolyzer is a device that produces hydrogen and oxygen from water. Water electrolysis can be considered a reverse process of a hydrogen-fueled fuel cell. Therefore, opposite to the electrochemical reaction occurring in a fuel cell, an electrolyzer converts the dc electrical energy into chemical energy stored in hydrogen. Water electrolysis is a long-established process, which started in the early part of the nineteenth century. Technically, the basic theory and reactions for electrolyzers are similar to those for fuel cells, but the reactions take place in the opposite direction. Different electrolytes can be used for electrolysis like in fuel cells; however, high-temperature electrolyzers are more challenging and still under R&D because steam has to be supplied to these types of electrolyzers, as opposed to liquid water, which can be used in low-temperature electrolyzers. The PEM and alkaline electrolyzer (both low-temperature) technologies are well established and available in the market [1]. However, the alkaline medium is preferred, and alkaline water electrolysis is the dominating technology today. Alkaline electrolyzers are more favorable than their PEM counterpart, mainly because they do not require expensive Platinum-based catalysts that are used in PEM electrolyzers. Further, the alkaline electrolyzer is a well-proven technology;

Modeling and Control of Fuel Cells, By M. Hashem Nehrir and Caisheng Wang

its large-scale production makes the unit cost lower than a similar PEM electrolyzer. Figure 5.1a shows a module of a pressurized alkaline electrolyzer unit using Inorganic Membrane Electrolysis Technology, IMET®, from Hydrogenics Corp (formerly Stuart Energy). IMET is capable of providing scalable H_2 generating capacity from 1 to 60 Nm^3/h with purity level greater than 99.9987%. Several modules could be put in series to form a stack, as shown in Fig. 5.1b. Figure 5.1c shows an assembled electrolyzer unit.

In this section, the principle of alkaline water electrolysis is summarized.

An alkaline electrolyzer uses potassium hydroxide (KOH) as electrolyte solution for transferring hydroxyl ions (OH^-). Figure 5.2 shows the schematic diagram of an alkaline electrolyzer. At the cathode, two water molecules are reduced electrochemically (using two electrons from the cathode) to produce one molecule of H_2 and two OH^- ions as follows:

$$2H_2O + 2e^- \rightarrow H_2\uparrow + 2OH^- \tag{5.1}$$

Under the external electric field, the OH ions are forced to move toward the anode through the porous diaphragm. At the anode, the two OH ions lose two electrons and are discharged into 1/2 molecule of O_2, one molecule of water, and two electrons, which migrate to the cathode through the dc electrical source according to the following chemical reaction.

$$2OH^- \rightarrow \tfrac{1}{2}O_2\uparrow + H_2O + 2e^- \tag{5.2}$$

By combining (5.1) and (5.2), the overall chemical reaction inside the electrolyzer is

$$H_2O \rightarrow H_2\uparrow + \tfrac{1}{2}O_2\uparrow \tag{5.3}$$

The current density of alkaline electrolyzer is normally less than 0.4 A/cm^2 and its energy conversion efficiency can range from 60 to 90%. Without auxiliary purification equipment, purities of 99.8% for H_2 can be achieved. Alkaline electrolysis technology can be implemented at a variety of scales from less than 1 kW to large industrial electrolyzer plants over 100 MW as long as there is a proper dc electricity supply available.

5.2 DYNAMIC MODELING OF ELECTROLYZERS

As explained in the previous section; electrolyzers use electricity to break water into hydrogen and oxygen. From the electrical circuit view point, an electrolyzer can be considered as a voltage-sensitive

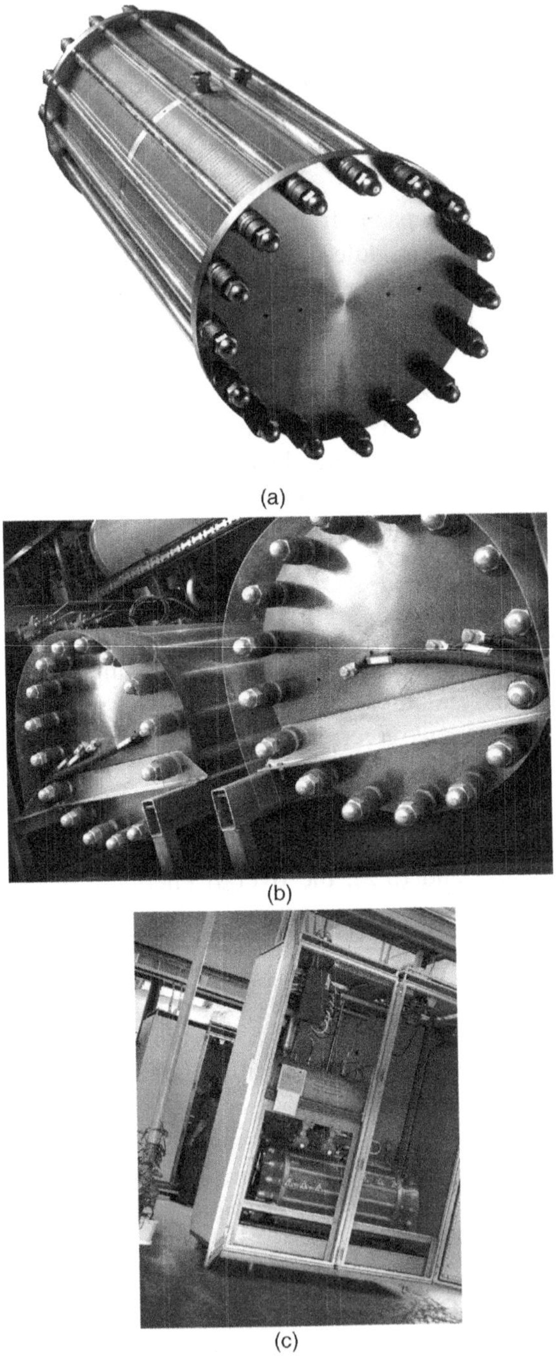

FIGURE 5.1 An alkaline electrolyzer unit in a HySTAT™-A Hydrogen Generator (a: module, b: stack, and c: the assembled unit) of Hydrogenics Corporation using IMET technology. *Source*: Courtesy of Hydrogenics Corporation.

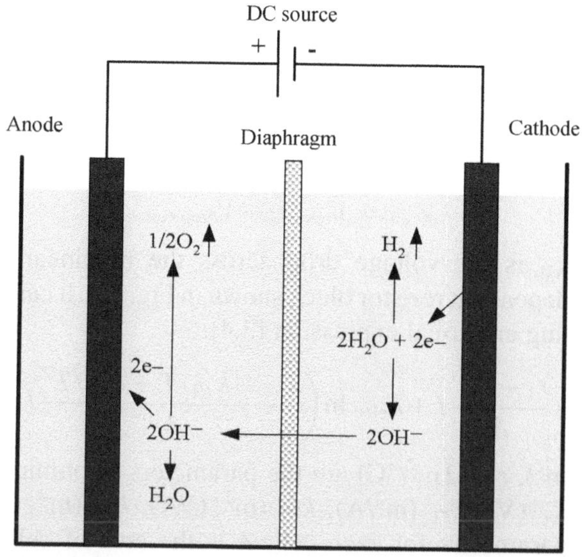

FIGURE 5.2 Schematic diagram of an alkaline electrolyzer.

nonlinear dc load. Within the electrolyzer rating, the higher the dc voltage applied to it, the larger is the load current, and the larger is its internal (reversible) voltage, hence larger amounts of hydrogen (and oxygen) produced.

5.2.1 Electrolyzer Steady-State (V–I) Characteristics

Figure 5.3 shows the equivalent circuit of an electrolyzer cell, which is similar to that of a battery operating in the charging mode. $V_{rev,cell}$ is the

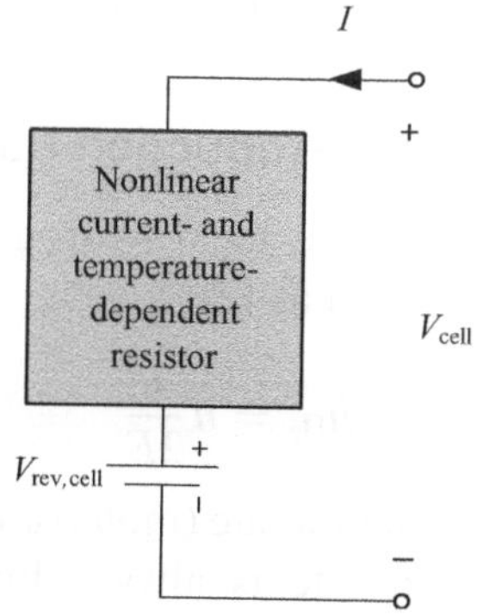

FIGURE 5.3 Electrolyzer equivalent circuit.

internal electrolyzer cell voltage, V_{cell} and I are the input dc voltage and current to the electrolyzer, and the nonlinear resistor is both current- and temperature-dependent representing the electrolyzer internal losses (voltage drops).

The internal voltage for each cell can be written as:

$$V_{cell} = V_{rev,cell} + V_{drop,cell} \tag{5.4}$$

where $V_{drop,cell}$ is the voltage drop across the nonlinear current- and temperature-dependent resistor block shown in Fig. 5.3. It can be expressed by the following empirical expression [3,4]:

$$V_{drop,cell} = \frac{r_1 + r_2 T}{A} I + k_{elec} \ln\left(\frac{k_{T1} + k_{T2}/T + k_{T3}/T^2}{A} I + 1\right) \tag{5.5}$$

where r_1 ($\Omega\,m^2$), r_2 ($\Omega\,m^2/°C$) are the parameters of ohmic resistance of each cell; k_{elec} (V), k_{T1} (m^2/A), k_{T2} ($m^2\,°C/A$), k_{T3} ($m^2\,°C^2/A$) are the overvoltage parameters for each cell; A is the area of cell (m^2); and T the cell temperature (°C).

The cell reversible voltage is determined by the Gibbs free energy due to change in the electrolyzer's electrochemical process and can be expressed by the empirical equation given below [3,4].

$$V_{rev,cell} = -\frac{\Delta G}{2F} = V_{rev}{}^{\circ} - k_{rev}(T - 25) \tag{5.6}$$

where $V_{rev}{}^{\circ}$ (V) is the reversible cell voltage at standard condition, and k_{rev} (V/°C) the empirical temperature coefficient of reversible voltage.

For an electrolyzer stack consisting of n cells in series, the terminal voltage of the stack can be expressed as

$$V_{elec} = nV_{cell} \tag{5.7}$$

The V–I characteristic for electrolyzer can be obtained from (5.4)–(5.7).

5.2.2 Modeling Hydrogen Production Rate

According to Faraday's Law, the ideal hydrogen production rate of an electrolyzer stack can be written as [2]:

$$\dot{n}_{H_2} = n\frac{I}{2F} \tag{5.8}$$

where, $\dot{n}_{H_2}$ is hydrogen production rate (mol/s) and n the number of cells in series. The actual hydrogen rate is always lower than the theoretical maximum value given by (5.8), which is due to parasitic current losses

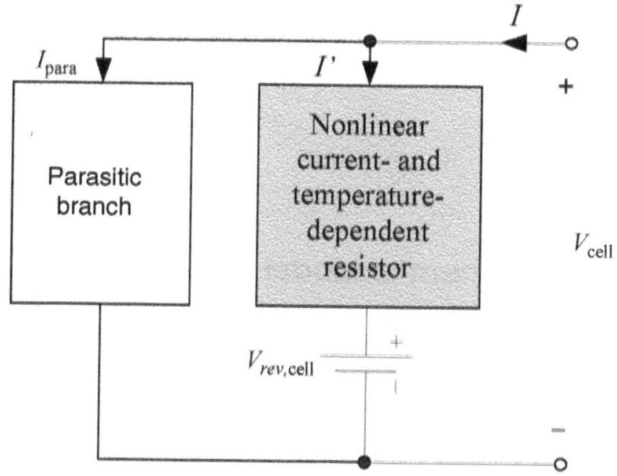

FIGURE 5.4 Electrolyzer equivalent circuit including parasitic losses.

(such as the leakage current through the cell insulation material). Figure 5.4 shows a modified equivalent circuit of the electrolyzer considering the parasitic losses. The total current has two parts, I' and I_{para}.

$$I = I' + I_{para} \tag{5.9}$$

where I' is the current actually used to generate hydrogen, and I_{para} the current due to parasitic current losses. The current efficiency or Faraday efficiency is defined as:

$$\eta_F = \frac{I'}{I} = 1 - \frac{I_{para}}{I} \tag{5.10}$$

The actual hydrogen generation rate then can be obtained as

$$\dot{n}_{H_2} = \eta_F \frac{nI}{2F} \tag{5.11}$$

The electrolyzer parasitic current is mainly determined by cell voltage. From Equation 5.10, η_F will increase as the electrolyzer current grows. Therefore, current efficiency is normally higher in higher current regions.

Similar to PEMFC and SOFC, as the electrolyzer internal temperature increases, the electrolyzer internal resistance will decrease. The decrease in the internal resistance will result in higher parasitic current and a lower current efficiency. However, the effect of temperature on current efficiency is small in high current density regions, and can be neglected [3]. Hence, we can consider η_F a function of current density. An empirical expression for η_F is as follows [4]:

$$\eta_F = \frac{(I/A)^2}{k_{f1} + (I/A)^2} k_{f2} \tag{5.12}$$

where k_{f1}, k_{f2} are empirical constants.

5.2.3 Electrolyzer Thermal Model

As we saw for PEMFC and SOFC, the operating temperature of an electrolyzer stack affects its performance. Therefore, a thermal model is necessary to predict electrolyzer stack operating temperature. Electrolyzer operating temperature can be estimated as a function of time from the following lumped thermal balance expression [4]:

$$C_{\text{elec}} \frac{dT}{dt} = \dot{Q}_{\text{gen}} - \dot{Q}_{\text{loss}} - \dot{Q}_{\text{cool}} \tag{5.13}$$

where T is the electrolyzer operating temperature (°C), C_{elec} the overall heat capacity of the electrolyzer stack (J/°C), $\dot{Q}_{\text{gen}}$ the heat rate (power) generated inside the electrolyzer stack, $\dot{Q}_{\text{loss}}$ the heat rate (power) loss, and $\dot{Q}_{\text{cool}}$ the heat rate (power) loss due to cooling.

The above heat rates can be written as follows:

$$\dot{Q}_{\text{gen}} = n(V_{\text{cell}} - V_{\text{th}})I \tag{5.14}$$

where V_{th} is the thermal voltage which can be expressed as [4],

$$V_{\text{th}} = -\frac{\Delta H}{2F} = -\left(\frac{\Delta G + T\,\Delta S}{2F}\right) = V_{\text{rev}} - \frac{T\,\Delta S}{2F} \tag{5.15}$$

$\dot{Q}_{\text{loss}}$ is the heat rate (power) loss expressed as,

$$\dot{Q}_{\text{loss}} = \frac{T - T_a}{R_{\text{T,elec}}} \tag{5.16}$$

where T_a is the ambient temperature (°C), and $R_{\text{T,elec}}$ the equivalent thermal resistance (°C/W). $\dot{Q}_{\text{cool}}$is the heat rate (power) loss due to cooling,

$$\dot{Q}_{\text{cool}} = C_{\text{cm}}(T_{\text{cm,o}} - T_{\text{cm,i}}) \tag{5.17}$$

where C_{cm} is the overall heat capacity of the flowing cooling medium (water) per second (W/°C), and $T_{\text{cm,o}}$, $T_{\text{cm,i}}$ are the outlet and inlet temperatures of the cooling medium (°C). Water is assumed to be the cooling medium for the electrolyzer.

Knowing the operating temperature (T) inside the electrolyzer, $T_{\mathrm{cm,o}}$ can be estimated as [3]:

$$T_{\mathrm{cm,o}} = T_{\mathrm{cm,i}} + (T - T_{\mathrm{cm,i}})\left[1 - \exp\left(\frac{-k_{\mathrm{HX}}}{C_{\mathrm{cm}}}\right)\right] \tag{5.18}$$

where k_{HX} is the effective heat exchange coefficient for the cooling process and can be estimated by the following empirical expression [4]:

$$k_{\mathrm{HX}} = h_{\mathrm{cond}} + h_{\mathrm{conv}} \cdot I \tag{5.19}$$

where h_{cond} is the coefficient of the conductive heat exchange (W/°C), and h_{conv} the coefficient of convective heat exchange (W/(°C A)).

5.3 ELECTROLYZER MODEL IMPLEMENTATION

A dynamic model for an electrolyzer stack can be obtained based on the mathematical expressions given in Sections 5.2.2 and 5.2.3. The simulation results reported in this section are for a 50 kW electrolyzer stack model consisting of 40 cells in series developed in MATLAB/SIMULINK. Figure 5.5 shows the block diagram of the electrolyzer model. The input quantities are the applied dc voltage V_{elec}, ambient temperature T_{a}, inlet temperature of the cooling water $T_{\mathrm{cm,i}}$, and the cooling water flow rate n_{cm}. The two output quantities are the hydrogen production rate $\bar{n}_{\mathrm{H_2}}$ and the stack temperature T. The V–I characteristic block (5.4–5.7) calculates the electrolyzer current. The electrolyzer operating temperature is estimated by the

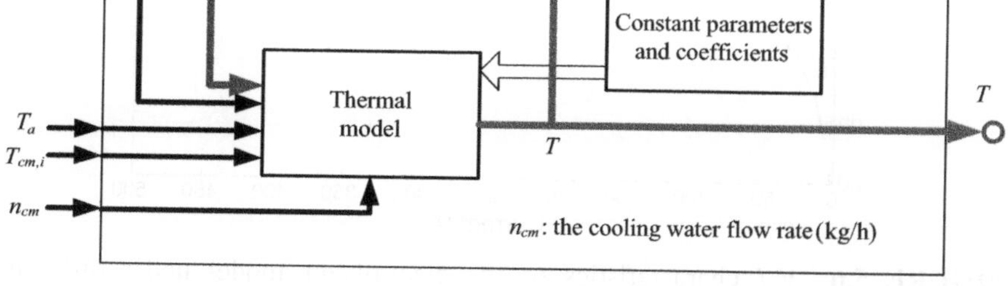

FIGURE 5.5 Block diagram of the electrolyzer model.

TABLE 5.1 Electrolyzer Model Parameters Used in the Simulation Study

r_1	$8.05 \times 10^{-5}\,\Omega\,m^2$
r_2	$-2.5 \times 10^{-7}\,\Omega\,m^2/°C$
k_{elec}	0.185 V
k_{T1}	$1.599\,m^2/A$
k_{T2}	$-1.302\,m^2\,°C/A$
k_{T3}	$421.3\,m^2\,°C^2/A$
n	40
k_{rev}	1.93×10^{-3} V/°C
$R_{T,elec}$	0.167°C/W
A	$0.25\,m^2$
k_{f1}	$2.5 \times 10^4\,A/m^2$
k_{f2}	0.96
h_{cond}	7.0 W/°C
h_{conv}	0.02 W/(°C A)
C_{elec}	6.252×10^5 J/°C
n_{cm} (see Fig. 5.5)	600 kg/h
C_{cm} (based on the value of n_{cm})	697.67 W/°C

thermal model block (5.13–5.19), and the hydrogen production rate is obtained based on the calculated electrolyzer current I and the current efficiency block (5.12). The model parameters used are given in Table 5.1 [4].

Figure 5.6 shows the V–I characteristic of the electrolyzer model under different cell operating temperatures. At the same current, the higher the

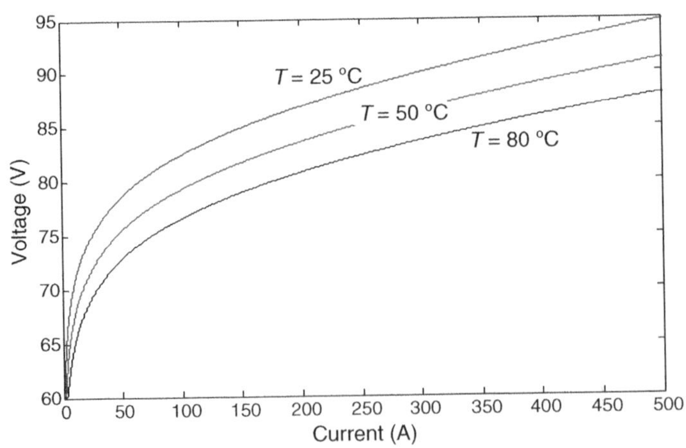

FIGURE 5.6 *V–I* characteristics of the electrolyzer model under different temperatures.

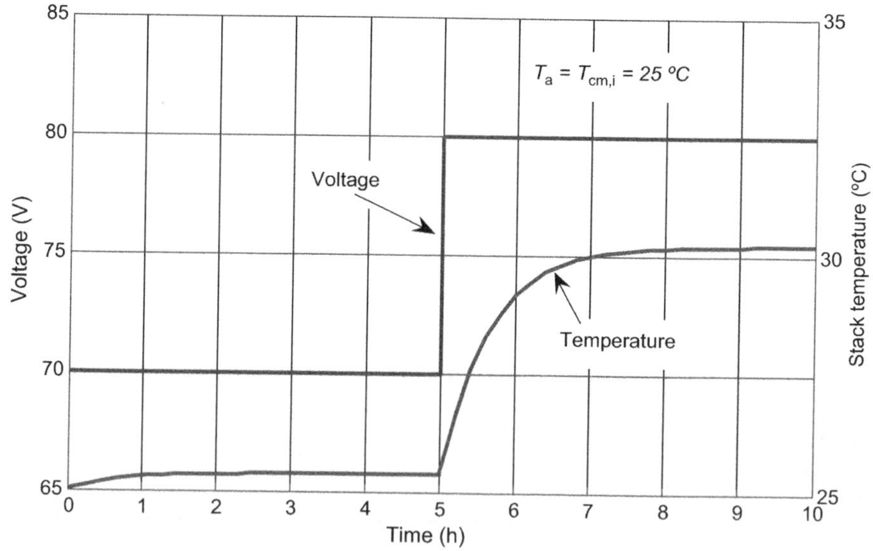

FIGURE 5.7 Temperature response of the model.

operating temperature, the lower the terminal voltage needed, and therefore a lower input electrical power is needed.

Figure 5.7 shows the electrolyzer model temperature response as a function of time for a step change in the electrolyzer input voltage. The ambient temperature and the inlet temperature of the cooling water are assumed to be 25°C. When the applied dc voltage steps up from 70 to 80 V, the stack temperature increases slowly from about 25.5 to 30.2°C.

REFERENCES

[1] M. Newborough, A report on electrolysers, future markets and the prospects for ITM Power Ltd's Electrolyser Technology, Online, http://www. h2fc.com/Newsletter/.

[2] J. Larminie and A. Dicks *Fuel Cell Systems Explained*, 2nd edn, Wiley, Chichester, UK, 2003.

[3] Ø. Ulleberg and S.O. Mørner TRNSYS simulation models for solar-hydrogen systems, *Solar Energy*, 59 (4–6), 271–279, 1997.

[4] Ø. Ulleberg, Modeling of advanced alkaline electrolyzers: a system simulation approach, *International Journal of Hydrogen Energy*, 28, 21–33, 2003.

CHAPTER 6

POWER ELECTRONIC INTERFACING CIRCUITS FOR FUEL CELL APPLICATIONS

6.1 INTRODUCTION

Power electronic (PE) interfacing circuits, also called power conditioning circuits, are normally necessary for a fuel cell (FC) based system to condition its output dc voltage and convert the dc to ac as shown in Fig. 1.4 in Chapter 1. Actually, nowadays, power electronic circuits are used almost everywhere, where there is a need to convert electric power from one type (dc or ac) to another. Power electronic interfacing circuits/devices deal with electric power conversion and control using power semiconductor switches that operate at high speeds. They come in various topologies and functions for a variety of applications such as motor drives, power adapters for computers and other electronic devices, uninterruptible power supplies (UPS), flexible ac transmission systems (FACTS) devices, and alternative energy power generation systems [1, 3]. In this chapter, the focus is on the modeling of power electronic circuits for FC systems. Depending on the actual FC applications, the power level of power electronic devices may vary from several watts to kilowatts and even megawatts, compared to microwatts to a few watts that are handled by low-power electronic circuits.

Modeling and Control of Fuel Cells, By M. Hashem Nehrir and Caisheng Wang

Power electronic interfacing circuits in a FC based system basically perform the following main functions:

- dc to dc conversion: Adapt one dc voltage (normally the FC output voltage) to another desired dc voltage.
- dc to ac inversion: Convert dc power into ac with desired magnitude and frequency.

A hybrid FC system may also need ac to dc rectification; an example of this case will be discussed in Chapter 9. Therefore, ac to dc conversion circuits are also discussed in this chapter.

High performance solid-state semiconductor switching devices, their circuit topologies, and how to control the high speed switches are the main three tasks of power electronics. In this chapter, we begin with an overview of various power electronic switching devices. Then we look at the different interfacing circuit topologies for different conversion tasks. We then develop small signal and state-space models for dc to dc converters and dc to ac inverters, which are important in simulation studies of these devices and for designing controllers for them. The details of controller design for dc to dc converters and dc to ac inverters will be addressed in Chapters 7 and 8.

An important issue in simulation studies of power electronic circuits is the simulation time step, which for accurate modeling often has to have a very small value. Depending on the actual switching frequency of the devices, the simulation time step needs to be in the order of μs or less for accurate switching simulation studies. As a result, a detailed model of a power electronic device together with its control switching signals will dramatically decrease the simulation speed. This will cause problems, especially when we need to carry out a long time (e.g., a 24 h) simulation study. Therefore, in addition to the state-space models of power electronic converters, effective average models of these circuits are also developed in this chapter. These effective average models can significantly improve the simulation speed and at the same time keep the desirable accuracy.

Throughout this chapter, it is assumed that the reader has some initial background in electric circuits, electronics, and control. The emphasis is on modeling of power electronic devices and circuits for computer simulation studies and controller design, but not on the detailed and complete discussion of power electronic circuit topologies and their design.

6.2 OVERVIEW OF BASIC POWER ELECTRONIC SWITCHES

Technological advances in power semiconductor devices and their widespread usage have reduced their costs significantly. Though the power electronic switching devices can sometimes be treated as ideal switches for the purpose of simplification, it is essential to know the characteristics of available devices before the right devices can be chosen and effective controllers can be designed for specific applications. Since the development of power semiconductor devices is a very dynamic area, only a brief summary of the most commonly available devices is given in this section.

Based on their controllability, currently available power semiconductor devices can be classified into the following three categories:

- Uncontrollable diodes: On/off switching of a diode is controlled by the circuit to which it is connected.
- Half-controllable thyristors: A thyristor can be turned on by a control signal only when it is already in the forward-biased mode. When in this mode, only the external circuit can turn the thyristor off.
- Controllable switches: On/off switching is determined by their control signals, not by the circuit.

Among various types of controllable switches, some commonly used devices include:

- Bipolar junction transistor (BJT)
- Metal-oxide semiconductor field effect transistor (MOSFET)
- Gate turn-off thyristor (GTO)
- Insulated gate bipolar transistor (IGBT)
- MOS-controlled thyristor (MCT)

The electrical terminal characteristics and switching capabilities of the above semiconductor devices are summarized in the remainder of this section.

6.2.1 Diode

Basically, a semiconductor diode is built on p–n junctions inside semiconductor materials or barriers between semiconductors and metals, like the Schottky diode [1]. Figure 6.1 shows the circuit symbol of a diode and its typical terminal i–v characteristic curve. Ideally, a diode only allows current flow in one direction, from its anode A to cathode K, as shown in

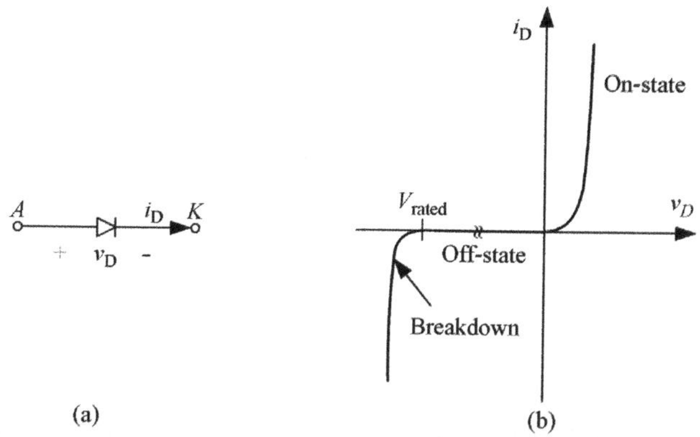

FIGURE 6.1 Circuit symbol of diode (a) and its *i–v* characteristic curve (b).

Fig. 6.1a. When the diode is forward biased, it starts to conduct with a small forward voltage drop (of normally less than 1 V) across it. This voltage drop can be neglected in some high voltage applications. When the diode is reverse biased, its current is almost zero, as shown in Fig. 6.1b; only a negligible leakage current can flow through it. When the reverse biased voltage exceeds the diode's rated breakdown voltage, current begins to flow in the reverse direction and increases very steeply. Under normal operation, the reverse voltage across the diode should not exceed its breakdown rating.

The most important steady-state parameters of a diode are its rated forward current, forward voltage drop, and rated breakdown voltage. Depending on the applications, the diodes chosen may vary widely in their rated currents and breakdown ratings (from tens of volts to several thousands of volts); their forward voltage drops are normally within 0.3 and 1 V.

6.2.2 Thyristor

The circuit symbol of a thyristor, shown in Fig. 6.2a is similar to that of a diode except it has a control gate G. A thyristor, also called silicon controlled rectifier (SCR), can block its current in either direction. Normally the rated reverse breakdown voltage (V_{rated_rb}) of a thyristor is the same as its rated forward breakdown voltage (V_{rated_fb}). Figure 6.2(b) shows a typical terminal *i–v* characteristic of a thyristor. The thyristor can be triggered to conduct current only when it is forward-biased and a positive signal is applied to its gate. When it conducts, the gate signal can be removed. However, when in its on-state, a thyristor cannot be turned off by applying a negative gate signal. It can only be turned off by its circuit.

Thyristors have a very good capability in handling high currents and voltages and have been widely used in high power applications, such as

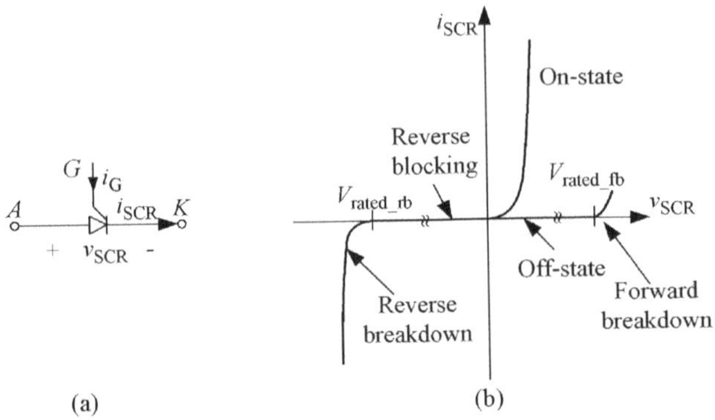

FIGURE 6.2 Circuit symbol of thyristor (a) and its i–v characteristic curve (b).

high voltage dc (HVDC) transmission systems. Their forward current and blocking voltage can go as high as several kilo-amperes and several kilo-volts, respectively. Depending on the application requirements, a variety of thyristors can be chosen to achieve a certain function. Their steady-state parameters such as current rating, blocking voltage, and forward voltage drop need to be considered in circuit design.

6.2.3 Bipolar Junction Transistor (BJT)

BJTs are current controlled switching devices. The circuit symbol of an NPN BJT and its steady-state i–v characteristics are shown in Fig. 6.3(a) and (b), respectively. The collector (C) and emitter (E) terminals of the BJT can be considered as the two terminals of a switch, which can be turned on and off by the base current I_B. When there is no base current, the switch is off, and i_C is only the negligible diffusion current. If a base current I_B is applied and is high enough to saturate the base–emitter junction of the BJT,

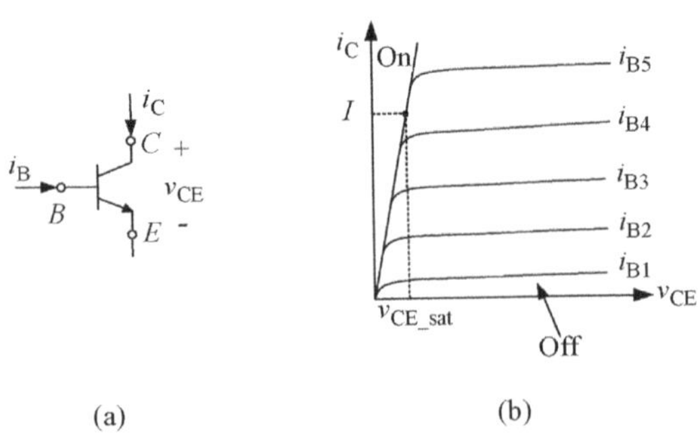

FIGURE 6.3 NPN BJT circuit symbol (a) and its terminal i–v characteristics at different base currents (b).

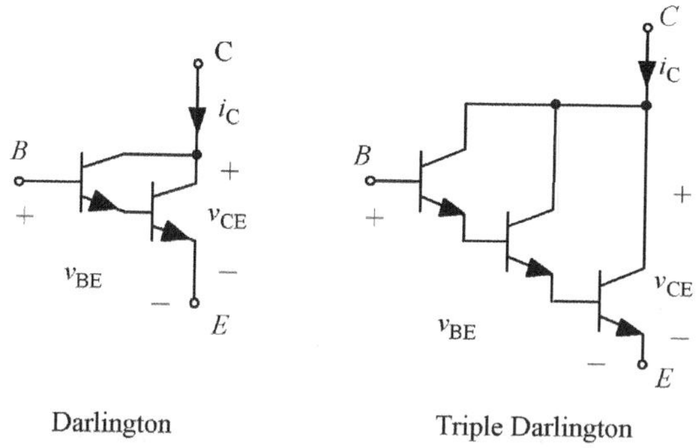

FIGURE 6.4 Darlington configuration.

then the switch is on. Under this condition, the voltage across the BJT (v_{CE_sat}) is normally within the 1–2 V range.

To keep a BJT on, the base current must be applied continuously. In order to reduce the value of base current required to keep the switch on, several BJTs can be connected together to form a Darlington or triple Darlington configuration, as shown in Fig. 6.4. Current flow in a BJT is unidirectional, that is for a NPN BJT, shown in Fig. 6.3, the current can only flow from terminal C to terminal E. In a PNP BJT, current flow is from its E terminal to C terminal.

6.2.4 Metal-Oxide Semiconductor Field Effect Transistor (MOSFET)

Opposite to BJTs, MOSFETs are voltage-controlled switching devices. The circuit symbol of an enhancement-mode N-channel MOSFET and its typical *i–v* characteristics at different gate-source voltages (v_{GS}) are shown in Fig. 6.5a and b, respectively.

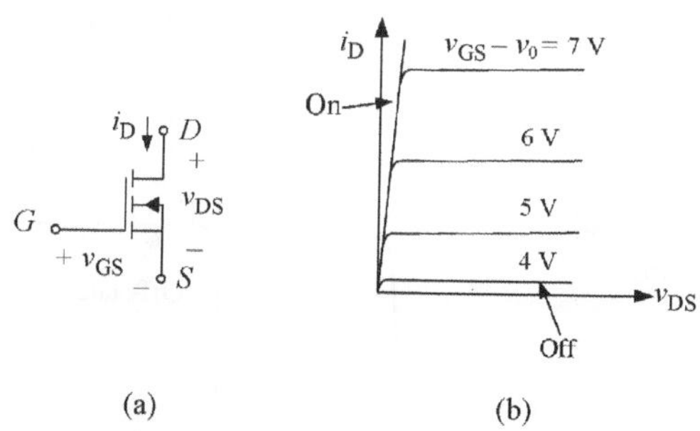

FIGURE 6.5 Enhancement-mode N-channel MOSFET circuit symbol (a) and typical *i–v* characteristics (b).

When v_{GS} is less than a threshold voltage (v_0), the switch is off, and, there is no conduction between the drain (D) terminal and source (S) terminal. Ideally, i_D is zero under this condition, but there actually is a negligible weak-inversion current, called sub-threshold leakage current. When v_{GS} is greater than the threshold voltage (v_0) and v_{DS} is less than the difference between v_{GS} and v_0 ($v_{GS} - v_0$), a conducting channel is created between terminals D and S, and the switch is on. Similar to BJTs, in order to keep the switch on, the MOSFET gate control signal needs to be continuous and has the right amplitude.

Since no current flows through the gate except during the switching transients and the switching times are very short, MOSFETs are particularly suitable for high-switching speed applications. However, they are not quite fit for high voltage and large current applications as high power rating is not a MOSFET's merit. Nevertheless, several MOSFETs of the same type can be connected in parallel to achieve a higher power/current rating if needed.

6.2.5 Gate Turn-Off Thyristor (GTO)

GTO is basically a fully controllable thyristor. Like a thyristor, a GTO can be turned on by a positive gate current signal, and the gate current is not required to be continuous. But unlike a thyristor, a GTO can be turned off by applying a negative gate current. The gate current does not require being continuous, but its rate of change (di_G/dt) needs to be very fast (around 100 A/μs), which limits the applications of GTOs to some extent.

The circuit symbol of a GTO and its terminal *i*–*v* characteristics are shown in Fig. 6.6a and b, respectively. GTOs have great capability of handling high voltages (in the order of kilovolts) and large currents (kilo-amperes). However, compared to MOSFETs, the switching

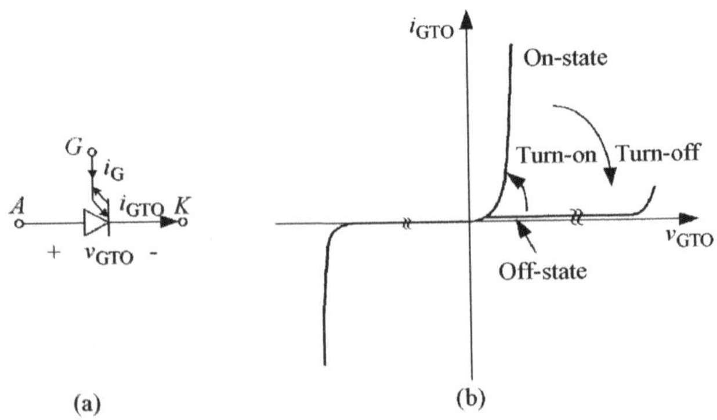

FIGURE 6.6 GTO circuit symbol (a) and terminal *i*–*v* characteristics (b).

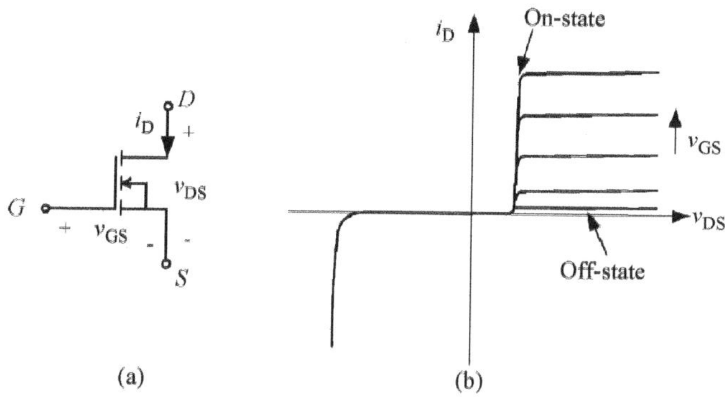

FIGURE 6.7 IGBT circuit symbol (a) and terminal *i–v* characteristics (b).

frequency of GTOs is lower, which confines their applications within slow switching speeds (a couple of kHz or below).

6.2.6 Insulated Gate Bipolar Transistor (IGBT)

IGBT is a voltage controlled switching device. Fig. 6.7a shows the circuit symbol of IGBT, and its *i–v* characteristics are given in Fig. 6.7b. As indicated by its name, an IGBT has an insulated control gate that can switch the device ON and OFF using a very small amount of energy. IGBT has some properties of bipolar transistors; like BJTs, the on-state voltage drop across the switch is small. It also has some similar properties of a GTO, including high power ratings and negative voltages blocking.

6.2.7 MOS-Controlled Thyristor (MCT)

MCT is a relatively new device for high power applications. The circuit symbol of a MCT and its terminal *i–v* characteristics are shown in Fig. 6.8a and b, respectively. Comparing Fig. 6.8b to Fig. 6.6b, we can see that a MCT has similar *i–v* characteristics as a GTO. A big difference between these two devices is that MCT is a voltage-controlled thyristor while GTO is a current-controlled device. Similar to MOSFET, a MCT has a voltage-controlled insulated gate that actually consists of two built-in MOSFETs: one is a p-channel and the other is an n-channel. Depending on which one is responsible for turning the device on (or off), there are two types of MCTs, P-MCT and N-MCT.

Due to the insulated gate, the driving circuit for a MCT is much simpler than that for a GTO. Moreover, MCT can be switched faster than a GTO. Although MCTs have the above advantages over GTOs, their power

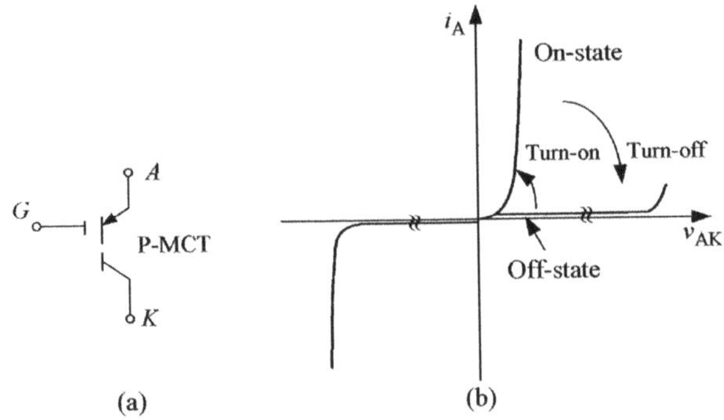

FIGURE 6.8 Circuit symbol of a p-channel MCT (a) and its *i–v* characteristics (b).

ratings are relatively lower. Their switching frequencies are also lower than MOSFETs'.

Power electronics is a dynamic area; in the last two decades; the performance of power electronic switching devices has improved dramatically and many new devices have emerged into the marketplace. In addition to MCTs, other emerging devices, such as power junction field effect transistor (JFET) and field controlled thyristor (FCT), may find their applications quickly and become popular in the future.

A comparative summary of the range of power handling capability, voltage, current, and switching frequency of major controllable power electronic devices discussed in this section is given in Table 6.1 on the basis of available technical data.

TABLE 6.1 Characteristics of Controllable Power Electronic Semiconductor Devices

Device	Power Capability[a]	Upper Voltage (kV)[b]	Upper Current (kA)[b]	Switching Frequency[c]
SCR	High	7	4	Low
BJT	Medium	1.4	0.4	Medium
MOSFET	Low	1	0.1	High
GTO	High	4.5	3	Low
IGBT	Medium	1.7	1.2	Medium
MCT	Medium	1.5	0.1	Medium

[a] Low (less than 10 kW); medium (between 10 and 100 kW); high (over 100 kW) [1].
[b] Voltage and current may not be able to reach the listed values at the same time.
[c] Low (normally less than 1 kHz); medium (between 1 and 100 kHz); high (above 100 kHz) [1].

The ac to dc rectification, dc to dc conversion, and dc to ac inversion circuits used in FC systems are discussed in the remainder of this chapter. The development of their corresponding state-space and simplified average models will be addressed as well.

6.3 ac/dc RECTIFIERS

6.3.1 Circuit Topologies

Basically, there are two types of ac/dc rectifiers: controllable and uncontrollable rectifiers. Uncontrollable rectifiers are normally referred to as diode (or diode bridge) rectifiers. Figure 6.9 shows a typical three-phase full diode bridge rectifier. L_s is the equivalent inductance at the ac side. C_{dc} is the filter capacitance connected at the dc side. Other filters can also be connected at the dc bus to further reduce ripples.

Assuming a balanced three-phase voltage applied to the rectifier, the average value of the output dc voltage V_{dc} is given as [1]

$$V_{dc} = \frac{3}{\pi}\sqrt{2}V_{LL} \tag{6.1}$$

where V_{LL} is the line–line RMS voltage applied to the rectifier.

The quality of the output dc voltage is fairly good. The voltage ripple is within ±5%, shown in Fig. 6.11. However, the input ac current, i_a for example, is distorted. The ac side inductor L_s can be increased to reduce the current distortion. However, increasing L_s will increase the voltage drop across the inductor. Another method to reduce the ac current distortion is to use a Y–Y and a Y–Δ connected transformer with the same line–line voltage ratio to form a 12-diode bridge rectifier, as shown in Fig. 6.10. Since there

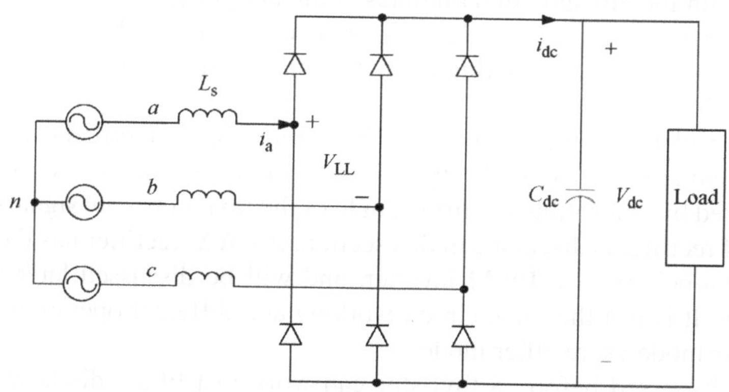

FIGURE 6.9 Three-phase full diode bridge rectifier.

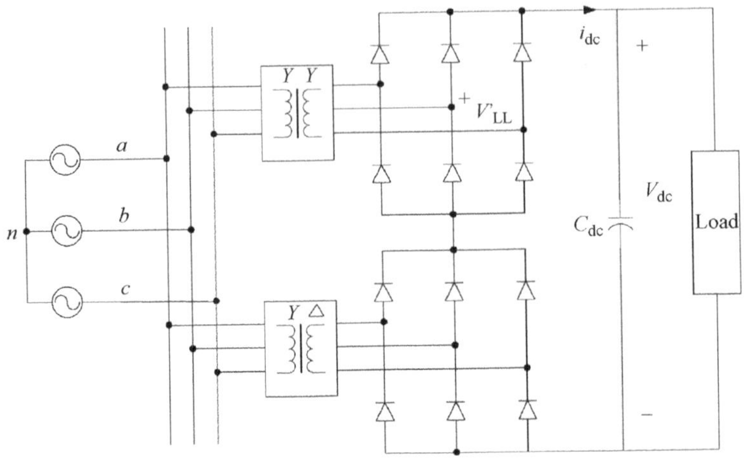

FIGURE 6.10 Three-phase 12-diode bridge rectifier with a Y–Y and a Y–Δ transformer.

is a 30° phase shift between the two transformer outputs, the transition period is reduced from 60 to 30° in one period (360°), as shown in Fig. 6.11. The average value of the dc output voltage now is

$$V_{dc} = \frac{6}{\pi}\sqrt{2}V'_{LL} \tag{6.2}$$

where V'_{LL} is the line–line RMS output voltage of each transformer.

The output dc voltage ripple now is reduced to within ±1.15%, and the harmonics of the ac side current are also much reduced. The total harmonic distortion (THD) of a 12-pulse rectifier is typically 12%, with the 11th and 13th harmonics dominant while the THD of a 6-pulse rectifier is about 30%, with the 5th and 7th harmonics dominant [1,4].

Based on the same principle, a 24-diode rectifier can be built using a zig-zag transformer group [4]. But, it is not commonly used in low power rating applications.

Controllable rectifiers are normally thyristor-based rectifiers and pulse width modulated (PWM) bridge rectifiers. Thyristor-based rectifiers are regulated by controlling the firing angle of the thyristors gate signals. This type of rectifier is discussed in this section. A PWM rectifier has the same circuit topology as a PWM inverter, and will be discussed later in this chapter. It is just the same circuit working at a different operation mode: inverter mode vs. rectifier mode.

As discussed before, a thyristor can work just like a diode with the exception that it starts conducting when it is forward biased and its

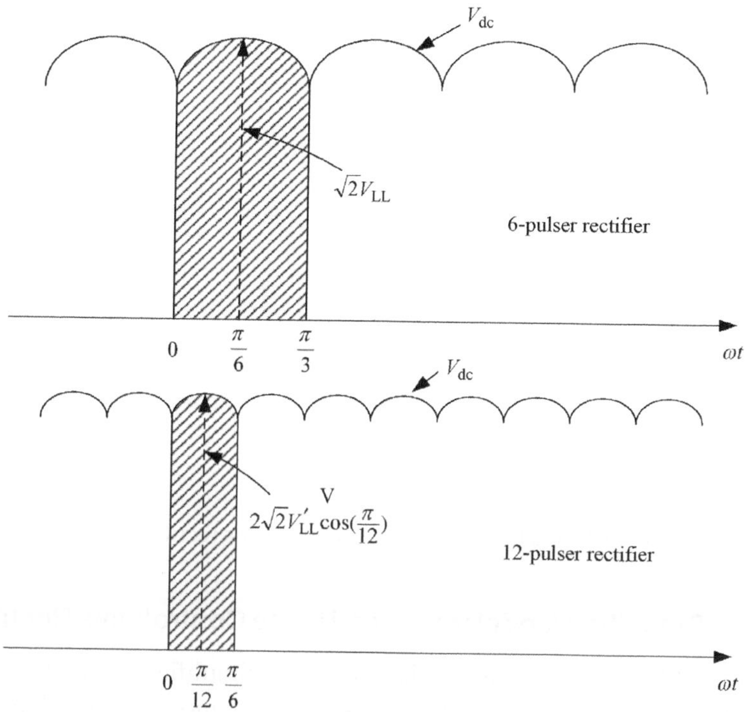

FIGURE 6.11 The dc output waveforms of the 6-pulse and 12-pulse diode rectifiers.

gate receives a positive current pulse; it continues conducting as long as it is forward biased. In Figs. 6.9 and 6.10, replacing the diodes by thyristors will give controlled rectifier circuits. Figure 6.12 shows a 6-pulse thyristor rectifier, whose output dc voltage can be regulated by controlling the firing angles of the thyristors. The firing angle α, shown in Fig. 6.12, is the angular delay between the time when a thyristor is forward biased and the time when a positive current pulse is applied to its gate. For a 6-pulse thyristor rectifier, the output dc voltage can be written as

$$V_{dc} = \frac{3}{\pi}\sqrt{2}V_{LL}\cos(\alpha) \tag{6.3}$$

where α is the firing angle.

From the above equation, the dc output voltage is a nonlinear function of firing angle. If $\cos(\alpha)$ is taken as a new control variable $g = \cos(\alpha)$, then the output voltage of the thyristor rectifier is proportional to the new control input, g.

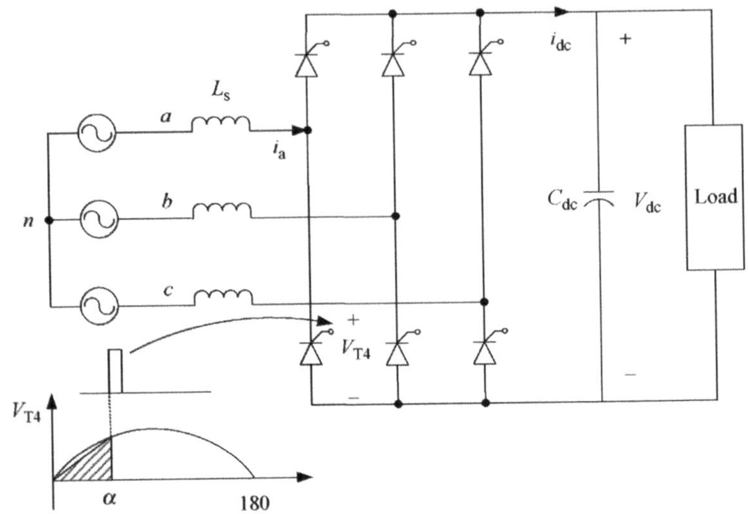

FIGURE 6.12 A 6-pulse thyristor rectifier.

6.3.2 Simplified Model for Three-Phase Controllable Rectifiers

Figure 6.13 shows the block diagram of a simplified ideal model for controllable three-phase 6-pulse rectifiers. This model is developed for long-time simulation studies. The "line–line voltage peak value" block calculates the peak value of the ac line–line voltage, $\sqrt{2}V_{LL}$. The input

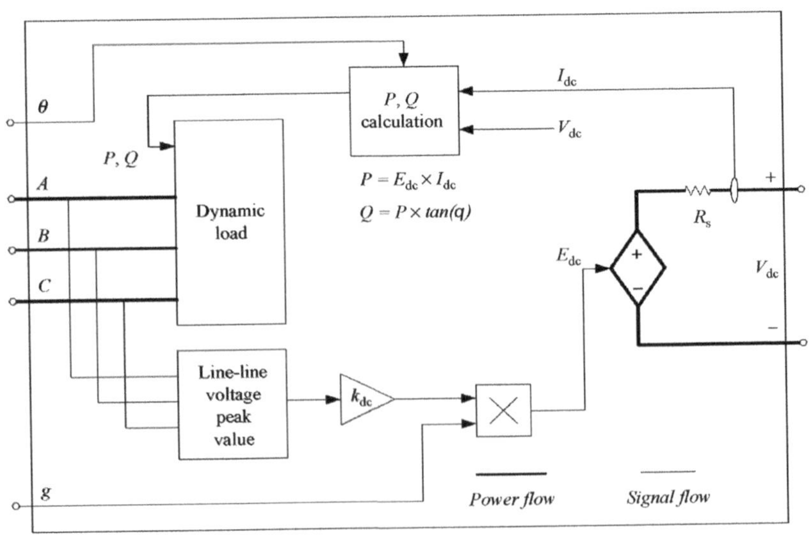

FIGURE 6.13 Block diagram of the simplified model for 3-phase controllable rectifiers.

quantity $g = \cos(\alpha)$, is the control variable to regulate the output dc voltage. The output dc voltage is calculated based on (6.1). For 6-pulse rectifiers, gain k_{dc} is $3/\pi$. Then the signal E_{dc} ($\sqrt{2}k_{dc}V_{LL}$) is used to control a controlled voltage source, which is the open-circuit dc output voltage of the model. The series resistor R_s is used to model the power and voltage losses within a real rectifier. The "P, Q Calculation" block calculates the amount of real and reactive power consumed at the ac side based on the dc power (real power) and the input quantity power angle θ, shown in Fig. 6.13. cos (θ) is the power factor of the rectifier. The "dynamic load" block consumes the real and reactive power assigned by its P & Q inputs. A built-in model of this dynamic load block can be found in MATLAB/SIMULINK [9].

Though the simplified model given in Fig. 6.13 is for 6-pulse controllable rectifiers, it can be easily modified to model other types of rectifiers. For instance, g can be set to 1 to represent a non-controllable 6-pulse diode rectifier.

Figures 6.14 and 6.15 compare the voltage and current responses of a simplified ideal rectifier model with the detailed 6-pulse and 12-pulse rectifier models with the same ac source and the same dc load. The detailed rectifier models are developed using built-in power electronic components from the SimPowerSystems Blockset in MATLAB/SIMULINK [9]. Figure 6.14 shows that the average output dc voltage of the ideal model is slightly higher than those of the detailed rectifier models. Figure 6.15 shows the ac side phase current of the different models. It is noted that the ac current is distorted for the 6-pulse rectifier model. However, the harmonic content is much reduced for the 12-pulse rectifier model, and the

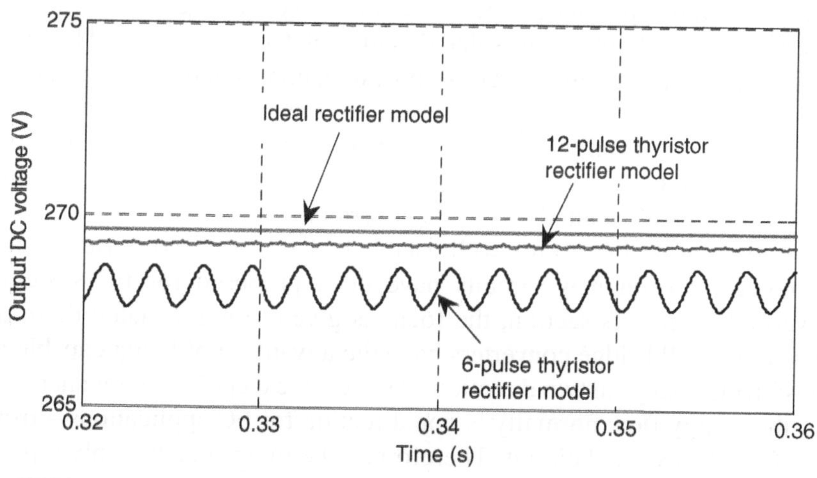

FIGURE 6.14 The dc Output voltages of the different rectifier models.

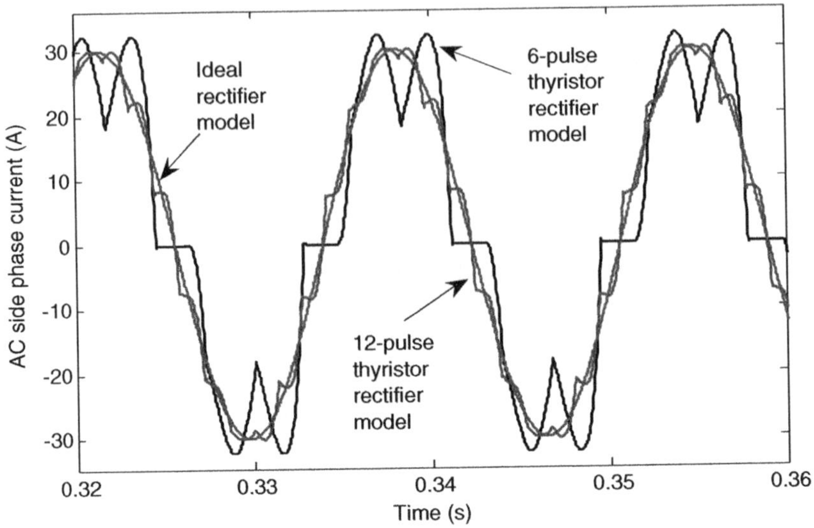

FIGURE 6.15 The ac side phase current of the different rectifier models.

response of this rectifier model is very close to that of the simplified rectifier model.

The simplified controllable rectifier model shown in Fig. 6.13 can be used in simulation studies of FC based hybrid energy systems. It can improve the simulation speed and at the same time keeps the desired accuracy.

6.4 dc TO dc CONVERTERS

There are many types of dc to dc converters. In general, dc to dc converters can be classified into non-isolated and isolated converters. The non-isolated converters include buck (step-down), boost (step-up), buck-boost, Cúk and full bridge converters. Isolated converters have an electrical isolation device (usually a high-frequency transformer) between their input and output. Flyback, forward, push–pull, and isolated half bridge and full bridge are typical isolated dc to dc converters.

In addition to the above well-established dc to dc converter circuit topologies, several new circuits have been proposed for FC systems. Nevertheless, in this section, the focus is given to non-isolated dc to dc converters. Full bridge converters have the advantage of being capable of transferring energy in two directions. However, except for the regenerative FC, the energy flow normally is one direction for FC applications—from the FC to the external circuit. Therefore, to be more specific, only typical boost converters and buck converters are discussed in this section.

6.4.1 Boost Converters

6.4.1.1 Circuit Topology Figure 6.16 shows the circuit diagram of a boost dc/dc converter. Let us only consider the continuous-conduction operation mode of the circuit, where the inductor current flows continuously, $i_{dd\_in} > 0$.

Figure 6.17 shows the steady-state inductor voltage and current waveforms. When the switching pulse is positive, switch S_{dd} (in Fig. 6.16) is on and diode D_{dd} is off. Then the voltage across the inductor is the same as the input voltage, V_{dd_in}. During the same time period, the inductor current increases with a slope V_{dd_in}/L_{dd}. When switch S_{dd} is off (the switching pulse is negative), diode D_{dd} is on, and the inductor voltage turns out to be $V_{dd\_in} - V_{dd\_out}$. During this time period (t_{off}), the inductor current will decrease at the slope of $(V_{dd\_out} - V_{dd\_in})/L_{dd}$. At steady state, the average value of the inductor current $i_{L_{dd}}$ is constant in one switching period T_s. Since $v_{L_{dd}} = L_{dd}(di_{L_{dd}}/dt)$, the integral of the inductor voltage $v_{L_{dd}}$ over one switching period is zero when the circuit is at steady state. Therefore, from Fig. 6.17, we can write

$$\int_{T_s} v_{L_{dd}}(t)dt = V_{dd\_in} \times t_{on} + (V_{dd\_in} - V_{dd\_out})t_{off} = 0 \tag{6.4}$$

Rearranging the above equation, we can get the output voltage at steady state as

$$V_{dd\_out} = \frac{1}{1-d} V_{dd\_in} \tag{6.5}$$

where $d = t_{on}/T_s$ is the duty ratio of the switching pulse.

The duty ratio d is always less than one for boost dc/dc converters; it cannot equal 1 because in that case switch S_{dd} in Fig. 6.16 would be always on, and the input would be short circuited all the time. The circuit, of

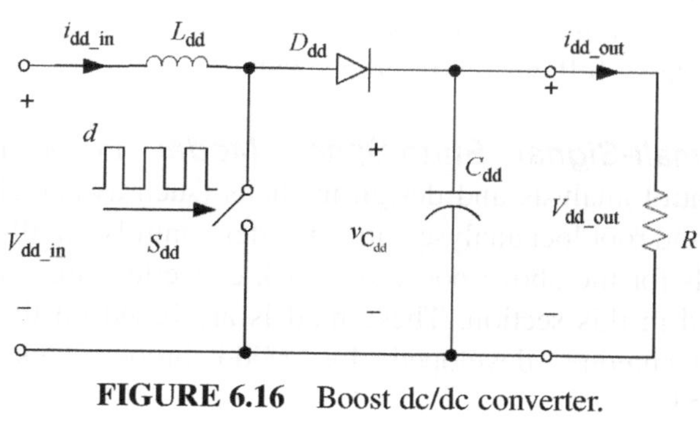

FIGURE 6.16 Boost dc/dc converter.

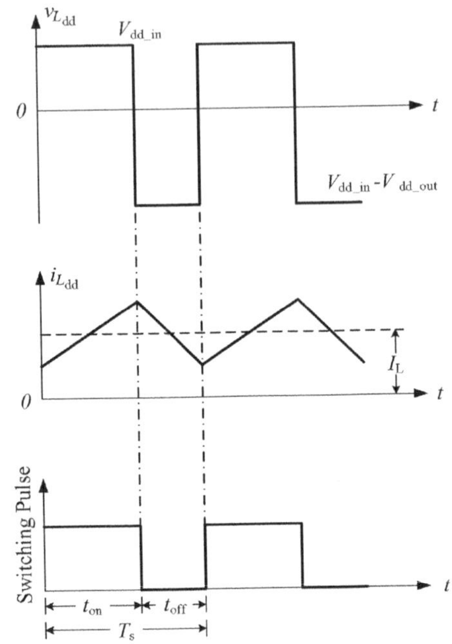

FIGURE 6.17 Inductor voltage and current under continuous-conduction mode.

course, would not work properly under this situation. For $0 \leq d < 1$, the converter output voltage is higher than the input voltage. In practical boost dc/dc converters, d normally is less than 0.85 [4].

Figure 6.18 shows a boost dc/dc converter with an output voltage regulation feedback control. The controller for the converter regulates the dc bus voltage within a desirable range. The output voltage is measured and compared with a reference value, and the error signal is processed through the PWM controller, which can be a simple proportional-integral (PI) controller. The output of the controller is used to control the pulse-width modulator to generate a PWM pulse series with the right duty ratios so that the output voltage follows the reference value.

6.4.1.2 Small-Signal State-Space Model In order to apply classical control analysis and design methods (such as Nyquist criterion, Bode plots, and root loci analyses) in converter controls, small-signal state-space models for the above boost and buck converters are needed, which are discussed in this section. These models are based on the state-space averaging technique developed by Middlebrook, Cúk and their colleagues [5].

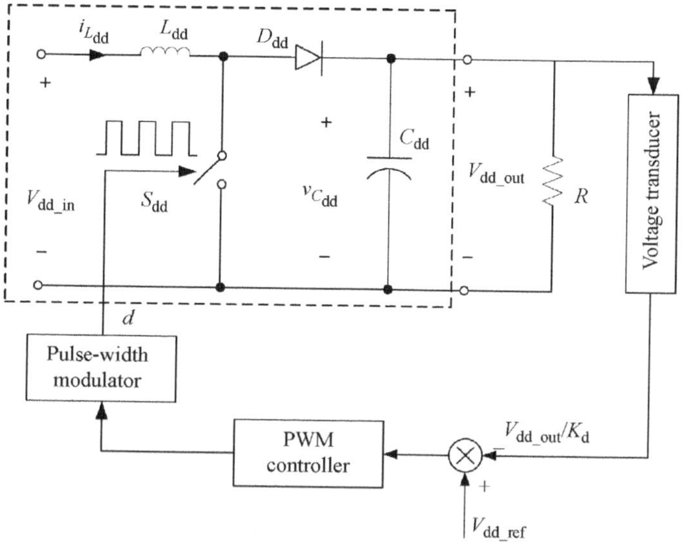

FIGURE 6.18 Boost dc/dc converter with feedback voltage control.

In Fig. 6.18, take $x_1 = i_{L_{dd}}$ and $x_2 = v_{C_{dd}}$ as state variables. Let $x = X + \tilde{x}$, $d = D + \tilde{d}$, $v_{dd\_in} = V_{dd\_in} + \tilde{v}_{dd\_in}$ and $v_{dd\_out} = V_{dd\_out} + \tilde{v}_{dd\_out}$. The symbol "$\sim$"is used to denote small perturbation signals, and state variable X denotes the system operating point.

When the switch S_{dd} is on and diode D_{dd} is off, the state-space representation of the main circuit can be written as

$$\begin{aligned} \dot{x} &= A_1 x + B_1 v_{dd\_in} \\ v_E &= C_1^T x \end{aligned} \tag{6.6}$$

where $x = \begin{bmatrix} x_1 & x_2 \end{bmatrix}^T$, $A_1 = \begin{bmatrix} 0 & 0 \\ 0 & \dfrac{-1}{RC_{dd}} \end{bmatrix}$, $B_1 = \begin{bmatrix} 1/L_{dd} \\ 0 \end{bmatrix}$, and $C_1^T = \begin{bmatrix} 0 & 1 \end{bmatrix}$. Symbol T represents the transpose operation of a matrix.

When switch S_{dd} is off and diode D_{dd} is on, the state-space equation of the circuit turns out to be

$$\begin{aligned} \dot{x} &= A_2 x + B_2 v_{dd\_in} \\ v_{dd\_out} &= C_2^T x \end{aligned} \tag{6.7}$$

where $A_2 = \begin{bmatrix} 0 & \dfrac{-1}{L_{dd}} \\ \dfrac{1}{C_{dd}} & \dfrac{-1}{RC_{dd}} \end{bmatrix}$, $B_2 = \begin{bmatrix} 1/L_{dd} \\ 0 \end{bmatrix}$, and $C_2^T = \begin{bmatrix} 0 & 1 \end{bmatrix}$

Therefore, the average state-space model of the main circuit at the operating point is

$$\begin{aligned} \dot{x} &= Ax + Bv_{\text{dd\_in}} \\ v_{\text{dd\_out}} &= C^{\text{T}}x \end{aligned} \tag{6.8}$$

where $A = [A_1 d + A_2(1-d)] = \begin{bmatrix} 0 & \dfrac{-(1-d)}{L_{\text{dd}}} \\ \dfrac{1-d}{C_{\text{dd}}} & \dfrac{-1}{RC_{\text{dd}}} \end{bmatrix}$, $B = [B_1 d + B_2(1-d)] = \begin{bmatrix} 1/L_{\text{dd}} \\ 0 \end{bmatrix}$, and $C^{\text{T}} = [C_1 d + C_2(1-d)]^{\text{T}} = [0 \quad 1]$

Then, the small signal model for the boost dc/dc converter can be written as

$$\begin{aligned} \dot{\tilde{x}} &= A_d \tilde{x} + B_d \tilde{d} + B_v \tilde{v}_{\text{dd\_in}} \\ \tilde{v}_{\text{dd\_out}} &= C^{\text{T}} \tilde{x} \end{aligned} \tag{6.9}$$

where $\tilde{x} = [\tilde{x}_1 \quad \tilde{x}_2]^{\text{T}}$, $A_d = \begin{bmatrix} 0 & \dfrac{-(1-D)}{L_{\text{dd}}} \\ \dfrac{1-D}{C_{\text{dd}}} & \dfrac{-1}{RC_{\text{dd}}} \end{bmatrix}$, $B_d = \begin{bmatrix} X_2/L_{\text{dd}} \\ -X_1/C_{\text{dd}} \end{bmatrix}$, $B_v = \begin{bmatrix} 1/L_{\text{dd}} \\ 0 \end{bmatrix}$, and X_1 and X_2 are the steady-state values of x_1 and x_2, respectively.

Using Equation (6.9) and only considering the perturbation of duty ratio $\tilde{d}$ (i.e., $\tilde{v}_{\text{dd\_in}} = 0$), the transfer function $T_{\text{vd}}(s) = [\tilde{v}_{\text{dd\_out}}(s)/\tilde{d}(s)]$ is obtained as

$$\begin{aligned} T_{\text{vd}}(s) &= C^{\text{T}}[SI - A_{\text{d}}]^{-1} B_{\text{d}} \\ &= \frac{1}{s[s + (1/RC_{\text{dd}})] + [(1-D)^2/L_{\text{dd}}C_{\text{dd}}]} \left[-\frac{X_1 s}{C_{\text{dd}}} + \frac{(1-D)X_2}{L_{\text{dd}}C_{\text{dd}}} \right] \end{aligned} \tag{6.10}$$

The developed transfer function will be used in controller design for boost converters, which will be covered in Chapters 7 and 8.

6.4.1.3 Average Model for Long-Time Simulation The transfer function given in (6.10) is a small signal model suitable for controller design; it is not suitable for long-time (such as 24 h) simulation studies with large variations in the circuit variables. An average model for boost

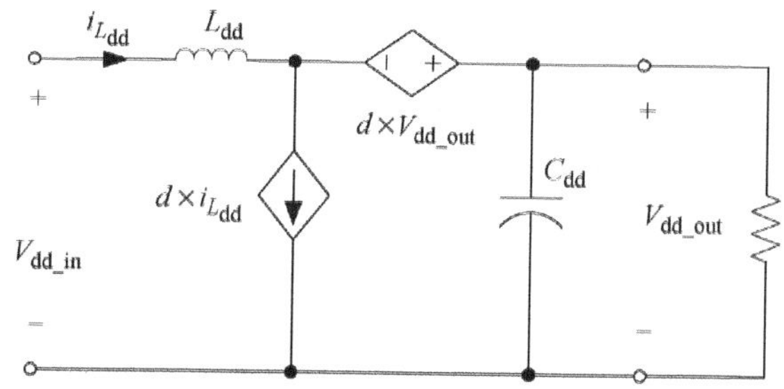

FIGURE 6.19 Average model for boost dc/dc converters.

dc/dc converter suitable for long-time simulation studies is needed for investigating system performance over a long period, as opposed to the small signal model, where the detailed cycle-to-cycle variations of values of the circuit voltages and currents are investigated.

Figure 6.19 shows the average model of the boost dc/dc converter given in Fig. 6.16. We can see that the power electronic switch is replaced by a current-controlled current source (CCCS) and the diode by a voltage-controlled voltage source (VCVS). The current magnitude of the CCCS and the voltage magnitude of the VCVS represent the average value of the current through the switch and the average voltage value across the diode, respectively. Simulation results for a boost dc/dc converter using its average model and detailed model are given in Figs. 6.20 and 6.21, respectively; the input to the converter is 110 V, and the duty ratio is fixed

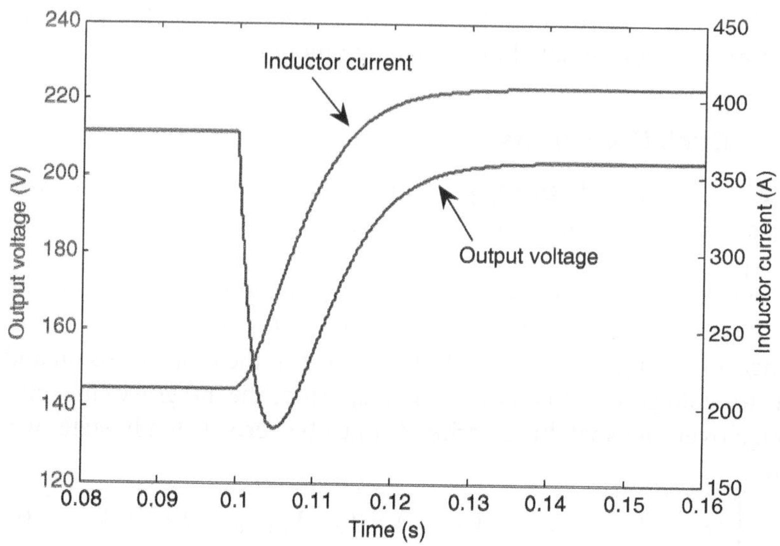

FIGURE 6.20 Simulation result of the average model for boost dc/dc converter.

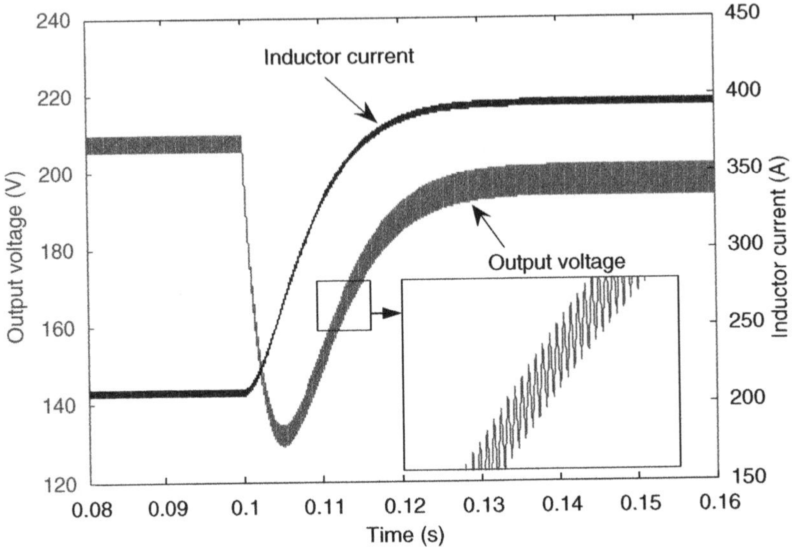

FIGURE 6.21 Simulation result of the detailed model for boost dc/dc converter.

at 0.5. The results show the converter input currents and output voltages of the two models when the load resistance (R in Figs. 6.16 and 6.19) is suddenly changed from 10 to 5 Ω. It is noted from these two figures that the average model can simulate the converter behavior at large time scale with reasonable accuracy. Figure 6.21 also shows the zoomed-in version of a small part of the converter output voltage, showing the actual voltage variations due to switching during a short time interval. These switching variations are not modeled in the average model.

6.4.2 Buck Converters

6.4.2.1 Circuit Topology A buck dc/dc converter with voltage feedback control is shown in Fig. 6.22. The circuit in the dashed rectangle is a typical buck dc/dc converter. Figure 6.23 shows the inductor voltage and current waveforms at steady state for continuous inductor current operation. When the switch S_{dd} is on, diode D_{dd} is off and the inductor voltage $V_{L_{dd}}$ is $V_{dd\_in} - V_{dd\_out}$. When S_{dd} is off, the diode turns on and the inductor voltage $V_{L_{dd}}$ changes to $-V_{dd\_out}$. Since the integral of the inductor voltage over one switching period T_s must be zero at steady state, we can write

$$\int_{T_s} v_{L_{dd}}(t)dt = (V_{dd\_in} - V_{dd\_out}) \times t_{on} - V_{dd\_out} \times t_{off} = 0 \qquad (6.11)$$

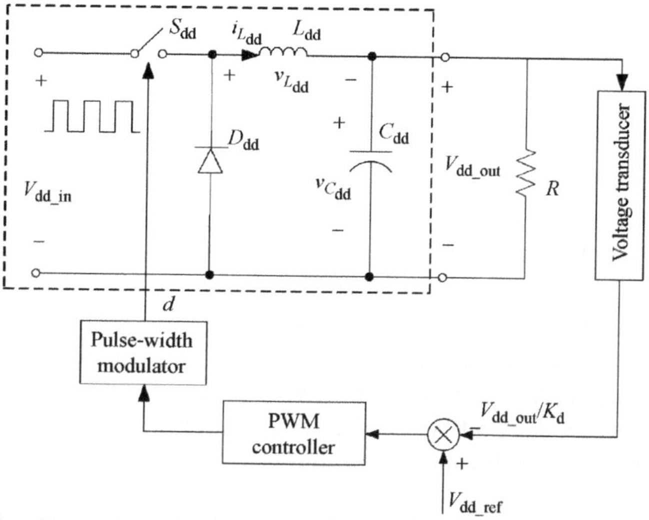

FIGURE 6.22 Buck dc/dc converter with voltage regulation.

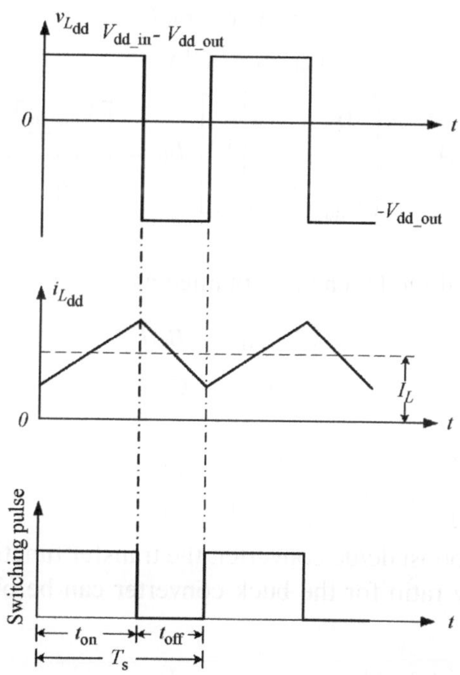

FIGURE 6.23 Inductor voltage and current waveforms for buck dc/dc converter.

Considering $T_s = t_{on} + t_{off}$ and $d = \frac{t_{on}}{T_s}$, the above equation can be written as

$$V_{dd\_out} = dV_{dd\_in} \tag{6.12}$$

Since $0 \leq d \leq 1$, the output voltage of a buck converter is always lower than its input voltage, making it a step-down dc/dc converter. Its output voltage can also be regulated by controlling the duty ratio (d) of the switching signal to regulate the dc bus voltage at a desired value. The voltage regulation feedback loop is shown in Fig. 6.22. The output voltage is measured and compared with the reference value, and the error signal is processed through a PWM controller, which controls the pulse-width modulator to generate a PWM pulse series with the right duty ratio so that the output voltage will follow the reference value.

6.4.2.2 Small-Signal State-Space Model for Buck dc/dc Converters Similar to the procedure discussed in the previous section for the boost converter, a small-signal state-space model of a buck converter (Fig. 6.22) can be obtained as follows [1,4]:

$$\begin{aligned} \dot{x} &= A_b x + B_b d \\ v_{dd\_out} &= C_b^T x \end{aligned} \tag{6.13}$$

where $x = \begin{bmatrix} i_{L_{dd}} \\ v_{C_{dd}} \end{bmatrix}$, $A_b = \begin{bmatrix} 0 & -\dfrac{1}{L_{dd}} \\ \dfrac{1}{C_{dd}} & -\dfrac{1}{RC_{dd}} \end{bmatrix}$, $B_b = \begin{bmatrix} \dfrac{V_{dd\_in}}{L_{dd}} \\ 0 \end{bmatrix}$, and $C_b^T = [0 \quad 1]$

The small signal model can be obtained as

$$\begin{aligned} \dot{\tilde{x}} &= A_b \tilde{x} + \tilde{B}_b \tilde{d} \\ \tilde{v}_{dd\_out} &= C_b^T \tilde{x} \end{aligned} \tag{6.14}$$

where $\tilde{x} = \begin{bmatrix} \tilde{i}_{L_{dd}} \\ \tilde{v}_{C_{dd}} \end{bmatrix}$ and $\tilde{B}_b = \begin{bmatrix} V_{dd\_in}/L_{dd} \\ 0 \end{bmatrix}$

Similar to the boost dc/dc converter, the transfer function for the output voltage over duty ratio for the buck converter can be obtained, as given below:

$$T_p(s) = \frac{\tilde{v}_{dd\_out}(s)}{\tilde{d}(s)} = \frac{V_{dd\_in}}{L_{dd}C_{dd}\{s^2 + [(s/RC_{dd}) + (1/L_{dd}C_{dd})]\}} \tag{6.15}$$

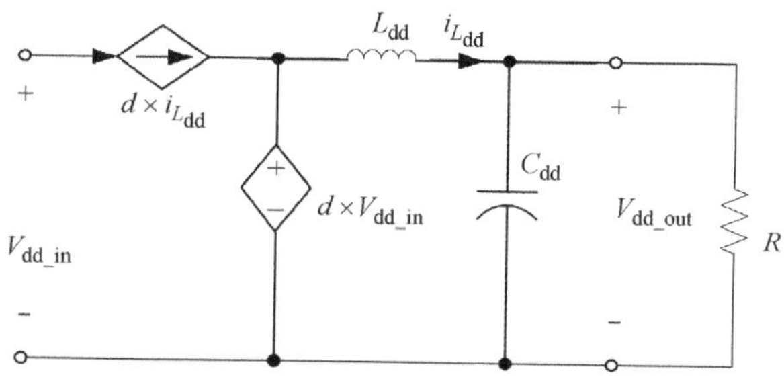

FIGURE 6.24 Average model for buck dc/dc converters.

6.4.2.3 Average Model for Long-Time Simulation Figure 6.24 shows the average model for buck converters. Similar to the average model for boost dc/dc converter (Fig. 6.19), the power electronic switch and diode are replaced by a CCCS and a VCVS, respectively. Simulation results for the inductor current and output voltage of the average and detailed circuit model of the buck dc/dc converter are shown in Figs. 6.25 and 6.26, respectively, when the load resistance (R in Figs. 6.22 and 6.24) is suddenly changed from 10 Ω to 5 Ω.The input voltage to the converter is 110 V, and the switching duty ratio is fixed at 0.5. A comparison of the results shown illustrates the effectiveness of the average model for buck dc/dc converters to represent the behavior of the converter with reasonable accuracy in long-time scale.

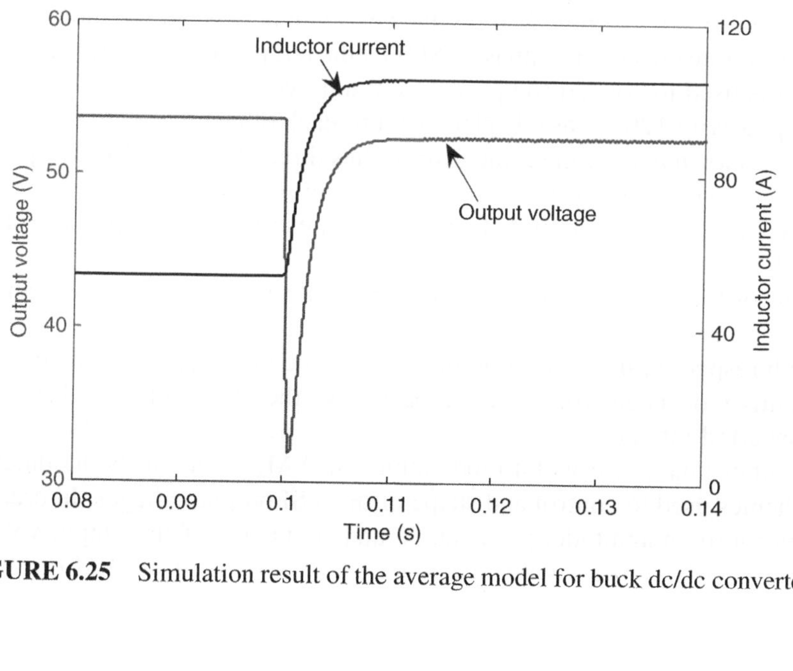

FIGURE 6.25 Simulation result of the average model for buck dc/dc converter.

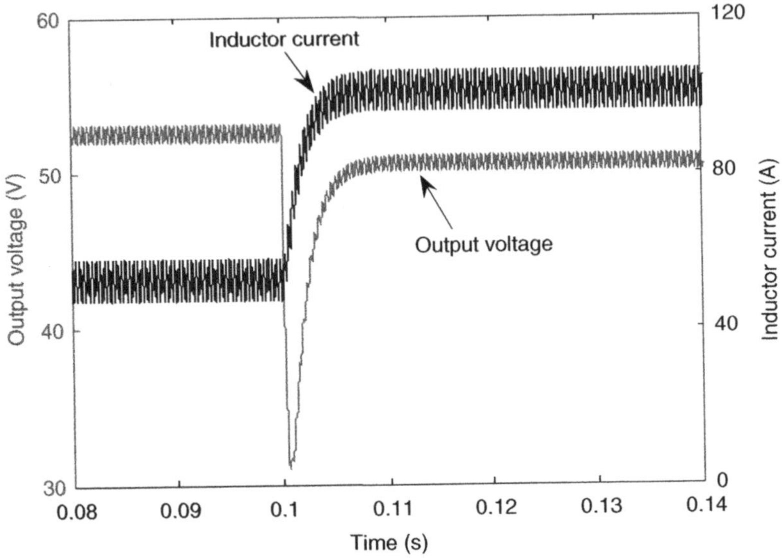

FIGURE 6.26 Simulation result of the detailed model for buck dc/dc converter.

6.5 THREE-PHASE dc/ac INVERTERS

6.5.1 Circuit Topology

Figure 6.27 shows the circuit topology of a typical 3-phase 6-switch PWM voltage source inverter (VSI) connected to an ac bus. At the moment, we concentrate on the three-phase VSI (i.e., the left part of Fig. 6.27) only. The VSI is used to convert the power from a dc voltage source to 3-phase ac outputs with 120° phase displacement to each other.

As shown in the figure, the inverter has three arms, one for each phase. Each arm has two power electronic switches, the upper switch (+) and the lower one (−). The output of each arm, for example the output of phase a with respect to the negative point on the dc side $v_{an'}$, depends on how we control the switches in arm a and on the dc voltage value. We can get either a positive or zero value output. The corresponding output voltage with respect to the neutral point (n) of the dc bus, v_{an}, can then be either positive or negative. That is, basically, the dc electricity has been converted into ac.

Sinusoidal pulse-width-modulation (SPWM) is one of the modulation schemes used to control and shape of the VSI output voltages. In order to control the magnitude, phase angle and frequency of the output voltage with a fairly constant voltage on the dc side of the VSI, SPWM is used to

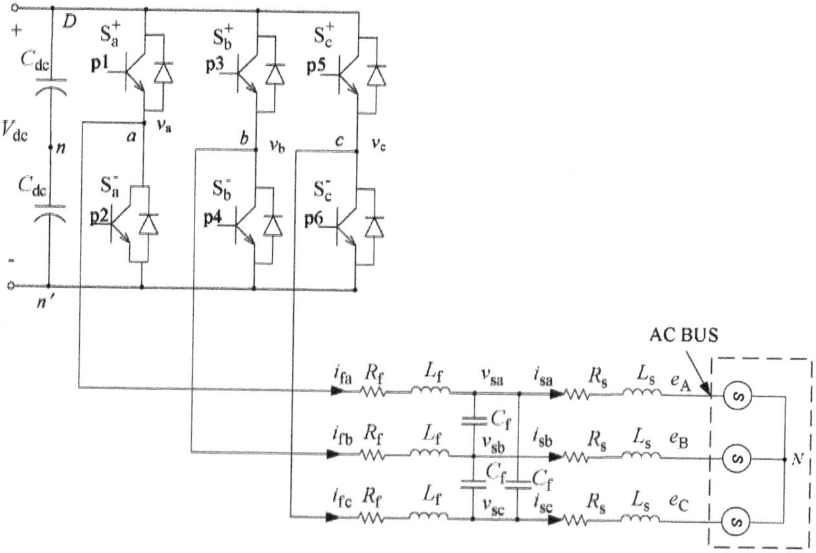

FIGURE 6.27 Three-phase dc/ac voltage source inverter.

generate appropriate switching pulses to control the six switches in the inverter. In a SPWM scheme, as shown in Fig. 6.28, three balanced sinusoidal control voltages (v_{actrl}, v_{bctrl}, and v_{cctrl}) are compared with the same triangular voltage (v_{tri}). The triangular waveform is at a switching frequency f_s, which is generally much higher than the frequency of the

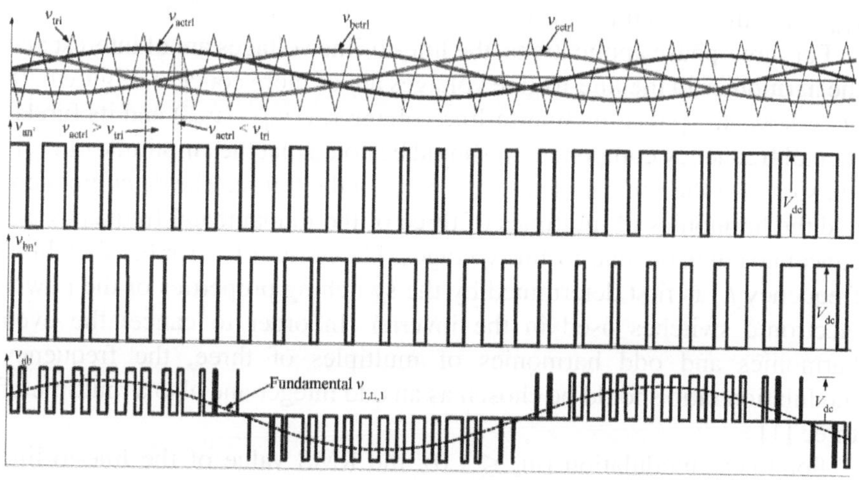

FIGURE 6.28 Three-phase PWM VSI waveforms.

control voltages and is called the carrier frequency. The three-phase sinusoidal control signals with the same frequency f_1 are used to modulate the duty ratios of the switching pulses (p1 to p6) for the six switches in Fig. 6.27.

The frequency modulation ratio is defined as

$$m_f = \frac{f_s}{f_1} \tag{6.16}$$

The amplitude modulation index is defined as

$$m_a = \frac{V_{p,ctrl}}{V_{p,tri}} \tag{6.17}$$

where $V_{p,ctrl}$ is the peak value of the control signal for each phase and $V_{p,tri}$ is the peak value of the triangular waveform.

Figure 6.28 shows how the output voltages are modulated by the SPWM scheme. Take phase a for example. When the control voltage v_{actrl} is larger than the triangular signal v_{tri}, the upper switch (S_a^+) is on and the lower switch (S_a^-) is off. As a result, the output voltage of phase a with respect to the negative terminal of the dc bus $v_{an'}$ is equal to V_{dc}. That is, when $v_{actrl} > v_{tri}$, $v_{an'} = V_{dc}$, and $v_{an'} = 0$ when $v_{actrl} < v_{tri}$. Similar argument holds for voltage $v_{bn'}$, that is, when $v_{bctrl} > v_{tri}$, $v_{bn'} = V_{dc}$, and $v_{bn'} = 0$ when $v_{bctrl} < v_{tri}$, as shown in Fig. 6.28. The output voltage of phase c is modulated similar to phases a and b. The waveform of the switching pulse (p1) for S_a^+ shares the same wave-shape (not necessarily the exact waveform) as $v_{an'}$, and the switching pulse p2 will be the inverse of p1.

For three-phase applications, the line-to-line voltages are of interest and the harmonics in the line-to-line voltages are of more concern. Figure 6.28 also shows the pulsating line-to-line voltage v_{ab} ($v_{an'} - v_{bn'}$) and its fundamental frequency component. It should be noted that the amplitude of v_{ab} is not shown at the same scale as $v_{an'}$ or $v_{bn'}$ in the figure. The dc component and the harmonics of multiples of three of the fundamental frequency are cancelled out in the line-to-line voltages. The choice of m_f (or the switching frequency f_s) is first determined by the switching properties of the power electronic switches used in the inverter. In order to cancel the even harmonics and odd harmonics of multiples of three, the frequency modulation ratio should be chosen as an odd integer and also a multiple of three [1].

For linear modulation ($m_a \leq 1.0$), the RMS value of the line-to-line voltage at fundamental frequency f_1 is:

$$V_{LL,1} = \frac{\sqrt{3}}{2\sqrt{2}} m_a V_{dc} \quad (m_a \leq 1.0) \tag{6.18}$$

As seen from Fig. 6.28, though some harmonics have been cancelled out in the line-to-line voltage v_{ab}, the voltage is still pulsating, not sinusoidal. To further reduce the harmonics in the output and obtain a better waveform, an output filter is normally needed, as shown in Fig. 6.27. R_f and L_f are the resistance and inductance of the filter inductor, and C_f is the filter capacitor. R_s and L_s are the resistance and inductance of the connecting line or a coupling inductor between the inverter and the system in the dashed rectangle shown in the figure. The ac system can be an ac load or an ac bus (e.g., a utility ac bus) at the same frequency.

6.5.2 State-Space Model

In this section, only the modeling of a three-phase inverter connected to a balanced ac system is considered. The neutral point of the ac system (point N in Fig. 6.27) is assumed to be at zero potential and is taken as the reference point. Define the following switching functions [6,7]:

$$d_1^* = \begin{cases} 1 & \text{when } S_a^+ \text{ is ON} \\ -1 & \text{when } S_a^- \text{ is ON} \end{cases}$$

$$d_2^* = \begin{cases} 1 & \text{when } S_b^+ \text{ is ON} \\ -1 & \text{when } S_b^- \text{ is ON} \end{cases} \tag{6.19}$$

$$d_3^* = \begin{cases} 1 & \text{when } S_c^+ \text{ is ON} \\ -1 & \text{when } S_c^- \text{ is ON} \end{cases}$$

Assuming there is a neutral point n at the dc side and the voltages across the two capacitors (C_{dc}) are identical, then $V_{Dn} = V_{nn'} = 0.5V_{dc}$, where V_{dc} is the dc bus voltage. The inverter output voltage of phase a (with respect to point n), v_{an}, is $0.5V_{dc}$ when S_a^+ is ON, and $-0.5V_{dc}$ when S_a^- is ON. Using the switching functions defined in (6.19), the voltages v_{an}, v_{bn}, v_{cn}, can therefore be written as:

$$\begin{cases} v_{an} = \dfrac{d_1^*}{2} V_{dc} \\ v_{bn} = \dfrac{d_2^*}{2} V_{dc} \\ v_{cn} = \dfrac{d_3^*}{2} V_{dc} \end{cases} \tag{6.20}$$

The inverter output phase potentials, v_a, v_b, and v_c, are

$$\begin{cases} v_a = v_{an} + v_n \\ v_b = v_{bn} + v_n \\ v_c = v_{cn} + v_n \end{cases} \tag{6.21}$$

where v_n is the voltage between the point n and the common reference point N.

Since the ac system is assumed to be balanced, $e_A + e_B + e_C = 0$, and $v_a + v_b + v_c = 0$. That is, $v_{an} + v_{bn} + v_{cn} + 3v_n = 0$. Therefore, v_n can be obtained as

$$v_n = -\frac{V_{dc}}{6}\sum_{k=1}^{3} d_k^* \tag{6.22}$$

For the circuit shown in Fig. 6.27, we can write the following dynamic equations:

$$\begin{cases} v_a = L_f \dfrac{di_{fa}}{dt} + R_f i_{fa} + v_{sa} \\ v_b = L_f \dfrac{di_{fb}}{dt} + R_f i_{fb} + v_{sb} \\ v_c = L_f \dfrac{di_{fc}}{dt} + R_f i_{fc} + v_{sc} \end{cases} \tag{6.23}$$

$$\begin{cases} i_{fa} = i_{sa} + C_f \dfrac{d(v_{sa} - v_{sb})}{dt} + C_f \dfrac{d(v_{sa} - v_{sc})}{dt} \\ i_{fb} = i_{sb} + C_f \dfrac{d(v_{sb} - v_{sa})}{dt} + C_f \dfrac{d(v_{sb} - v_{sc})}{dt} \\ i_{fc} = i_{sc} + C_f \dfrac{d(v_{sc} - v_{sa})}{dt} + C_f \dfrac{d(v_{sc} - v_{sb})}{dt} \end{cases} \tag{6.24}$$

$$\begin{cases} R_s i_{sa} + L_s \dfrac{di_{sa}}{dt} = v_{sa} - e_A \\ R_s i_{sb} + L_s \dfrac{di_{sb}}{dt} = v_{sb} - e_B \\ R_s i_{sc} + L_s \dfrac{di_{sc}}{dt} = v_{sc} - e_C \end{cases} \tag{6.25}$$

Writing the above equations into the state-space form, we have

$$\dot{X}_{abc} = A_{abc} X_{abc} + B_{abc} U_{abc} \tag{6.26}$$

where $X_{abc} = [i_{fa}, \; i_{fb}, \; i_{fc}, \; v_{sa}, \; v_{sb}, \; v_{sc}, \; i_{sa}, \; i_{sb}, \; i_{sc}]^T$

$$U_{abc} = [V_{dc}, \; e_A, \; e_B, \; e_C]^T,$$

$$A_{abc} = \begin{bmatrix} -\frac{R_f}{L_f} & 0 & 0 & \frac{-1}{L_f} & 0 & 0 & 0 & 0 & 0 \\ 0 & -\frac{R_f}{L_f} & 0 & 0 & \frac{-1}{L_f} & 0 & 0 & 0 & 0 \\ 0 & 0 & -\frac{R_f}{L_f} & 0 & 0 & \frac{-1}{L_f} & 0 & 0 & 0 \\ \frac{1}{3C_f} & 0 & 0 & 0 & 0 & 0 & \frac{-1}{3C_f} & 0 & 0 \\ 0 & \frac{1}{3C_f} & 0 & 0 & 0 & 0 & 0 & \frac{-1}{3C_f} & 0 \\ 0 & 0 & \frac{1}{3C_f} & 0 & 0 & 0 & 0 & 0 & \frac{-1}{3C_f} \\ 0 & 0 & 0 & \frac{1}{L_s} & 0 & 0 & -\frac{R_s}{L_s} & 0 & 0 \\ 0 & 0 & 0 & 0 & \frac{1}{L_s} & 0 & 0 & -\frac{R_s}{L_s} & 0 \\ 0 & 0 & 0 & 0 & 0 & \frac{1}{L_s} & 0 & 0 & -\frac{R_s}{L_s} \end{bmatrix}$$

and

$$B_{abc} = \begin{bmatrix} \left(\frac{d_1^*}{2} - \frac{1}{6}\sum_{k=1}^{3} d_k^*\right) \Big/ L_f & 0 & 0 & 0 \\ \left(\frac{d_2^*}{2} - \frac{1}{6}\sum_{k=1}^{3} d_k^*\right) \Big/ L_f & 0 & 0 & 0 \\ \left(\frac{d_3^*}{2} - \frac{1}{6}\sum_{k=1}^{3} d_k^*\right) \Big/ L_f & 0 & 0 & 0 \\ 0 & 0 & 0 & 0 \\ 0 & 0 & 0 & 0 \\ 0 & 0 & 0 & 0 \\ 0 & -\frac{1}{L_s} & 0 & 0 \\ 0 & 0 & -\frac{1}{L_s} & 0 \\ 0 & 0 & 0 & -\frac{1}{L_s} \end{bmatrix}$$

From the above state-space model, the inverter currents, and the system currents and voltages can be obtained as a function of time knowing the ac system and the inverter dc side voltages.

6.5.3 *abc/dq* Transformation

The *abc/dq0* transformation transfers a stationary 3-coordinate (*abc*) system to a rotating 3-coordinate (*dq0*) system. For a balanced three-phase system, the *0*-coordinate is zero, and then the three-phase ac stationary system is transformed into a 2-coordinate (*dq*) rotating dc system. The constant (dc) *dq* signals can be used to achieve zero tracking error control. Due to this merit, the *abc/dq0* transformation has been widely used in PWM converter and rotating machine control [2,6–8].

The *abc/dq0* transformation can be expressed as [2]

$$V_{dq0} = T_{abc/dq0} V_{abc} \tag{6.27}$$

where $V_{dq0} = \begin{bmatrix} V_d \\ V_q \\ V_O \end{bmatrix}$, $V_{abc} = \begin{bmatrix} V_a \\ V_b \\ V_c \end{bmatrix}$, $T_{abc/dq0} =$

$$\frac{2}{3}\begin{bmatrix} \sin(\theta) & \sin\left(\theta - \frac{2\pi}{3}\right) & \sin\left(\theta + \frac{2\pi}{3}\right) \\ \cos(\theta) & \cos\left(\theta - \frac{2\pi}{3}\right) & \cos\left(\theta + \frac{2\pi}{3}\right) \\ \frac{1}{2} & \frac{1}{2} & \frac{1}{2} \end{bmatrix}, \text{ and } \theta = \int_0^t \omega(\xi)\, d\xi + \theta(0)$$

In the above equations, θ is the angular position, ω the angular speed, and ξ the dummy variable of integration. For *abc* variables with constant frequency f and zero initial phase, θ can be written as $\theta = 2\pi f t$. The transformation equation (6.27) holds for other physical three-phase vectors, such as current, flux linkage, or electric charge.

For a balanced three-phase system, the value of the "*0*" component is zero, and the *abc/dq0* transformation reduces to *abc/dq* transformation, as given below.

$$T_{abc/dq} = \frac{2}{3}\begin{bmatrix} \sin(\theta) & \sin\left(\theta - \frac{2\pi}{3}\right) & \sin\left(\theta + \frac{2\pi}{3}\right) \\ \cos(\theta) & \cos\left(\theta - \frac{2\pi}{3}\right) & \cos\left(\theta + \frac{2\pi}{3}\right) \end{bmatrix} \tag{6.28}$$

We can see that the transformation matrix of $T_{abc/dq}$ is a sub-matrix of $T_{abc/dq0}$ by deleting the row and column for the *0*-component.

The inverses *dq0-abc* transformation can be expressed as [2]

$$T_{dq0/abc} = (T_{abc/dq0})^{-1} = \begin{bmatrix} \sin(\theta) & \cos(\theta) & 1 \\ \sin\left(\theta - \frac{2\pi}{3}\right) & \cos\left(\theta - \frac{2\pi}{3}\right) & 1 \\ \sin\left(\theta + \frac{2\pi}{3}\right) & \cos\left(\theta + \frac{2\pi}{3}\right) & 1 \end{bmatrix} \tag{6.29}$$

The corresponding inverses *dq-abc* transformation is

$$T_{dq/abc} = \begin{bmatrix} \sin(\theta) & \cos(\theta) \\ \sin\left(\theta - \frac{2\pi}{3}\right) & \cos\left(\theta - \frac{2\pi}{3}\right) \\ \sin\left(\theta + \frac{2\pi}{3}\right) & \cos\left(\theta + \frac{2\pi}{3}\right) \end{bmatrix} \tag{6.30}$$

6.5.4 *dq* Representation of the State-Space Model

Using the *abc/dq* transformation explained in the previous section, the system state-space equation given in (6.26) can be represented in the rotating *dq* frame as

$$\dot{X}_{dq} = A_{dq}X_{dq} + B_{dq}U_{dq} \tag{6.31}$$

where

$$X_{dq} = \begin{bmatrix} i_{dd} \\ i_{dq} \\ v_{sd} \\ v_{sq} \\ i_{sd} \\ i_{sq} \end{bmatrix} = \begin{bmatrix} T_{abc/dq} & 0 & 0 \\ 0 & T_{abc/dq} & 0 \\ 0 & 0 & T_{abc/dq} \end{bmatrix} \begin{bmatrix} i_{da} \\ i_{db} \\ i_{dc} \\ v_{sa} \\ v_{sb} \\ v_{sc} \\ i_{sa} \\ i_{sb} \\ i_{sc} \end{bmatrix}, \quad U_{dq} = \begin{bmatrix} V_{dc} \\ e_d \\ e_q \end{bmatrix},$$

and

$$\begin{bmatrix} e_d \\ e_q \end{bmatrix} = T_{abc/dq} \begin{bmatrix} e_A \\ e_B \\ e_C \end{bmatrix}$$

In the above equation, ***0*** is a matrix with the same dimension as the matrix $T_{abc/dq}$, that is 2×3. All elements of the ***0*** matrix are zero.

The system matrix A_{dq} and the input matrix B_{dq} are

$$A_{dq} = \begin{bmatrix} -\frac{R_f}{L_f} & \omega & \frac{-1}{L_f} & 0 & 0 & 0 \\ -\omega & -\frac{R_f}{L_f} & 0 & \frac{-1}{L_f} & 0 & 0 \\ \frac{1}{3C_f} & 0 & 0 & \omega & \frac{-1}{3C_f} & 0 \\ 0 & \frac{1}{3C_f} & -\omega & 0 & 0 & \frac{-1}{3C_f} \\ 0 & 0 & \frac{1}{L_s} & 0 & -\frac{R_s}{L_s} & \omega \\ 0 & 0 & 0 & \frac{1}{L_s} & -\omega & -\frac{R_s}{L_s} \end{bmatrix}$$

$$B_{dq} = \begin{bmatrix} \frac{f_1(t, d_1^*, d_2^*, d_3^*)}{L_f} & 0 & 0 \\ \frac{f_2(t, d_1^*, d_2^*, d_3^*)}{L_f} & 0 & 0 \\ 0 & 0 & 0 \\ 0 & 0 & 0 \\ 0 & -1 & 0 \\ 0 & 0 & -1 \end{bmatrix}$$

f_1 and f_2 in the input matrix are obtained as

$$f_1(t, d_1^*, d_2^*, d_3^*) = \frac{1}{3}\left[\sin(\omega t)\left(d_1^* - \frac{1}{3}\sum_{k=1}^{3} d_k^*\right) + \sin\left(\omega t - \frac{2\pi}{3}\right)\left(d_2^* - \frac{1}{3}\sum_{k=1}^{3} d_k^*\right) + \sin\left(\omega t + \frac{2\pi}{3}\right)\left(d_3^* - \frac{1}{3}\sum_{k=1}^{3} d_k^*\right)\right]$$

$$f_2(t, d_1^*, d_2^*, d_3^*) = \frac{1}{3}\left[\cos(\omega t)\left(d_1^* - \frac{1}{3}\sum_{k=1}^{3} d_k^*\right) + \cos\left(\omega t - \frac{2\pi}{3}\right)\left(d_2^* - \frac{1}{3}\sum_{k=1}^{3} d_k^*\right) + \cos\left(\omega t + \frac{2\pi}{3}\right)\left(d_3^* - \frac{1}{3}\sum_{k=1}^{3} d_k^*\right)\right]$$

$T_{abc/dq}$ is the *abc/dq* transformation matrix given in Equation 6.28.

The *dq* variables in Equation 6.31 can be used in the inverter controller design. Then, they can be transferred back into the *abc* variables for appropriate control purposes. Examples of using *abc/dq* and its inverse transformation in the inverter controller design are covered in Chapters 7, 8 and 9.

6.5.5 Ideal Model for Three-Phase VSI

Detailed modeling of dc/ac inverters requires detailed switching models of the power electronic switches. In order to properly model these high-frequency switching devices, the simulation time step has to be very small (in the order of μs or even smaller). Such small simulation time steps are not suitable for long-time simulation studies, where an ideal (or simplified) inverter model can be useful.

Figure 6.29 shows an ideal lossless model for a three-phase VSI, which can be used for long time simulation studies. The input to the model is V_{dc}, and the outputs are the three phase voltages, $V_a(t)$, $V_b(t)$ and $V_c(t)$. Also, there are three input quantities, the desired output ac frequency f, the ac voltage amplitude index (similar to the modulation index in a real VSI control) m, and the initial phase of the three-phase output voltages ϕ_0. The "*abc* signal formation" block gives the base signals for the three phases, $v_a(t)$, $v_b(t)$, and $v_c(t)$, as follows:

$$\begin{cases} v_a(t) = m\sin(2\pi ft + \phi_0) \\ v_b(t) = m\sin(2\pi ft + 2\pi/3 + \phi_0) \\ v_c(t) = m\sin(2\pi ft + 4\pi/3 + \phi_0) \end{cases} \tag{6.32}$$

The output voltage values, $V_a(t)$, $V_b(t)$, and $V_c(t)$, are calculated by multiplying the base signal values by the pulsating peak value of each phase output, which is $0.5V_{dc}$ for linear modulation ($m \leq 1.0$) [see the discussion below Equation 6.19]. For example, $V_a(t) = 0.5V_{dc} \times v_a(t)$.

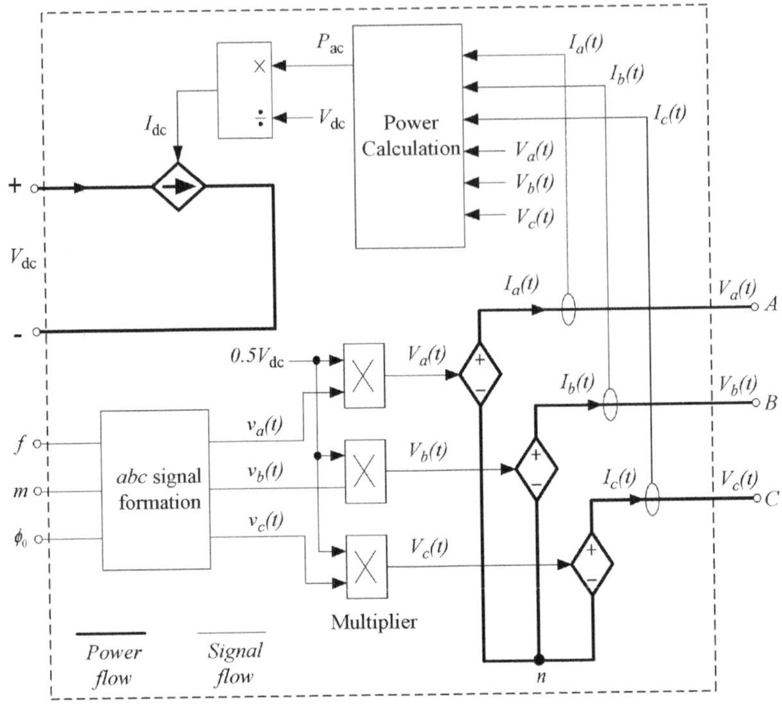

FIGURE 6.29 Ideal model for a three-phase VSI.

Then, $V_a(t)$, $V_b(t)$, and $V_c(t)$ are used to control the three controlled voltage sources, which are the three-phase output voltages of the inverter model. The ac power is calculated through the "Power Calculation" block as

$$P_{ac} = I_a(t) \times V_a(t) + I_b(t) \times V_b(t) + I_c(t) \times V_c(t) \tag{6.33}$$

where $I_a(t)$, $I_b(t)$, and $I_c(t)$ are the instantaneous values of the inverter current.

The load current of the input dc voltage source (I_{dc}) is determined by the power consumed at the ac side as

$$I_{dc} = P_{ac}/V_{dc} \tag{6.34}$$

For the purpose of comparison, a 208 V three-phase VSI is simulated to supply 100 kW to a load both by the ideal model and the detailed switched model. Figure 6.30 shows the output ac phase voltage of the two models. It is noted that though there are harmonics in the voltage obtained from the switched model, its output voltage waveform is very close to the output from the ideal model when a properly designed filter is used at the output of the switched model. Figure 6.31 shows the output ac power to the load from the two models. The ideal model delivers a pure constant power of 100 kW

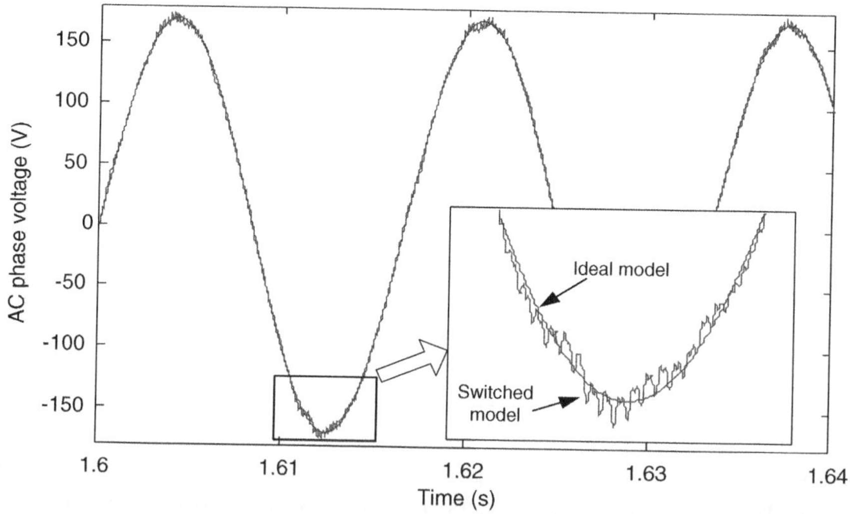

FIGURE 6.30 The ac phase output voltages of the ideal VSI model and the switched VSI model.

to the load. On the contrary, the detailed switched model delivers the same amount of average power (100 kW) but with ripples. Nevertheless, the ideal model is capable of simulating a VSI with a good accuracy and can be used for long-time simulation studies of power electronically interfaced

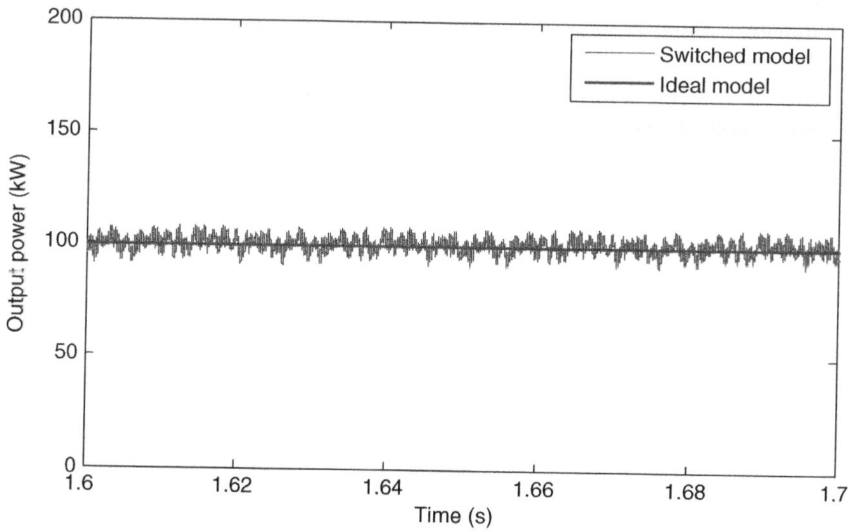

FIGURE 6.31 Output power of the ideal VSI model and the switched VSI model.

DGs. This model is used in a hybrid alternative energy system study discussed in Chapter 9.

REFERENCES

[1] N. Mohan, T.M. Undeland, and W.P. Robbins, *Power Electronics—Converters, Applications, and Design*, Wiley, Huboken, NJ, 2003.

[2] P.C. Krause, O. Wasynczuk, and S.D. Sudhoff, *Analysis of Electric Machinery*, IEEE Press, Piscataway, NJ, 1994.

[3] M.R. Patel, *Wind and Solar Power Systems*, CRC Press, Boca Raton, FL, 1999.

[4] D.W. Hart, *Introduction to Power Electronics*, Prentice-Hall, Englewood Cliffs, NJ, 1997.

[5] R.D. Middlebrook, Small-signal modeling of pulse-width modulated switched-mode power converters, *Proceedings of the IEEE*, 76 (4), 343–354, 1988.

[6] H. Mao, Study on Three-Phase High-Input-Power-Factor PWM-Voltage-Type Reversible Rectifiers and Their Control Strategies, Ph.D. Dissertation, Zhejiang University, 2000.

[7] M. Tsai and W.I. Tsai, Analysis and design of three-phase ac-to-dc converters with high power factor and near-optimum feedforward, *IEEE Transactions on Industrial Electronics*, 46 (3), 535–543, 1999.

[8] C.T. Rim, N.S. Choi, G.C. Cho, and G.H. Cho, A complete DC and AC analysis of three-phase controlled-current PWM rectifier using circuit D-Q transformation, *IEEE Transactions on Power Electronics*, 9 (4), 390–396, 1994.

[9] http://www.mathworks.com/access/helpdesk/help/pdf_doc/physmod/powersys/powersys.pdf.

CHAPTER 7

CONTROL OF GRID-CONNECTED FUEL CELL POWER GENERATION SYSTEMS

7.1 INTRODUCTION

As discussed in the previous chapters, fuel cells (FCs) show great promise to be an important distributed generation (DG) source of the future due to many advantages they have, such as high efficiency, zero or low emission (of pollutant gases), and flexible modular structure [1]. Fuel cell DG (FCDG) systems can either be connected to a utility grid or installed in a remote area to supply stand-alone power.

Grid-connected FCDGs can be strategically placed in a power system (normally at the distribution level) for grid reinforcement, deferring, or eliminating the need for system upgrades, and improving system integrity, reliability, and efficiency. At the same time, however, certain operational requirements are imposed on FCDGs. For example, they should be able to deliver a preset amount of real and reactive power to the grid, or be able to follow a time-varying load profile [2,6,7]. Therefore, proper controllers need to be designed for a FCDG to make its performance characteristics as desired [2–8].

In this chapter, control design strategies are discussed for grid-connected FC systems. Two types of fuel cells, PEMFC (polymer electrolyte membrane) and SOFC (solid oxide fuel cell), which both are

Modeling and Control of Fuel Cells, By M. Hashem Nehrir and Caisheng Wang

likely to be used in DG applications, will be addressed. The 500-W PEMFC model developed in Chapter 3, the 5-kW SOFC model presented in Chapter 4, and the models for boost dc/dc converters and a three-phase inverter discussed in Chapter 6, will be used in this chapter for simulation study and controller design. Simulation models will be developed for both a PEMFC DG system and a SOFC DG system in the MATLAB/ SIMULINK environment. Simulation studies are carried out to investigate real and reactive power control and load-following performance of FCDG systems, as well as their performances under severe electrical faults.

7.2 GRID-CONNECTED SYSTEM CONFIGURATION

A FCDG normally is interfaced to the power grid through a set of power electronic devices. The interface is very important as it affects the operation of the fuel cell system as well as the power grid. Various power electronic circuits have been proposed in recent work to interface different energy sources with a utility grid [9–16]. Direct current/direct current (dc/dc) converters are needed to regulate the fuel cell output voltage and adapt it to a desired voltage to be supplied to an ac/dc converter (inverter) and at the same time smooth the fuel cell output current [17,18]. The inverter converts the regulated dc bus voltage to a desired ac voltage, which can be synchronized with the grid. Pulse-width modulated (PWM) voltage source inverters (VSI) are widely used for this purpose. The real and reactive power flow from the FCDG to the power grid can be controlled through these devices [9,11,15,16].

The schematic diagram of a grid-connected FCDG system is shown in Fig. 7.1. The system consists of a 480-kW FC power plant, which is made of parallel connection of several fuel cell arrays. A boost converter is used to adapt the output voltage of each fuel cell array to a 480 V dc bus. The dc bus voltage (dc/dc converter output) is mainly determined by the inverter ac output voltage and the voltage drop across the smoothing LC (inductor–capacitor) filter. The FC power plant in the figure can either be PEMFC or SOFC type. A lower bound on the dc bus voltage (V_{dc}) can be determined from the following equation (given for unity power factor loads), which should be satisfied between the dc side and ac side voltages of the inverter [15]:

$$\frac{\sqrt{3}}{2\sqrt{2}} m_a V_{dc} \geq \sqrt{(V_{dc,LL})^2 + 3(\omega L_f I_{ac})^2} \tag{7.1}$$

where $V_{ac,LL}$ is the line–line RMS (root-mean-square) voltage on the inverter ac side, L_f is the filter inductance, I_{ac} is the maximum possible

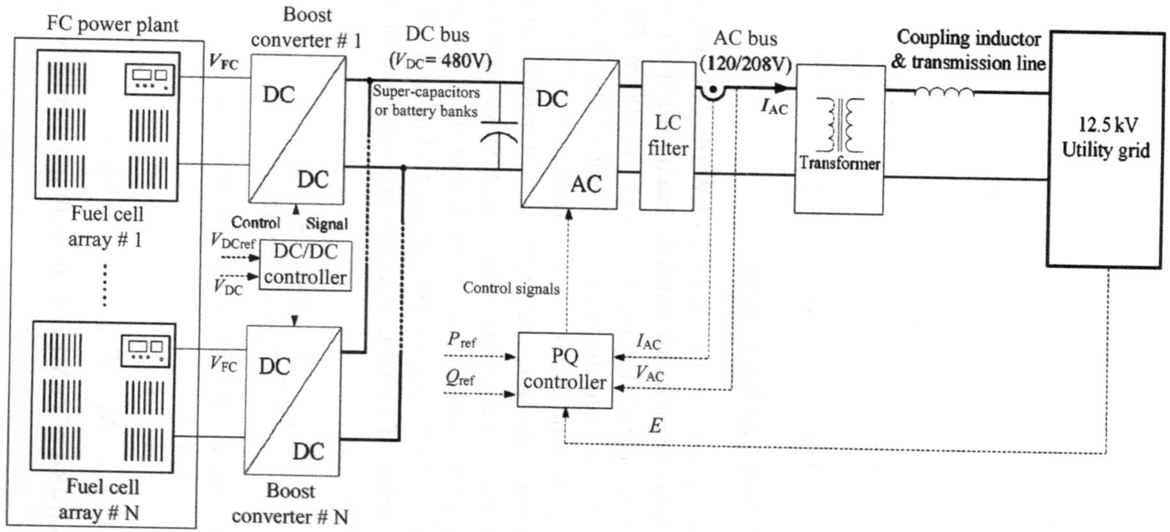

FIGURE 7.1 Block diagram of a fuel cell distributed generation system.

RMS value of the ac load current, and m_a is the modulation index of the inverter. Linear pulse-width modulation (PWM) is used for the inverter; $m_a \leq 1.0$ in the study in this chapter.

For a boost converter, the higher the duty ratio (or the larger the voltage difference between the input and output), the lower is its efficiency [16]. On the contrary, the output voltage of the converter will not be boosted to a distinctly higher voltage if the duty ratio is too low. A duty ratio around 55% can be used for the dc/dc converters at the fuel cells rated operating point. The approximate desired input voltage for the dc/dc converter, or the output voltage of each FC array, V_{FC}, can then be obtained, according to Equation 6.5 in Chapter 6, as $(1-0.55)\ V_{dc} = 216$ V.

In the next two subsections, the design of the PEMFC and SOFC arrays will be presented, from the 500-W PEMFC model developed in Chapter 3 and the 5-kW SOFC model developed in Chapter 4, respectively.

7.2.1 PEMFC Unit Configuration

The PEMFC power plant consists of ten 48-kW PEMFC arrays connected in parallel for a total capacity of 480 kW. Each 48-kW PEMFC array is composed of a series-parallel combination of the 500-W PEMFC stacks developed in Chapter 3. According to the *V–I* characteristics of the PEMFC, given in Fig. 3.14, when the PEMFC load current is over 23 A the fuel cell is in the concentration zone. Fuel cell operation should be avoided in this zone [19,20]. In order to leave some safe margin between the fuel cell linear operating region (ohmic zone) and concentration zone, its rated operating point is chosen at 20 A and output voltage (V_{PEMFC}) of about 27 V. Therefore, the number of PEMFC stacks to be connected in series to get a voltage of 216 V is

$$N_s = \frac{V_{FC\ array}}{V_{PEMFC}} = \frac{216}{27} = 8 \tag{7.2}$$

The number of series PEMFC stacks needed to get a 48-kW PEMFC array can be obtained as follows:

$$N_p = \frac{P_{array}}{N_s \times P_{stack}} = \frac{48\ \text{kW}}{8 \times 0.5\ \text{kW}} = 12 \tag{7.3}$$

Therefore, each 48-kW PEMFC array is composed of $8 \times 12 = 96$,500-W stacks.

7.2.2 SOFC Unit Configuration

The 480-kW SOFC power plant can be built using twelve 40-kW SOFC arrays connected in parallel. The series-parallel configuration for each array can be determined similar to designing the PEMFC arrays, discussed above. According to the V–I and P–I characteristics shown in Figs. 4.5 and 4.7 for a 5-kW SOFC stack, the stack is in the concentration zone when its current is over 110 A. Beyond this load current, the stack output voltage will decrease sharply as the load current increases. This situation should also be avoided [19]. In order to leave some safe margin, the SOFC stack rated operating point is chosen at 100 A and an output voltage (V_{SOFC}) of about 55 V. Therefore, the number (N_s) of series 5-kW SOFC stacks needed to get a voltage of 220 V is

$$N_s = \frac{V_{\text{FC array}}}{V_{\text{SOFC}}} = \frac{220}{55} = 4 \tag{7.4}$$

Therefore, four 5-kW SOFC stacks are connected in series to compose a 20-kW SOFC unit. The number (N_p) of the 20-kW SOFC units to be connected in parallel to compose a 40-kW SOFC array is 2:

$$N_p = \frac{P_{\text{array}}}{N_s \times P_{\text{stack}}} = \frac{40\,\text{kW}}{4 \times 5\,\text{kW}} = 2 \tag{7.5}$$

Each 40-kW SOFC array is therefore composed of $4 \times 2 = 8$, 5-kW SOFC stacks. The system configuration, ratings, and parameters of the PEMFC and SOFC DG systems are given in Table 7.1.

TABLE 7.1 Configuration Parameters of the Proposed System

FC power plant	PEMFC power plant	216 V/480 kW Ten 48-kW FC arrays connected in parallel
	PEMFC array	216 V/48 kW, consisting of 8 (series) × 12 (parallel) = 96 500-W PEMF stacks
	SOFC power plant	216 V/480 kW Twelve 40-kW FC arrays connected in parallel
	SOFC array	220 V/40 kW, consisting of 4 (series) × 2 (parallel) = 8 5-kW SOFC stacks.
	Boost dc/dc converter	200 V/480 V, 50 kW each
	3-phase dc/ac inverter	480 V dc/208 V ac, 500 kW
	LC filter	$L_f = 0.15$ mH, $C_f = 306.5$ μF
	Step-up transformer	$V_n = 208$ V/12.5 kV, $S_n = 500$ kW $R_1 = R_2 = 0.005$ PU, $X_1 = X_2 = 0.025$ PU
	Coupling inductor	$X_c = 50\ \Omega$
	Transmission line	0.5 km ACSR 6/0 $R = 2.149$ Ω/km, $X = 0.5085$ Ω/km
	dc bus voltage	480 V
	ac bus voltage	120 V/208 V

A fast-response energy storage device, such as a super-capacitor or a battery bank, is connected to the dc bus to provide storage capability and fast dynamic response to load transients. A three-phase six-switch inverter interfaces the dc bus to a 120/208 V ac power system. An LC filter is connected to the output of the inverter to reduce the harmonics introduced by the inverter. A 208 V/12.5 kV step-up transformer connects the fuel cell power system to the utility grid through a coupling inductor and a short transmission line. The coupling inductor is needed to control the real and reactive power flow between the FCDG system and the utility grid and to limit power disturbance and fault currents.

The controllers for the boost dc/dc converters are designed to keep the dc bus voltage within ±5% of the desired values. Therefore, the input to the three-phase inverter can be considered as a fairly constant voltage source. The inverter controller controls the real and reactive (P, Q) power flows to the utility grid. P, Q power flows follow their respective reference values, which can either be set as fixed values or to follow a certain load demand. The dc/dc converter controller and the PQ controller for the inverter, shown in Fig. 7.1, will be discussed in the following section.

7.3 CONTROLLER DESIGNS FOR dc/dc CONVERTERS AND THE INVERTER

The output voltage of a fuel cell is a function of its load, as shown in Figs. 3.14 and 4.5. The boost dc/dc converter adapts the fuel cell output voltage to the dc bus voltage, and the voltage controller helps regulate the output voltage within a ±5% tolerance band under normal operation. A conventional PI controller can be used for the boost dc/dc converters. In practice, a load-sharing controller (not discussed in this book) can be applied on the converters, connected in parallel, to achieve a uniform load distribution among them [27]. Also, a dq-transformed two-loop current control scheme is presented for the inverter to control the real and reactive power delivered from the fuel cell power system to the grid. The controller design for the dc/dc converters and the inverter of the FCDG system are given in this section, using the models developed for these devices, discussed in Chapter 6.

7.3.1 Circuit and Controller Design for the Boost dc/dc Converter

7.3.1.1 Circuit Design The main components of a dc/dc converter can be determined by its technical specifications, such as rated and peak

TABLE 7.2 Specifications of the Boost dc/dc Converter

Specification	Value
Input voltage, V_i	200 V
Output voltage, V_o	480 V
Rated power	50 kW
Input current ripple, $\Delta i_L/I_L$	$\leq$20%
Output voltage ripple $\Delta V_o/V_o$	$\leq$5%

voltages and currents, and tolerable input current and output voltage ripples [15,16]. Based on the FC system configuration discussed in Section 7.2, the technical specifications of each dc/dc converter in the system of Fig. 7.1 are summarized in Table 7.2. The input voltage is determined as 200 V so that the converter will have a little wider input voltage range than the designed input voltage, selected as 216 V. Another reason for choosing a slightly lower converter-rated voltage is because of degradation effect of a fuel cell stack over its operating time. The output voltage of a fuel cell will drop slightly over time due to its performance degradation [19,20]. In order to draw a relatively smooth current from each fuel cell array, the ripple of the inductance current at the rated operating point is set to be less than 20%. The output voltage ripple is set to a typical value, less than 5%. A boost dc/dc converter with the topology shown in Fig. 6.16 will be designed to meet these specifications.

Switching Frequency Since the rated power of the boost converter will be 50-kW, IGBT (Insulated Gate Bipolar Transistor) is a good choice for the switching device in the circuit [15]. The switching frequency of IGBTs is normally less than 50 kHz. In this design, the switching frequency (f_s) is chosen as 5 kHz.

Inductance L_{dd} The equivalent resistive load at the rated operating point is

$$R - \frac{V_{\mathrm{dd\_out}}^2}{P_{\mathrm{dd,N}}} = \frac{480^2}{50 \times 10^3} = 4.608\ \Omega \tag{7.6}$$

The rated duty ratio is

$$D = 1 - \frac{V_{\mathrm{dd\_in}}}{V_{\mathrm{dd\_out}}} = 0.5833 \tag{7.7}$$

Then, the rated input (inductor) current can be obtained as

$$\bar{I}_{L_{\mathrm{dd}},\mathrm{N}} = \frac{V_{\mathrm{dd\_in}}}{(1-D)^2 R} = \frac{200}{(1-0.5833)^2 \times 4.608} \approx 250\ \mathrm{A} \tag{7.8}$$

Based on the requirement specified for the input current ripple (to be less than 20%), we can get

$$\frac{\Delta i_{L_{\rm dd}}}{\bar{I}_{L_{\rm dd},\rm N}} = \frac{D(1-D)^2 R}{f_{\rm s} L_{\rm dd}} \le 20\% \Rightarrow L_{\rm dd} \ge \frac{5D(1-D)^2 R}{f_{\rm s}} \tag{7.9}$$

That is,

$L_{\rm dd} \ge \frac{5\times 0.5833\times(1-0.5833)^2\times 4.608}{5\times 10^3} = 0.4667$ mH. $L_{\rm dd}$ is chosen as 1.2 mH for this design.

Capacitance C_{dd} The output voltage ripple percentage can be calculated as

$$\frac{\Delta V_{\rm o}}{V_{\rm o}} = \frac{D}{RC_{\rm dd} f_{\rm s}} \le 5\% \Rightarrow C_{\rm dd} \ge \frac{D}{0.05Rf_{\rm s}} = \frac{0.5833}{0.05\times 4.608\times 5000} = 506.337\,\mu\text{F} \tag{7.10}$$

$C_{\rm dd}$ is selected as 1000 μF in this design.

Switch S According to the selected component parameters, the maximum current through the switch at the rated operating point is

$$I_{S_{\rm dd},\max} = I_{L_{\rm dd},\max} = 250 \times 1.1 = 275 \text{ A}$$

The peak switch voltage at the rated operating point is

$$V_{S_{\rm dd},\max} = V_{\rm dd\_out} = 480 \text{ V}$$

Power Diode D The peak forward current through diode D at the rated operating point is

$$I_{D_{\rm dd},\max} = 275 \text{ A}$$

The peak inverse voltage across the diode at the rated operating point is

$$V_{D_{\rm dd},\max} = V_{\rm dd\_out} = 480 \text{ V}$$

The component values for each 50 kW dc/dc converter (used for each FC array) are summarized in Table 7.3; based on those values, a controller can be designed to regulate the converter output voltage.

7.3.1.2 Controller Design The circuit of a boost dc/dc converter with an output voltage regulation loop is shown in Fig. 6.18. Figure 7.2

TABLE 7.3 System Parameters FOR the Boost dc/dc Converter

L_{dd}	1.2 mH
C_{dd}	1000 μF
D	0.5833
R	4.608 Ω
$\bar{I}_{L_{dd},N}$ (X_1)	250 A
$V_{dd\_out}(X_2)$	480 V
k_{di}	50
k_{dp}	0.5

shows a transfer-function block diagram for the circuit given in Fig. 6.18 when it is linearized around its operating point. In that figure, the PWM controller is the compensator that needs to be designed so that the dc/dc converter provides the desired specifications. The pulse-width modulator is controlled/modulated by the controller output (v_c) to generate a pulse series with right duty ratios. The control voltage $v_c(t)$ is compared with a repetitive saw-tooth waveform $v_{tri}(t)$ at the switching frequency f_s, as shown in Fig. 7.3.

By assuming a small sinusoidal perturbation in the control voltage, the transfer function of a pulse-width modulator can be derived as [15,16]:

$$T_{PWM}(s) = \frac{1}{V_{tri}} \tag{7.11}$$

where V_{tri} is the saw-tooth waveform amplitude of the pulse-width modulator.

For the boost dc/dc converter shown in Fig. 6.18, its transfer function (output voltage over duty ratio) $T_p(s)$ is obtained as (see Eq. 6.10 in Chapter 6):

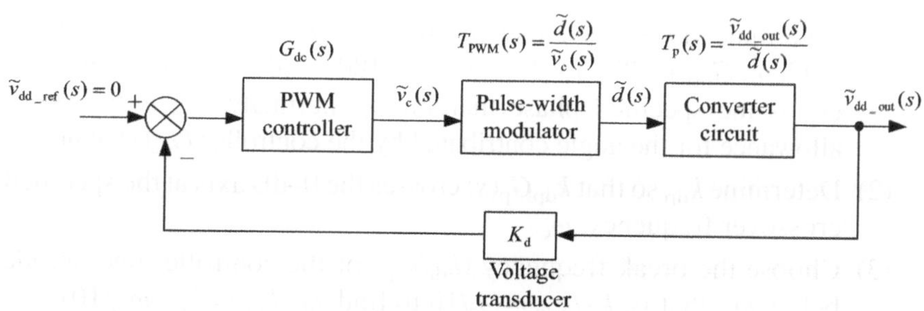

FIGURE 7.2 Block diagram of the control loop for the boost dc/dc converter.

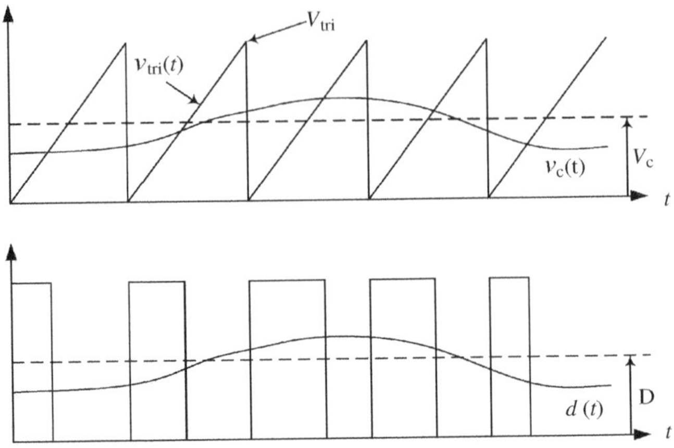

FIGURE 7.3 Pulse-width modulator.

$$T_{\rm p}(s) = \frac{\tilde{v}_{\rm dd\_out}(s)}{\tilde{d}(s)} = \frac{1}{s[s+(1/RC_{\rm dd})]+[(1-D)^2/L_{\rm dd}C_{\rm dd}]}\left[-\frac{X_1 s}{C_{\rm dd}}+\frac{(1-D)X_2}{L_{\rm dd}C_{\rm dd}}\right] \tag{7.12}$$

where D is the rated duty ratio of the switch, X_1 the average current through the inductor $L_{\rm dd}$, and X_2 the converter output voltage at the rated operating point.

The constant $K_{\rm d}$ of the voltage transducer in Fig. 7.2 is set to 1/480 (480 V being the desired voltage at the dc bus) to normalize the output dc voltage. The PWM controller in the figure is chosen as a PI controller with the transfer function:

$$G_{\rm dc}(s) = \frac{k_{\rm dp}s + k_{\rm di}}{s} = k_{\rm dp}\left(1 + \frac{k_{\rm di}}{k_{\rm dp}}\frac{1}{s}\right) \tag{7.13}$$

The transfer function given in (7.12) can be used to design the PI current controller as follows [21]:

(1) Determine the desired crossover frequency $\omega_{\rm c}$, where the phase angle of plant, $G_{\rm p}(s) = T_{\rm p}(s) \times T_{\rm PWM}(s)$ is $-180° + \phi_{\rm m} + \phi_{\rm c}$ (see Fig. 7.2). $\phi_{\rm m}$ is the specified phase margin. $\phi_{\rm c}$, normally, 5°, is the phase allowance for the angle contributed by the controller $G_{\rm dc}(s)$ at $\omega_{\rm c}$.

(2) Determine $k_{\rm dp}$ so that $k_{\rm dp}G_{\rm p}(s)$ crosses the 0–dB axis at the specified crossover frequency, $\omega_{\rm c}$.

(3) Choose the break frequency ($k_{\rm di}/k_{\rm dp}$) of the controller one decade below $\omega_{\rm c}$, that is, $k_{\rm di}/k_{\rm dp} = \omega_{\rm c}/10$ to find $k_{\rm di}$, $k_{\rm di} = k_{\rm dp}(\omega_{\rm c}/10)$.

(4) Plot $G_{\rm dc}(s)G_{\rm p}(s)$ to verify that all specifications are met.

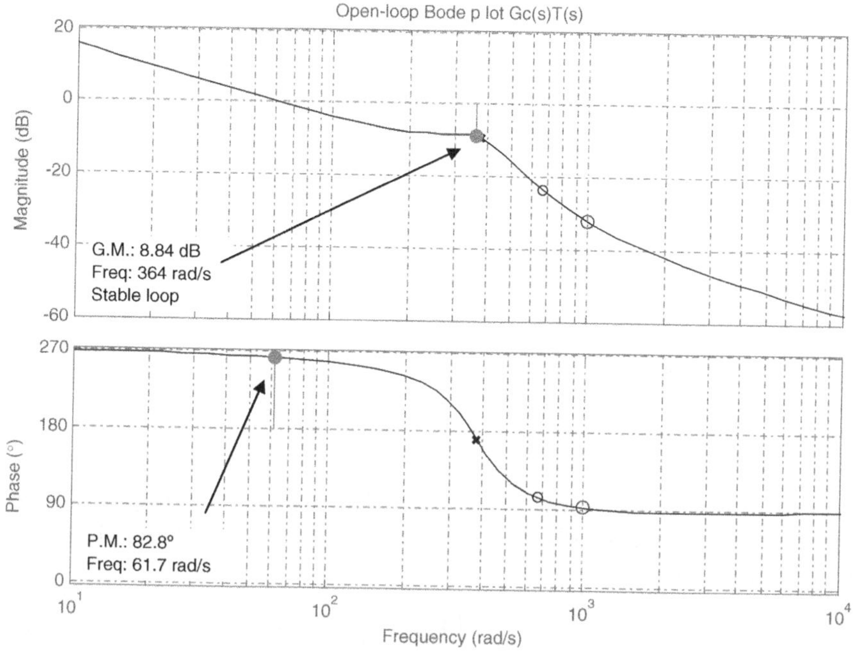

FIGURE 7.4 Bode diagram of the control loop for the boost dc/dc converter.

The controller parameters can be adjusted from the values obtained through the above design procedure based on actual circuit testing or simulation study. The actual parameters of the *PI* voltage controller for each dc/dc converter in Fig. 7.1 are listed in Table 7.3 as well. Figure 7.4 shows the open-loop Bode diagram (magnitude and phase) for the dc/dc converter with the PI controller. It is noted from the figure that the system is stable and has a phase margin of $267.2° - 180° = 82.8°$.

7.3.2 Controller Design for the Three-Phase VSI

In order to meet the requirements for interconnecting a fuel cell system to a utility grid and control the real and reactive power flow between them, it is necessary to shape and control the inverter output voltage in amplitude, angle, and frequency [22]. In this section, a PWM controller is designed for the inverter shown in Fig. 6.27 to satisfy voltage regulation as well as to achieve real and reactive power control.

Though the state-space Equation 6.31 in Chapter 6 is sufficient for theoretical analysis and design, it is too complicated to be applied directly in actual controller design. In order to simplify the analysis and design, a control scheme consisting of an outer voltage regulator with an inner current control loop, presented in Ref. [23], is adapted for inverter control, as shown in Fig. 7.12, discussed later in this section. With the help of "*abc*/*dq*" transformation (discussed in Chapter 6), the inner current control loop is designed to respond faster than the outer voltage control loop so that these two loops can be treated separately [23,24]. The controller output signals in the *dq* frame are then converted back to *abc* coordinates through "*dq*/*abc*" transformation to produce the proper pulses for the inverter switches. The details of the controller design for the current control loop, voltage control loop, and the overall control scheme are discussed in the following subsections.

7.3.2.1 Current Control Loop The ramp comparison method for one phase of the three-phase inverter, shown in Fig. 7.5, is used to generate the six PWM pulses. Based on the state-space averaging technique discussed in Chapter 6 [15,23], the average value of the switching functions in a switching period is

$$(d_k^*)_{\text{avg}} = \frac{i_{\text{ck}}}{V_{\text{m\_tri}}}, \quad k = 1,2,3 \tag{7.14}$$

where i_{ck} is the modulating (control) signal and $V_{\text{m\_tri}}$ is the amplitude of the carrier signal. The gain of the pulse-width modulator, K_{PWM} can then be obtained as

$$K_{\text{PWM}} = \frac{V_{\text{dc}}}{2V_{\text{m\_tri}}} \tag{7.15}$$

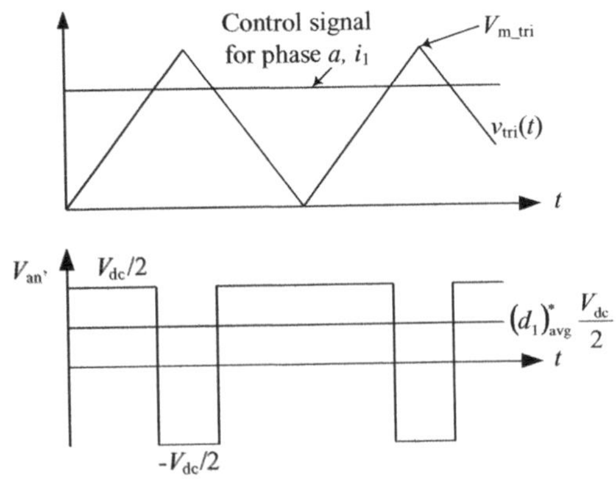

FIGURE 7.5 Sinusoidal pulse-width modulator for one phase of the three-phase inverter.

Converting the control signals from *abc* coordinates into the *dq* frame yields:

$$\begin{bmatrix} i_{cd} \\ i_{cq} \\ i_{c0} \end{bmatrix} = T_{abc/dq0} \begin{bmatrix} i_{c1} \\ i_{c2} \\ i_{c3} \end{bmatrix} \tag{7.16}$$

where i_{cd} and i_{cq} are the control signals in *dq* frame and $i_{c0} = 0$ for a balanced system. $T_{abc/dq0}$ is the *abc/dq*0 transformation matrix [28], also given in Equation 6.27 in Chapter 6.

Only unity power factor operation of the inverter, that is, zero value for the *q*-axis component) is discussed. The effect of the capacitor C_f in the low pass LC filter (see Fig. 6.27) on the power frequency[1] current is very small. Therefore, C_f can be neglected when designing the current loop controller [23]. Based on the state-space Equation 6.31, the current control loop in *d*-axis (frame) can be represented in the diagram shown in Fig. 7.6, where, for example, i_{sd} is the *d*-axis component value of the *abc/dq0* transformation of the *abc* signals $\{i_{sa}, i_{sb}, i_{sc}\}$, and v_d and e_d are the *d*-axis components of $\{v_a, v_b, v_c\}$ and $\{e_A, e_B, e_C\}$ in Fig. 6.27, respectively. i_{sd_ref} is generated from the outer voltage control loop, which will be discussed next.

In the above figure, $G_{ACR}(s)$ is the current controller, which is chosen as a generic PI controller:

$$G_{ACR}(s) = \frac{k_{cp}s + k_{ci}}{s} \tag{7.17}$$

The expression $1/(R_s + L_s s)$ is the model of the overall equivalent admittance of the LC filter inductor, power transformer, coupling inductor and transmission line, and k_c is the current transducer ratio. The parameters of the current control loop are summarized in Table 7.4.

The transfer function of the current control loop can be obtained as

$$T_c(s) = \frac{i_{sd}(s)}{i_{sd\_ref}(s)} = \frac{P_{1c}\Delta_{1c}}{\Delta_c} \tag{7.18}$$

where $P_{1c} = G_{ACR} K_{PWM} \frac{1}{(R_s + L_s s)}$, $\Delta_{1c} = 1$, and

$$\Delta_c = 1 + G_{ACR} \frac{K_{PWM}}{k_c} \frac{1}{(R_s + L_s s)}$$

[1]Power frequency is 60 Hz in North America and 50 Hz in Europe.

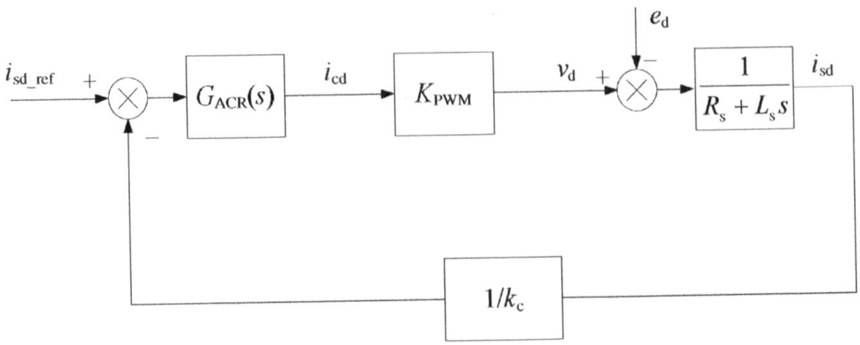

FIGURE 7.6 Block diagram of the *d*-axis component of the current control loop of the inverter.

Using the parameters listed in Table 7.4, the Bode plots (magnitude and phase) for the current control loop are given in Fig. 7.7. It is noted from this figure that the system is stable with a phase margin of 88.2°.

The unit step response of the current control loop is shown in Fig. 7.8. It is noted that the overshoot is about 2.5% and the equivalent dominant time constant is about 0.45 ms. Therefore, the overall current control loop can be approximated by a simple lag with gain k_c as

$$T_c(s) = \frac{k_c}{1 + \tau_i s} \tag{7.19}$$

where $\tau_i = 0.45$ ms is the equivalent time constant of the current control loop and k_c is the current transducer ratio.

The unit step response of $T_c(s)$is also shown in Fig. 7.8 when k_c is set to 1. This approximated overall transfer function will be used in the following section to design the voltage control loop.

7.3.2.2 Voltage Control Loop Using the state-space inverter model given in Equation 6.31 in Chapter 6, the voltage control loop

TABLE 7.4 Parameters of the Current Control Loop the Inverter

V_{dc}	480 V
V_{tri_m}	1 V
R_s	0.006316 Ω
L_s	0.1982 mH
k_c	1388
k_{ci}	250
k_{cp}	2.5

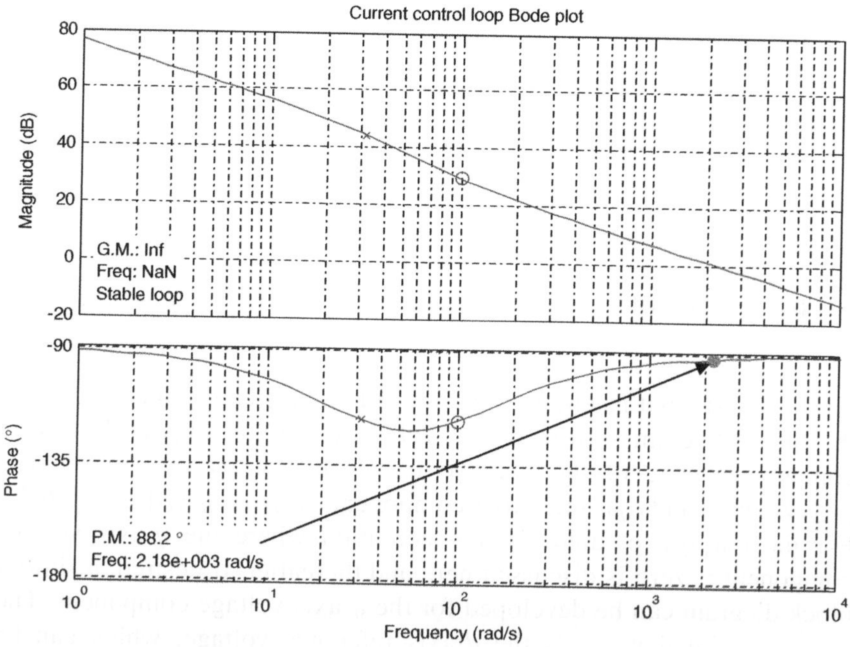

FIGURE 7.7 Bode plot of the current control loop.

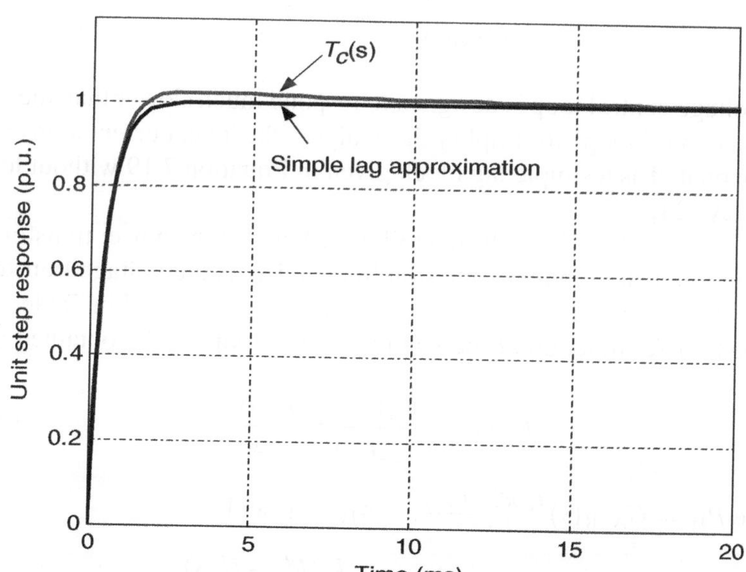

FIGURE 7.8 Unit step response of the current control loop and its approximation.

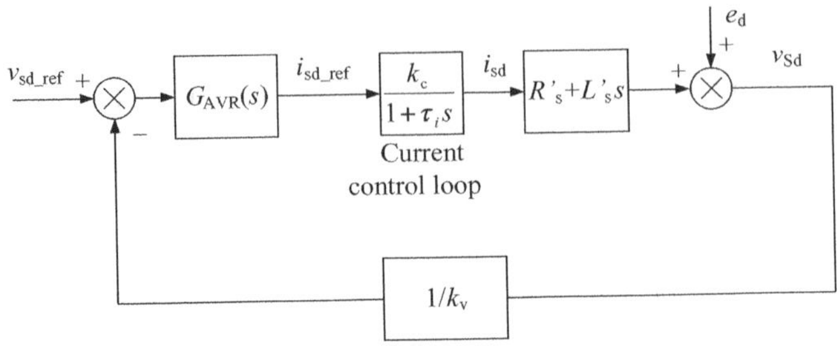

FIGURE 7.9 Block diagram of the voltage control loop for the inverter.

for the inverter with a circuit topology in Fig. 6.27 can be represented by the block diagram shown in Fig. 7.9. In this figure, v_{sd} is the d-axis component value of the *abc/dq0* transformation of the *abc* signals $\{v_{sa}, v_{sb}, v_{sc}\}$, and e_d is the d-axis components of $\{e_A, e_B, e_C\}$ in Fig. 6.27. For the purpose of analysis, the condition where the q-axis voltage component is zero is discussed here and thereafter. If desired, a similar block diagram can be developed for the q-axis voltage component. The reference signal v_{sd_ref} is the d-axis reference voltage, which can be determined by (7.26).

In Fig. 7.9, $G_{AVR}(s)$ is the voltage regulator, which similar to the current controller, is chosen as a PI controller.

$$G_{AVR}(s) = \frac{k_{vp}s + k_{vi}}{s} \tag{7.20}$$

The voltage control loop is designed to respond much slower than the inner current control loop. To simplify the analysis, the inner current loop can be approximated as a simple lag block given in Equation 7.19 without losing accuracy [24].

$R'_s + L'_s\, s$ in Fig. 7.9 is the overall impedance of the power transformer, coupling inductor, and transmission line, and k_v is the voltage transducer ratio. The parameters of the voltage control loop are listed in Table 7.5.

The transfer function of the voltage control loop can be obtained as

$$T_v(s) = \frac{v_{sd}(s)}{v_{sd\_ref}(s)} = \frac{P_{1v}\Delta_{1v}}{\Delta_v} \tag{7.21}$$

where $P_{1v} = G_{AVR}(s)\frac{k_c(R'_s + L'_s\, s)}{1+\tau_i s}$, $\Delta_{1v} = 1$, and

$$\Delta_v = 1 + G_{AVR}(s)\frac{k_c(R'_s + L'_s\, s)}{k_v(1+\tau_i s)}$$

TABLE 7.5 Parameters of the Voltage Control Loop the Inverter

R'_{s}	0.005316 Ω
L'_{s}	0.0482 mH
τ_{I}	0.45 ms
k_{c}	1388
k_{v}	169.8
k_{vi}	25
k_{vp}	0.2

The Bode plot of the voltage control loop is shown in Fig. 7.10. It is noted that the system is stable with phase margin of 91°.

Control of the real and reactive power flow through the inverter can be achieved by the following:

Consider a voltage source $V_{\mathrm{s}}\angle\delta$ connected to a utility grid $E\angle 0°$ through a coupling impedance $R + jX$, as shown in Fig. 7.11. The real and reactive powers delivered to the utility grid are [25]

$$P = \frac{EV_{\mathrm{s}}}{Z}\cos(\theta_z - \delta) - \frac{E^2}{Z}\cos(\theta_z) \tag{7.22}$$

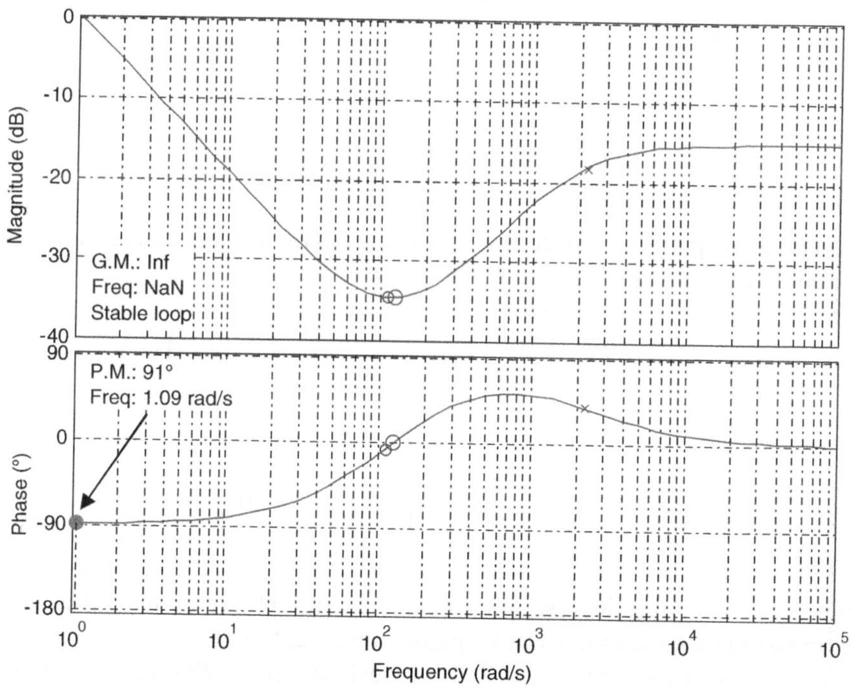

FIGURE 7.10 Bode plot of the voltage control loop.

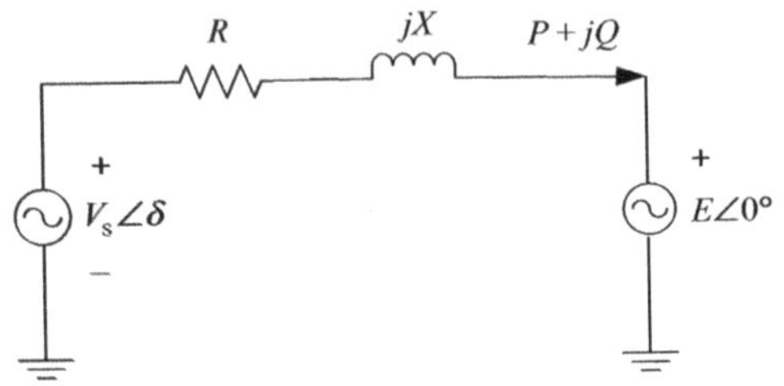

FIGURE 7.11 Power flow between a voltage source and utility grid.

$$Q = \frac{EV_s}{Z}\sin(\theta_z - \delta) - \frac{E^2}{Z}\sin(\theta_z) \tag{7.23}$$

where $Z = \sqrt{R^2 + X^2}$ and $\theta_z = \tan^{-1}(X/R)$.

From (7.22) and (7.23), it is clear that the real and reactive powers delivered to the utility grid are completely determined by the amplitude and angle of the sending-end voltage source, in the case of FCDG, the output voltage of the inverter. If the desired values of real and reactive power are known, the values of V_s and δ can be determined from (7.22) and (7.23):

$$V_s = \left[\frac{Z^2}{E^2}(P^2 + Q^2) + E^2 + 2PZ\cos(\theta_z) + 2QZ\sin(\theta_z)\right]^{\frac{1}{2}} \tag{7.24}$$

$$\delta = \theta_z - \cos^{-1}\left(\frac{ZP}{EV_s} + \frac{E}{V_s}\cos(\theta_z)\right) \tag{7.25}$$

The corresponding *dq0* component values of the voltage can be obtained through *abc/dq* transformation as follows:

$$\begin{bmatrix} V_d \\ V_q \\ V_0 \end{bmatrix} = T_{abc/dq0}\begin{bmatrix} v_a(t) \\ v_b(t) \\ v_c(t) \end{bmatrix} = T_{abc/dq0}P^{-1}\left(\begin{bmatrix} V_a \\ V_b \\ V_c \end{bmatrix}\right) = T_{abc/dq0}P^{-1}\left(\begin{bmatrix} V_s\angle\delta \\ V_s\angle(\delta - 120^\circ) \\ V_s\angle(\delta + 120^\circ) \end{bmatrix}\right) \tag{7.26}$$

where P^{-1} is the inverse phasor transform, which obtains the time expression of a variable from its phasor representation.

Based on Equations 7.24–7.26, a desired set of *PQ* values can be converted into *dq* reference voltages to control the inverter.

7.3.2.3 Overall Power Control System for the Inverter The overall control system block diagram for the inverter is given in Fig. 7.12. In this figure, the "*dq* reference signal computation" block, which is based on Equations 7.22–7.26, calculates the magnitude and angle of the filtered output voltages of the inverter and then converts them into *dq* voltage reference signals. The "*abc*/*dq* transformation" block takes the (*abc* coordinates) current and voltage values from the voltage and current meters and converts them into *dq* values. The outer voltage controller takes the error signals between the actual output voltage in the *dq* frame ($V_{d,q}$) and the reference voltage ($V_{d,q(\mathrm{ref})}$) and generates the current reference signals ($I_{d,q(\mathrm{ref})}$) for the current control loop. The inner current controller produces the *dq* control signals, which are converted back into the control signals in *abc* coordinates through the "*dq*/*abc* transformation" block. These signals are used to control the sinusoidal pulse-width modulator (SPWM) to produce the proper pulses for the inverter switches (Fig. 6.27), which control the inverter output voltage.

In general, the controller and the inverter are capable of operation in four quadrant. In other words, both real and reactive power can be controlled to flow in either direction (from or to the utility grid) based on the actual need. However, only one direction of real power from the FC system to the grid is discussed in this chapter since the FC stack considered in this chapter is not

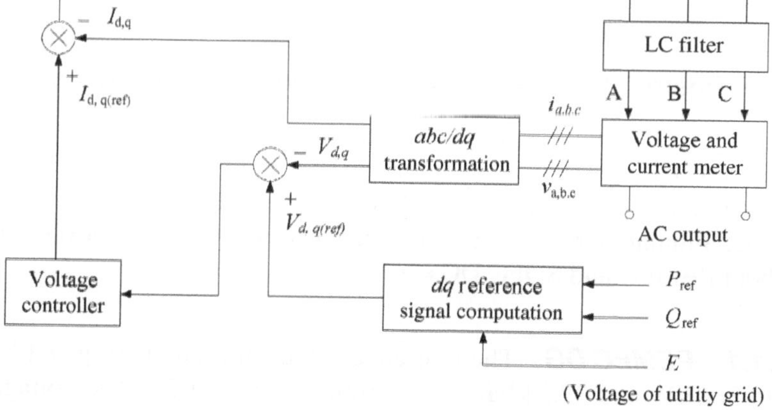

FIGURE 7.12 Block diagram of the overall power controller for the inverter.

a regenerative FC. Nevertheless, the FCDG system can be used for reactive power compensation at the same time when the real power is being delivered to the grid, which can be seen from the simulation studies given in the next section.

The controllers given in this section for power electronic converters are designed on the basis of small-signal linearized models; there is no guarantee that the designed controllers will work well over a large operating range because of the nonlinearity of power electronic devices. However, the majority of controllers for power electronic devices are linear time invariant (LTI) that are designed on the basis of linearized models. The effectiveness of LTI controllers can be checked through simulations on switched (nonlinear) models, rather than the linearized ones used for the initial controller design [29]. We have verified the validity of the proposed control schemes over a large operating range through simulations on nonlinear converter/inverter models. The simulation results will be discussed in the following section.

7.4 SIMULATION RESULTS

The individual fuel cell stack models developed in Chapters 3 and 4, and the controllers designed in this chapter were used to develop simulation models for both a PEMFC DG system and a SOFC DG system. The models were built in MATLAB/SIMULINK using SimPowerSystems block-set. Each system was tested under several different operating conditions to investigate its power management and load-following capabilities as well as its stability during electrical faults. Sample simulation results of the scenarios studied follow.

7.4.1 Desired P and Q Delivered to the Grid—Heavy Loading

Normally, when the electric power demand on a utility is heavy, the connected DGs are set to deliver more real power to the grid. The DGs may also be required to deliver reactive power to the grid to help boost the grid voltage. Simulation results for such a scenario are given in the next sections for both PEMFC and SOFC DGs.

7.4.1.1 PEMFC DG The reference values of P and Q (Fig. 7.12) are set to 360 kW and 32.3 kVar with a ramp loading of 2 s. The voltage of the utility grid was set to $E = 0.98\ \angle 0°$ PU. Using (7.24) and (7.25), the desired magnitude and angle of the filtered output voltage of the inverter turn out to be $V_s = 1.05$ PU, $\delta = 8.2678°$. Converting the values from the *abc* reference frame into the *dq* coordinates yields $V_{d(\text{ref})} = 1.0391$ PU and $V_{q(\text{ref})} = 0.151$ PU. A three-phase ac circuit breaker connects the inverter to the utility grid at $t = 0.08$ s. Figure 7.13 shows the real and reactive power delivered from the PEMFC DG to the grid when the DG reaches its steady-state operation from the initial loading. The steady-state values of the fuel cell P and Q agree with their reference values very well. The *dq* components of the inverter output voltage also reach their corresponding reference values, as shown in Fig. 7.14.

The output voltage and current responses of each PEMFC array for this case are shown in Fig. 7.15. When the FCDG system reaches a steady state, the fuel cell array output current ripple is about 10% and the output voltage ripple is less than 3.3%. These relatively small variations of the current and voltage are indicative of the healthy operation of fuel cells [26]. The output voltage of the boost dc/dc converter (dc bus voltage) for this case is shown in Fig. 7.16. After loading, the dc bus voltage increases to its reference value (480 V), although the PEMFC array terminal voltage falls well below its no-load value under this heavy loading condition, as shown in Fig. 7.15. The voltage ripple at the dc bus is about 1.25%, which is within the acceptable range of ±5%.

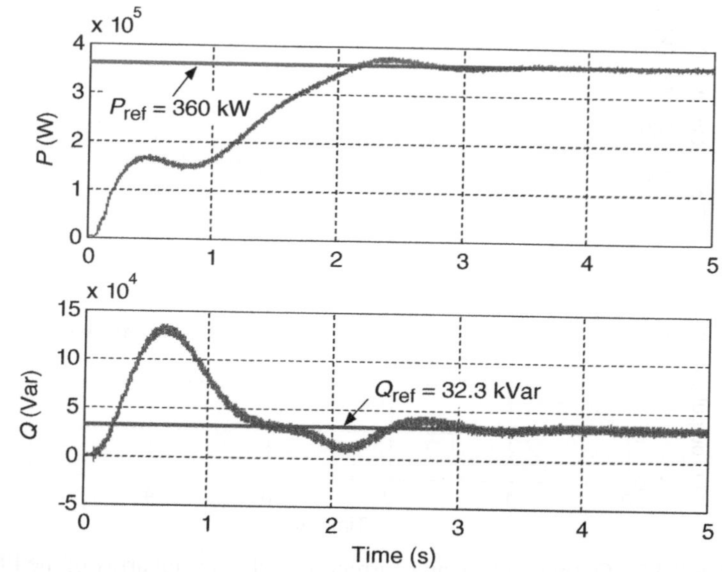

FIGURE 7.13 PEMFC DG: P and Q delivered to the grid under heavy loading.

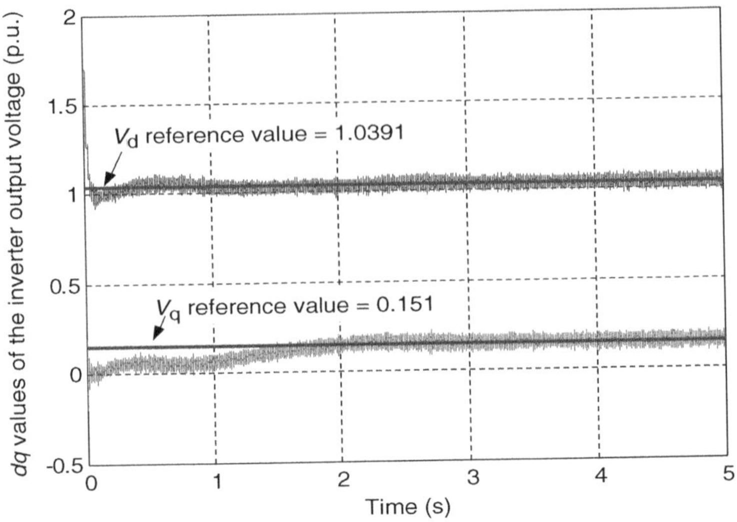

FIGURE 7.14 PEMFC DG: *dq* values of the inverter output voltage under heavy load.

7.4.1.2 SOFC DG Under heavy loading condition, the reference values of P and Q (Fig. 7.12) are set to 450 kW and 84.5 kVar, respectively, with a ramp loading in 2 s. The voltage of the utility grid is set to $E = 0.98 \angle 0°$ PU. Figure 7.17 shows the reference and actual values of the

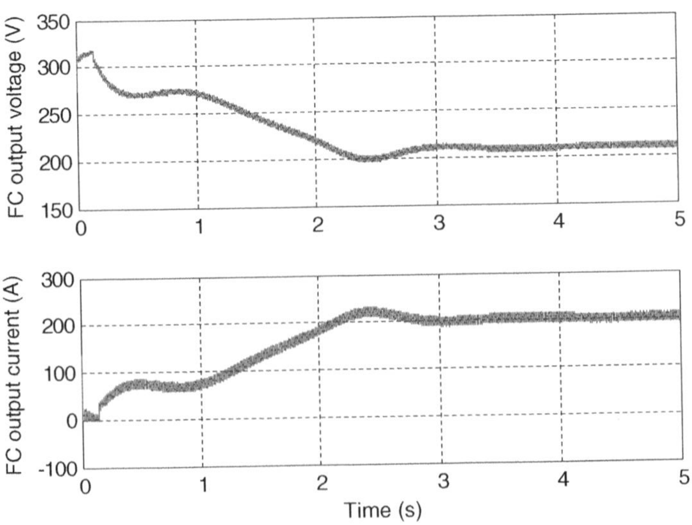

FIGURE 7.15 Output voltage and current of each fuel cell array of the PEMFC DG under heavy loading.

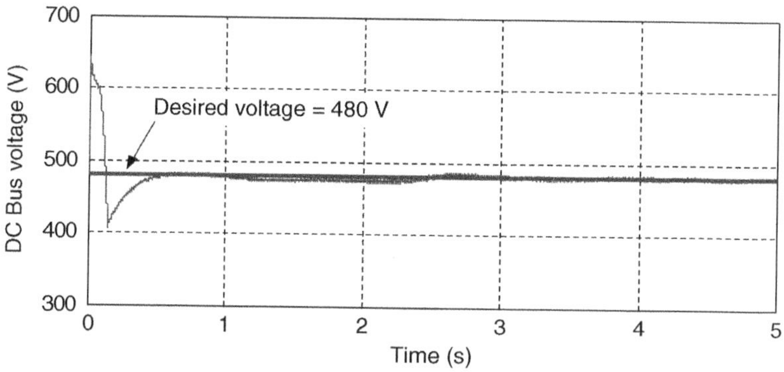

FIGURE 7.16 PEMFC DG: dc bus voltage waveform under heavy load.

real and reactive power delivered from the SOFC DG to the grid. The final values of P and Q delivered to the grid by the SOFC DG reach their reference values after a disturbance period of 3–4 s.

The output voltage and current responses of each 40-kW SOFC array for the above heavy loading case are shown in Fig. 7.18. Although the SOFC array output voltage drops as a result of loading; the array output current ripple is about 10% and its output voltage ripple is around 3%. Again, these relatively small current and voltage ripples are indicative of the healthy operation of FCDG system [30]. The dc bus voltage (output voltage of the boost dc/dc converter) for this case is shown in Fig. 7.19. The dc bus

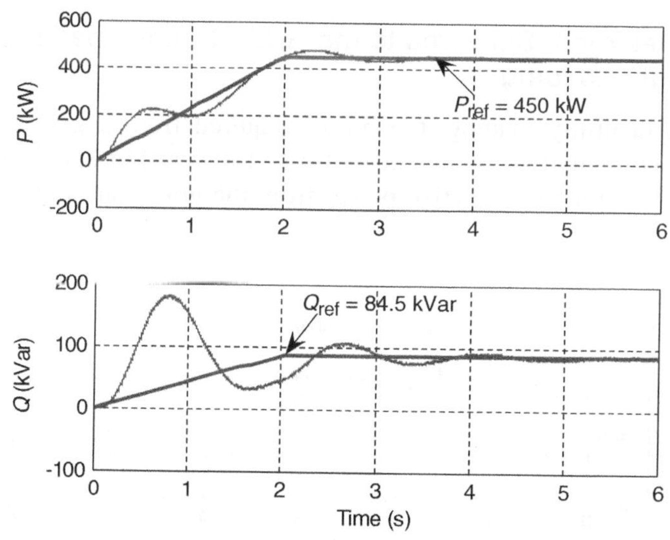

FIGURE 7.17 Real and reactive power delivered to the grid by the SOFC DG under heavy loading.

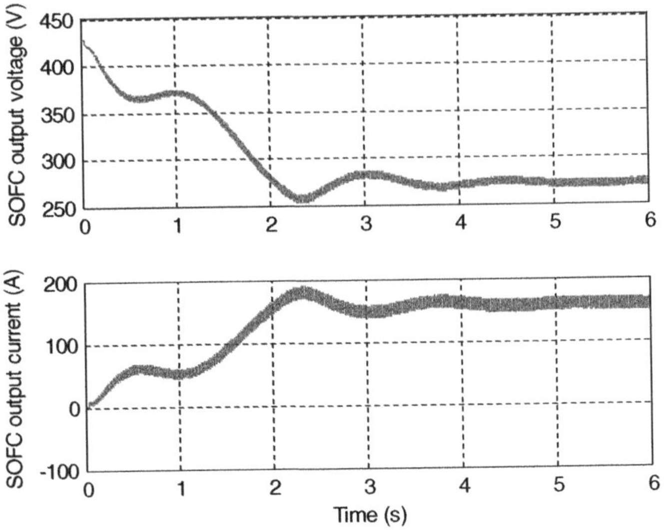

FIGURE 7.18 SOFC DG: Output voltage and current of each FC array under heavy load.

voltage comes up to its reference value (480 V) though the fuel cell terminal voltage is much lower than its no-load value under this heavy loading condition (Fig. 7.18). The voltage ripple at the dc bus is about 1.5%, which is within the acceptable range.

7.4.2 Desired P Delivered to the Grid, Q Consumed from the Grid: Light Loading

Under light utility loading, the power required from a grid-connected FCDG is normally low (compared to its rated power). Further, the DG may be set to consume the reactive power from the grid, that is $Q<0$. These

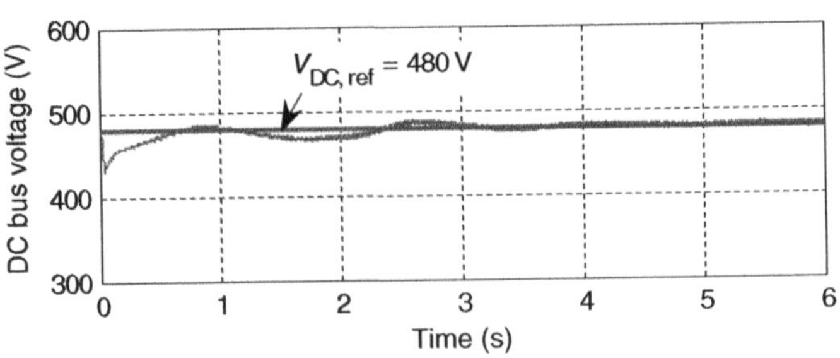

FIGURE 7.19 SOFC DG: dc bus voltage under heavy loading.

scenarios are examined in the next section for both PEMFC and SOFC DGs.

7.4.2.1 PEMFC DG For light loading of the PEMFC DG, the reference values of P and Q in Fig. 7.12 are set to $P_{ref} = 100\,\text{kW}$ and $Q_{ref} = -31.8\,\text{kVar}$ with an initial step-change from no-load to the reference load. The voltage of the utility grid is set to $E = 1.0\,\angle 0°$ PU. Using Equations 7.24 and 7.25, the desired amplitude and angle of the filtered output voltage of the inverter turn out to be $V_s = 1.0\,\angle 2.648°$ PU. The dq reference voltages (in Fig. 7.12) are then obtained by finding the dq components of V_s using Equation 7.26: $V_{d(ref)} = 0.99893$ PU and $V_{q(ref)} = 0.0463$ PU.

Figure 7.20 shows the real and reactive power responses of the PEMFC DG. After the initial transients, the FCDG comes to steady state delivering 100 kW of real power to the grid and absorbing 31.8 kVar of reactive power from the grid ($Q = -31.8\,\text{kVar}$), the reference values of P and Q, respectively.

Figure 7.21 shows the dc output voltage and current of each fuel cell array under light loading. In this case the fuel cell array output current percentage ripple is higher than that of the heavy loading case. These ripples are due to the electronic switches in the dc/dc converters; although the ripple magnitudes are about the same as the heavy loading case, the percent ripple is higher mainly due to lighter loading. On the contrary, the

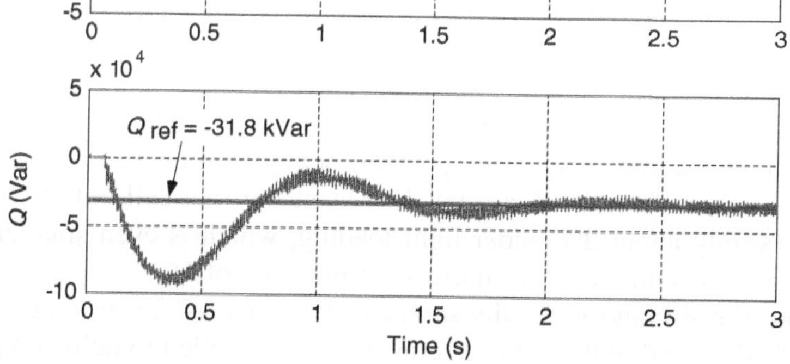

FIGURE 7.20 PEMFC DG: P and Q delivered to the grid: light loading.

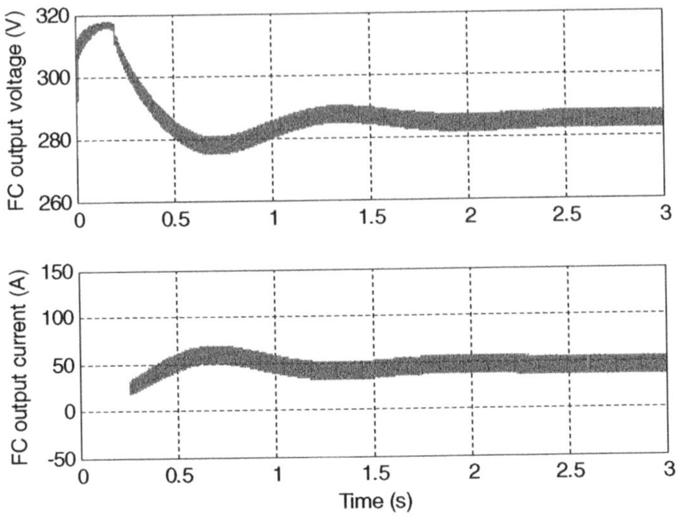

FIGURE 7.21 PEMFC DG: Output voltage and current of each fuel cell array under light loading.

output voltage ripple of each fuel cell array is only about 2%, which is even less than that of the heavy loading case. This is due mainly to a higher steady-state fuel cell output voltage under light loading.

7.4.2.2 SOFC DG Similar to the case of PEMFC DG, under light loading the SOFC DG delivers less real power to the grid (compared to heavy loading condition). It may also be set to absorb reactive power from the grid, that is $Q<0$, as shown in the simulation results for this scenario given in Figs. 7.22 and 7.23. The reference values of P and Q are set as 250 kW and −53.1 kVar with a ramp loading in 2 s. The voltage of the utility grid is again set to $E = 1.0 \angle 0°$ PU. Figure 7.22 shows that after the initial transients, the actual value of real power delivered from the SOFC DG to the grid and the reactive power absorbed from the grid by the SOFC DG reach their respective reference values.

The fuel cell array output voltage and current are given in Fig. 7.23. The current ripple percentage, which is about 25%, of the SOFC output current under light loading is again larger than heavy loading due to the lower average value of the load current. However, the fuel cell output voltage ripple is only about 2% under light loading, which is even smaller than heavy loading and well within the tolerable limit of 5%.

From the simulation results shown in Figs. 7.13–7.23, it is noted that both PEMFC and SOFC DG systems can be controlled to deliver a preset amount of real and reactive power. In practice, however, to get the

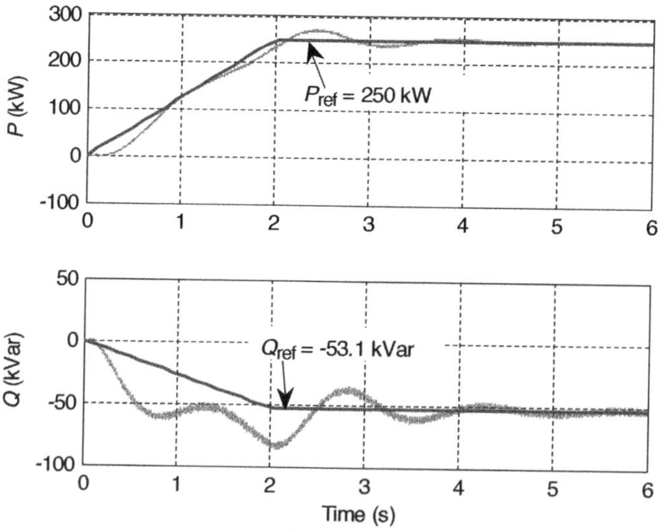

FIGURE 7.22 SOFC DG: P and Q delivered to the grid under light loading.

maximum benefit from fuel cells, their normal operating points are set close to their rated power and their load cycling is not desired.

7.4.3 Load-Following Analysis for Fuel Cells

In this section, load-following capability of fuel cells when connected to a microgrid (Bus 1 in Fig. 7.24), to which the main utility grid is also connected, is investigated. The microgrid could supply some critical loads, where it is essential that the power to such loads be uninterrupted. Two scenarios will be investigated. In Case 1, the utility grid is set to supply a fixed amount of power to the microgrid, and the remainder of the load power demand will be supplied by the FCDG. In Case 2, the FCDG is set to deliver a fixed amount of power to the microgrid, and the remainder of the load power demand will be supplied by the utility grid. Simulation results are shown for the PEMFC DG only.

7.4.3.1 Fixed Power Supply from the Grid In this case, a proper load-following controller must be designed to ensure that only scheduled power is delivered from the grid to the microgrid, and that the fuel cell system follows the remainder of load demand. In the deregulated power market, the ancillary service costs due to load-following operations can be as high as 20% of the total ancillary service costs [7]. Figure 7.24 shows the system configuration for which the load-following study was carried out for

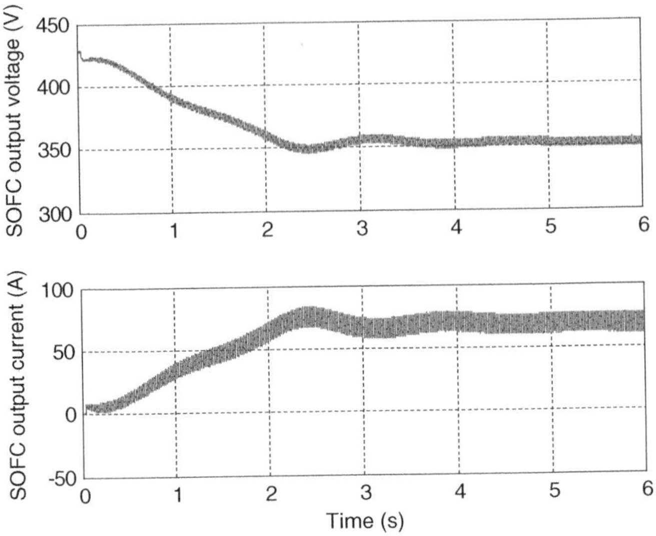

FIGURE 7.23 SOFC DG: Output voltage and current of each fuel cell array under light load.

a PEMFC DG. Bus 1 in the figure can be considered as a microgrid, to which the PEMFC DG and the power grid are connected to supply power to the load. The power delivered from the grid (P_{Grid}) is kept constant ($P_{Grid,sched}$), and the FCDG is required to follow the remainder of the load demand at the microgrid. A load-following controller is placed between the microgrid and the FCDG to achieve the desired load following scenario. The power supplied to the microgrid by the main grid (P_{Grid}) is measured and compared

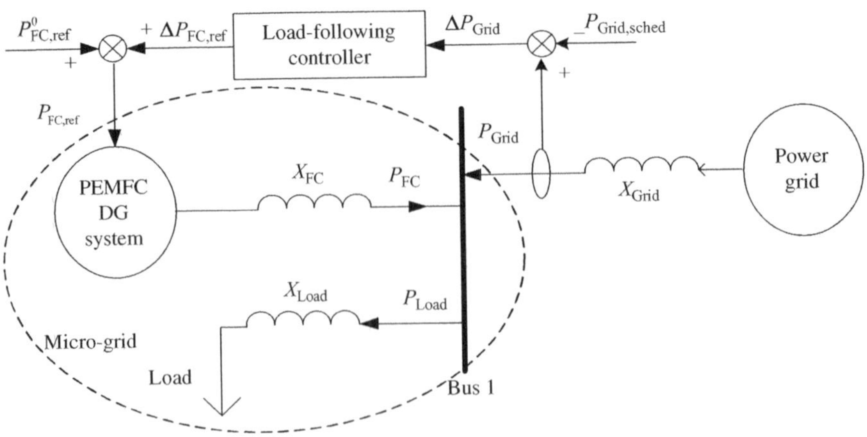

FIGURE 7.24 System for the PEMFC DG load-following study.

with the scheduled power from the grid. The error (ΔP_{Grid}) is then fed through the load-following controller (e.g., a PI controller) to generate the adjustment needed ($\Delta P_{FC,ref}$) to the initial power reference value ($P^0_{FC,ref}$). The summation of $P^0_{FC,ref}$ and $\Delta P_{FC,ref}$ will be the new power reference ($P_{FC,ref}$) for the FCDG system. As the load at the microgrid varies, $P_{FC,ref}$ is adjusted accordingly to ensure that the FCDG will compensate for the load variations. In practice, an upper and a lower limit should be imposed on $\Delta P_{FC,ref}$ so that the power adjustment to the FCDG is within its safe operating range.

In a case study, the results of which are shown in Fig. 7.25, P_{Grid} is set to 100 kW (1 PU). Initially, the load at the microgrid is 2 PU; therefore the FCDG delivers the remaining 1 PU power to the load. At $t = 0.3$ s, the load demand increases to 3 PU and at $t = 4.1$ s drops back to 2 PU. During the transient periods, the power grid responds to the initial load changes, but the FCDG picks up the changes in load after the transient period and keeps the grid power at its scheduled value.

7.4.3.2 Fixed Power Supply from the FCDG In this case (Fig. 7.26), the FC power plant is scheduled to supply a fixed amount of power to the microgrid while the grid will pick up the load variations. The power delivered from the FCDG is set to 200 kW (2 PU). Again, the load suddenly increases to 3 PU at $t = 0.3$ s. and changes back to its original

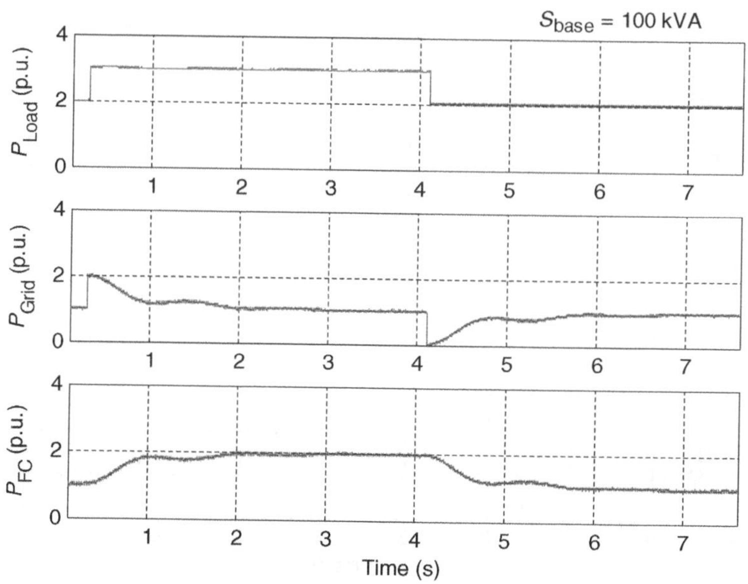

FIGURE 7.25 Power curves of the load-following study with fixed grid power.

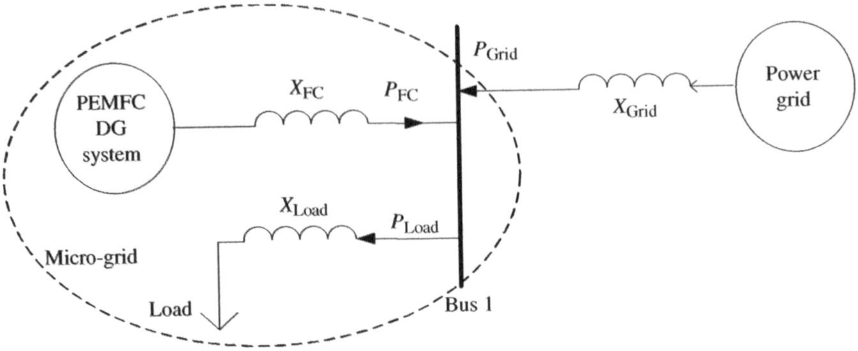

FIGURE 7.26 System for the PEMFC DG supplying fixed amount of power.

value at 4.1 s. The simulation results for this case study are shown in Fig. 7.27. Because the FCDG is scheduled to deliver a fixed power to the load and the grid absorbs the variations in load demand, no transient is observed in the grid or the fuel cell power response. The grid responses to the load fluctuations almost simultaneously.

7.4.4 Fault Analysis

It is important to know whether a FCDG system will remain stable under an electrical fault. In this section, the stability of the PEMFC DG under

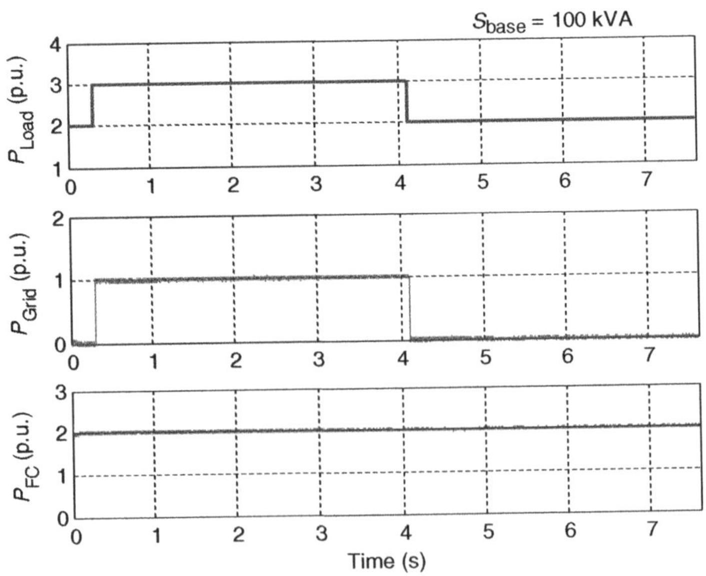

FIGURE 7.27 Power curves of the load-following study with fixed FC power.

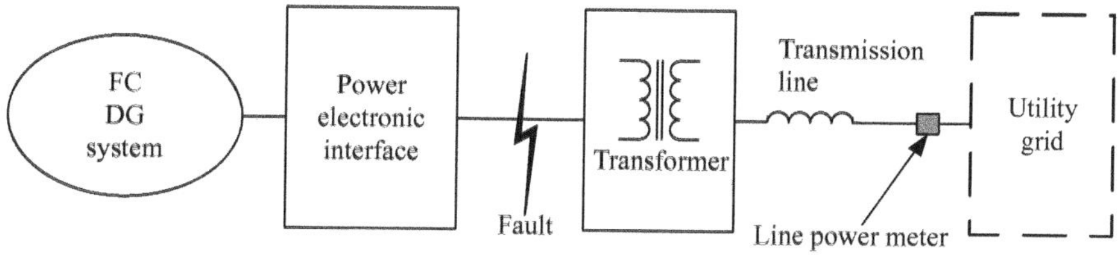

FIGURE 7.28 The fuel cell DG system under faulted.

electrical faults is investigated. The FCDG system, shown in Fig. 7.28, delivers 2 PU to the utility grid before the fault. A simulated three-phase fault occurs at $t = 0.7$ s at the low voltage side of the transformer connecting the FCDG to the utility grid. The fault lasts for five cycles (0.0833 s) and is cleared at $t = 0.7833$ s.

The power flow through the transmission line is shown in Fig. 7.29. During the fault, the transmission line power changes direction, flowing from the grid toward the fault, since the utility grid also supplies power to the fault. As shown in the figure, the fuel cell system remains stable after the disturbance. Immediately after the fault is cleared, power flow changes direction in the transmission line flowing toward the grid again, and there is a rush of power delivered to the grid. This is due to the large phase difference between the utility grid and the inverter output voltage when the

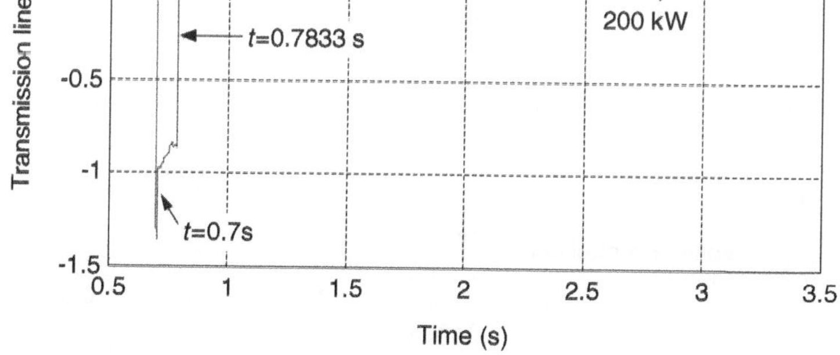

FIGURE 7.29 Power flow of the transmission line in the faulted fuel cell DG system.

fault is cleared. A soft-starting procedure and a reconnection controller can be added to the system to limit the peak power when the system recovers from a fault.

One advantage of the two-loop (voltage and current) control for the inverter (Section 7.3.2), used in the above simulation study, is its capability to limit fault currents. To show this advantage, the previous three-phase fault is simulated on the system of Fig. 7.28 with voltage control only and with the original two-loop control for the inverter for comparison. The output power of the PEMFC DG under these two inverter control scenarios is shown in Fig. 7.30. It is noticed that the system does not remain stable when only voltage control is applied for inverter control. Since there is no current limit control for this case, the fuel cell output current increases sharply due to the fault and exceeds the current corresponding to the fuel cell maximum power point, moving the fuel cell operating point to the concentration zone. As a result, the fuel cell output power drops sharply making the FC system unstable. On the other hand, when the two-loop control strategy is applied for inverter control, the fuel system remains stable under the same electrical disturbance, and the system returns to its pre-fault state after the disturbance.

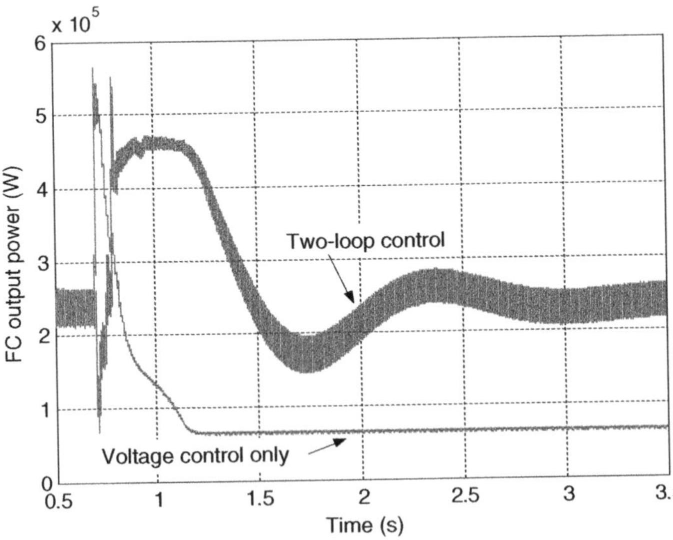

FIGURE 7.30 Fuel cell output power for a three-phase fault applied on the transmission line in the PEMFC DG system shown in Fig. 7.29.

7.5 SUMMARY

Control of grid-connected FC DG systems were investigated in this chapter. The PEMFC and SOFC models, the model for boost dc/dc converter and for a three-phase voltage source inverter (Chapter 6) were used to develop a simulation model for a grid-connected FC system. Controller designs for the dc/dc converter and the three-phase inverter were given using linearized small-signal converter/inverter models. Conventional PI voltage feedback controllers were used for the dc/dc converters to regulate the dc bus voltage, and a *dq* transformed two-loop current control scheme was used on the inverter to control the real and reactive power delivered from the fuel cell system to the utility grid.

A MATLAB/SIMULINK model of the proposed FC DG system was implemented using the SimPowerSystems block-set. The validity of the proposed control schemes over a large operating range was verified through simulations of the switched nonlinear converter/inverter models. Simulation results of the case studies given in the chapter show that the real and reactive power delivered from the fuel cell system to the utility grid can be controlled as desired while the dc bus voltage is maintained within its prescribed range. The results also show that the fuel cell system is capable of load following and can remain stable under severe electrical faults. Further, it is noticed that a two-loop inverter control scheme has the advantage over a voltage only control scheme for the inverter on fault protection and system stability.

REFERENCES

[1] C. Wang, M.H. Nehrir, and S.R. Shaw, Dynamic models and model validation for PEM fuel cells using electrical circuits, *IEEE Transactions on Energy Conversion*, 20 (2), 442–451, 2005.

[2] C.J. Hatziadoniu, A.A. Lobo, F. Pourboghrat, and M. Daneshdoost, A simplified dynamic model of grid-connected fuel-cell generators, *IEEE Transactions on Power Delivery*, 17 (2), 467–473, 2002.

[3] M.D. Lukas, K.Y. Lee, and H. Ghezel-Ayagh, Development of a stack simulation model for control study on direct reforming molten carbonate fuel cell power plant, *IEEE Transactions on Energy Conversion*, 14 (4), 1651–1657, 1999.

[4] J. Padullés, G.W. Ault, and J.R. McDonald, An integrated SOFC plant dynamic model for power system simulation, *Journal of Power Sources*, 495–500, 2000.

[5] K. Sedghisigarchi and A. Feliachi, Dynamic and transient analysis of power distribution systems with fuel cells—Part I: Fuel-cell dynamic model, *IEEE Transactions on Energy Conversion*, 19 (2), 423–428, 2004.

[6] K. Sedghisigarchi and A. Feliachi, Dynamic and transient analysis of power distribution systems with fuel cells—Part II: Control and stability enhancement, *IEEE Transactions on Energy Conversion*, 19 (2), 429–434, 2004.

[7] Y. Zhu and K. Tomsovic, Development of models for analyzing the load-following performance of microturbines and fuel cells, *Journal of Electric Power Systems Research*, 62, 1–11, 2002.

[8] Z. Miao, M.A. Choudhry, R.L. Klein, and L. Fan, Study of a fuel cell power plant in power distribution system—Part I: Stability control, Proceedings of the IEEE PES General Meeting, Denver, CO, June 2004.

[9] D. Candusso, L. Valero, and A. Walter, Modelling, control and simulation of a fuel cell based power supply system with energy management, Proceedings of the 28th Annual Conference of the IEEE Industrial Electronics Society (IECON 2002), Vol. 2, 1294–1299, 2002.

[10] R. Naik, N. Mohan, M. Rogers, and A. Bulawka, A novel grid interface, optimized for utility-scale applications of photovoltaic, wind-electric, and fuel-cell systems, *IEEE Transactions on Power Delivery*, 10 (4), 1920–1926, 1995.

[11] W. Shireen and M.S. Arefeen An utility interactive power electronics interface for alternate/renewable energy systems, *IEEE Transactions on Energy Conversion*, 11 (3), 643–649, 1996.

[12] G. Spiazzi, S. Buso, G.M. Martins, and J.A. Pomilio, Single phase line frequency commutated voltage source inverter suitable for fuel cell interfacing, 2002 IEEE 33rd Annual IEEE Power Electronics Specialists Conference, Vol. 2, 734–739, 2002.

[13] K. Ro and S. Rahman, Control of grid-connected fuel cell plants for enhancement of power system stability, *Renewable Energy*, 28 (3), 397–407, 2003.

[14] H. Komurcugil and O. Kukrer, A novel current-control method for three-phase PWM AC/DC voltage-source converters, *IEEE Transactions on Industrial Electronics*, 46 (3), 544–553, 1999.

[15] N. Mohan, T.M. Undeland, and W.P. Robbins, Power Electronics—Converters, Applications, and Design, Wiley, Hoboken, NJ. 2003.

[16] D.W. Hart, Introduction to Power Electronics, Prentice Hall, 1997.

[17] M.H. Todorovic, L. Palma, and P. Enjeti, Design of a wide input range DC-DC converter with a robust power control scheme suitable for fuel cell power conversion, 19th Annual IEEE Applied Power Electronics Conference, California, 2004.

[18] S.M.N. Hasan, S. Kim, and I. Husain, Power electronic interface and motor control for a fuel cell electric vehicle, 19th Annual IEEE Applied Power Electronics Conference, California, 2004.

[19] *Fuel Cell Handbook, 6th edn*, EG&G Services, Parsons Inc., DEO, Division of Fossil Energy, National Energy Technology Laboratory, 2004.

[20] J. Larminie and A. Dicks, *Fuel Cell Systems Explained*, John Wiley & Sons, Ltd., New York, 2001.

[21] J. Van de Vegte, *Feedback Control Systems*, 3rd edn, Prentice-Hall, Upper Saddle River, NJ, 1994.

[22] IEEE Std 1547, *IEEE Standard for Interconnecting Distributed Resources with Electric Power Systems*, 2003.

[23] M.Tsai and W.I. Tsai, Analysis and design of three-phase AC-to-DC converters with high power factor and near-optimum feedforward, *IEEE Transactions on Industrial Electronics*, 46 (3), 535–543, 1999.

[24] H. Mao, Study on three-phase high-input-power-factor PWM-voltage-type reversible rectifiers and their control strategies, Ph. D. Dissertation (in Chinese), Zhejiang University, 2000.

[25] J.D. Glover and M.S. Sarma, *Power System Analysis and Design*, 4th edn, Wadsworth Group, Brooks/Cole, Florence, KY, 2007.

[26] R.S. Gemmen, Analysis for the effect of inverter ripple current on fuel cell operating condition, *Transactions of the ASME—Journal of Fluids Engineering*, 125 (3), 576–585, 2003.

[27] R. Wu, T. Kohama, Y. Kodera, T. Ninomiya, and F. Ihara, Load-current-sharing control for parallel operation of DC-to-DC converters, IEEE PESC'93, Seattle, WA, 101–107, June 1993.

[28] P.C. Krause, O. Wasynczuk, and S.D. Sudhoff, *Analysis of Electric Machinery*, IEEE Press, New York, 1995.

[29] J.G. Kassakian, M.F. Schlecht, and G.C.Verghese, *Principles of Power Electronics*, Addison-Wesley, Reading, MA, 280–293, 1991.

[30] M.K. Acharya, C.L. Haynes, R. Williams Jr., M.R. von Spakovskym, D.J. Nelson, D.F. Rancruel, J. Hartvigsen, and R.S. Gemmen, SOFC performance and durability: resolution of the effects of power-conditioning systems and application loads, *IEEE Transactions On Power Electronics*, 19 (5), 1263–1278, 2004.

CHAPTER 8

CONTROL OF STAND-ALONE FUEL CELL POWER GENERATION SYSTEMS

8.1 INTRODUCTION

Stand-alone fuel cell power generation systems have applications such as remote area power (where the FC system is part of an isolated microgrid with other generation sources and loads), backup power, or transportation. The load following controllers designed for grid-connected FC systems (Chapter 7) can also be implemented for stand-alone FC systems to extract a desired amount of power from the system. However, in such applications, the system needs to have sufficient storage capacity to handle power variations and disturbances, such as load transients.

FCs are good energy sources to provide reliable power at steady state, but they may not respond to electrical load transients as fast as desired. This is mainly due to their slow internal electrochemical and thermodynamic responses and slow responses of their mechanical accessory systems [1]. Load transients can cause a low-reactant (fuel starvation) condition inside the FCs, which is considered to be harmful and will shorten their life [2]. In order to overcome this weakness, FCs can be combined with other

Modeling and Control of Fuel Cells, By M. Hashem Nehrir and Caisheng Wang

energy sources with fast dynamics, such as battery or super-capacitor, to form a hybrid FC system [3–6].

In this chapter, the design of a load transient mitigation technique is presented for stand-alone FC-battery power generation systems. In the designed control strategy, the battery supplies the transient load, and the FC provides the steady-state load. Meanwhile, the battery voltage is controlled to remain within a desired range, and the FC output current ripple is also limited to within an acceptable range. The former is essential for proper operation of the FC-battery system, and the latter is important to FC's healthy operation and life time [2,7,8].

Both PEMFC and SOFC show great potential for stand-alone power generation applications; the dynamic models for PEMFC and SOFC, developed in Chapters 3 and 4 [1,9], are used in this chapter to investigate the effectiveness of the designed load transient mitigation controller. A validated circuit model for lead–acid batteries, reported in [10], is used in the study. It should be mentioned that depending on the application, other fast energy storage devices such as super-capacitors or superconducting magnetic energy storage (SMES) devices [11,12] can also be used in the design of the load transient mitigation controller.

8.2 SYSTEM DESCRIPTION AND CONTROL STRATEGY

Figure 8.1 shows the schematic diagram for the proposed stand-alone FC-battery power generation system [13]. The basic structure of the stand-alone FC system is similar to the grid-connected FC system described in Chapter 8. A boost dc/dc converter is used to adapt the FC output voltage to the battery voltage, and a series inductor is used for current continuity. The bottom part of the figure shows the load transient mitigation controller and the battery charge/discharge controller, which are explained in the next section The main component of the load transient mitigation controller is a low-pass filter, through which the sharp load transients are filtered and the filtered signal is used as a reference current for the current controller.

A 2-kW PEMFC unit and a 5-kW SOFC unit are used in the study. The 2-kW PEMFC unit was built based on the 500-W PEMFC model developed in Chapter 3. The unit consists of four 500-W PEMFC stacks connected in series. For the SOFC-based system, the FC unit is a single 5-kW SOFC stack, using the model developed in Chapter 4. The system load can either be dc or ac (interfaced through an inverter).

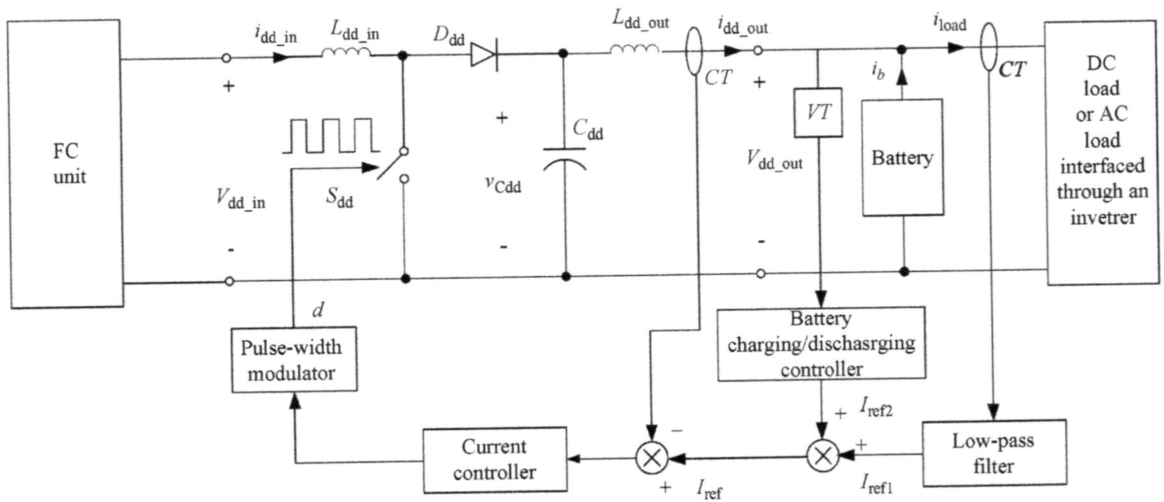

FIGURE 8.1 Schematic diagram of a FC-battery hybrid system with load transient mitigation control.

8.3 LOAD TRANSIENT MITIGATION CONTROL

Load mitigation control is achieved through a current controller for the dc/dc converter. The purpose of the proposed control strategy is to control the FC to only supply steady-state power, while the battery will supply the transient power to the load. The load current (I_{load}), measured through a current transducer (CT), is fed back through a low-pass filter to filter out the high-frequency transients. The sum of the filtered load current signal (I_{ref1}) and the output signal (I_{ref2}) from the battery charge/discharge controller is taken as the reference signal (I_{ref}) to control the dc/dc converter. The output current of the dc/dc converter (i_{dd_out}) is compared with the reference signal I_{ref}, and the error signal is processed by a current controller to control the duty ratio of the switch in the converter. The current controller can be a simple PI compensator [13].

The peak-to-peak ripple of the converter input current (FC output current, i_{dd_in} in Fig. 8.1) should also be controlled within a desired range (10% current ripple is used as acceptable range in this study) for healthy operation of the FC [2,7,8]. The main components of the dc/dc converter can be selected on the basis of technical specifications, such as rated and peak voltage and current, input current ripple, output voltage ripple, etc., using the classic design procedure, as discussed in Chapter 7 and in references [15,17]. The component values for a 5-kW dc/dc converter model built for use in this study are listed in Table 8.1. The value of the coupling inductor (L_{dd_out}) is selected so that the combination of this inductor with the capacitor (C_{dd}) yields a low resonant frequency to smooth out the converter output current (i_{dd_out}). A resonant frequency of less than 10 Hz was assumed to be acceptable in this study.

An approximate state-space model of the converter at the rated operating point can be derived using the averaging technique discussed in Chapter 6 [15,18]. Using the converter model, a PI current controller ($k_p + k_i/s$) can be designed using the classic Bode-plot and root-locus method as was done in Chapter 7 [16]. The parameters of the PI current controller are also listed in Table 8.1.

TABLE 8.1 Parameters of the dc/dc Converter

L_{dd_in}	12 mH
C_{dd}	2500 μF
L_{dd_out}	120 Mh
f_s (switch frequency)	5 kHz
k_i	20
k_p	0.02

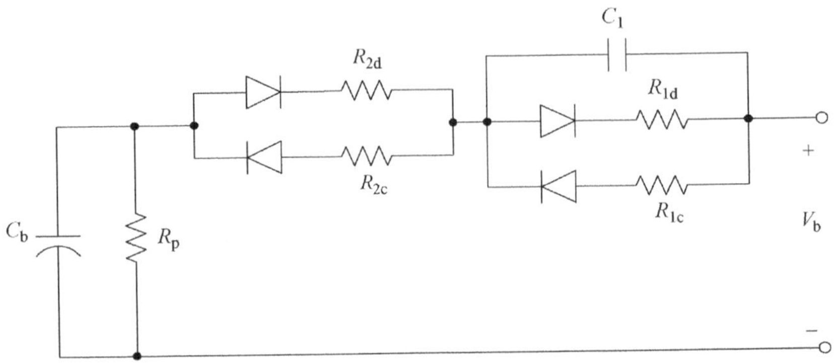

FIGURE 8.2 Equivalent circuit model for battery reported in [10]. C_b: battery capacitance; R_p: self-discharge resistance, or insulation resistance; R_{2c}: internal resistance for charge; R_{2d}: internal resistance for discharge; R_{1c}: overvoltage resistance for charge; R_{1d}: overvoltage resistance for discharge; C_1: overvoltage capacitance.

8.3.1 Circuit Model for Lead–Acid Batteries

A validated electrical circuit model for lead–acid batteries, shown in Fig. 8.1, was used in this study [10]. The diodes in the model are all ideal and are used to select different resistances for charging and discharging states of battery. The model capacitances and resistance values are defined below in Fig. 8.2; they are functions of battery current and depend on battery temperature and state of charge (SOC) [10,14]. The nominal values used for 0.5-kWh battery model are given in Table 8.2. Assuming the battery voltage will remain within the range of interest ($100 \pm 5\%$), the difference between the charge and discharge resistances is negligible [10], and they are assumed constant within the$100 \pm 5\%$ voltage range used. The model shown in Fig. 8.2 is then simplified into the circuit model given in Fig. 8.3. This simplified model is used in the load transient mitigation controller design and simulation study. The battery model used can be replaced with a more advanced storage system model.

TABLE 8.2 Parameters of a 220 V, 0.5 kWh Battery Model

C_b	300 F
R_p	25 MΩ
R_2	0.075 Ω
C_1	500 F
R_1	0.1 Ω

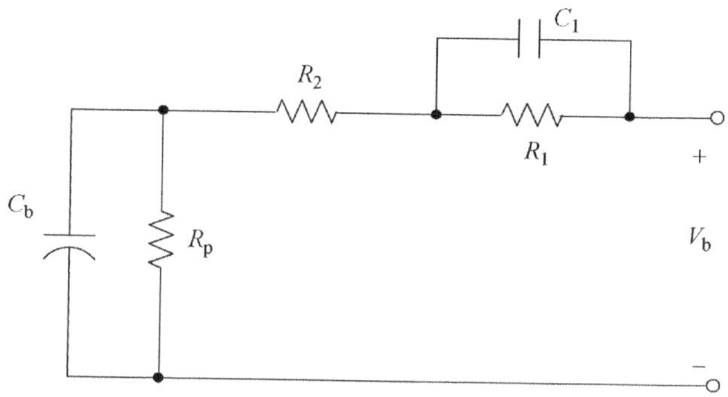

FIGURE 8.3 Simplified circuit model for lead–acid batteries.

8.3.2 Battery Charge/Discharge Controller

The battery in the FC system given in Fig. 8.1 is designed to buffer transient power. A 220 V, 0.5-kWh battery model (Table 8.2) is used in the study. The actual battery size needed for each application depends on the type of load transients applied to the system.

The block diagram for the battery charge/discharge controller used to keep the output voltage of the battery within ±5% of its nominal value (for its healthy operation) is shown in Fig. 8.4.

In Fig. 8.4, V_{ref} is the battery nominal voltage (220 V). The regulation constant K_c in the figure is determined by assuming that the 5% charge/discharge voltage deviation can be achieved in a certain time duration T. In our study, T is assumed to be a half hour (1800 s). Therefore, K_c which has a unit of siemens, can be written as:

$$K_c = \frac{0.05C_b}{1800} \tag{8.1}$$

where C_b is the battery capacitance.

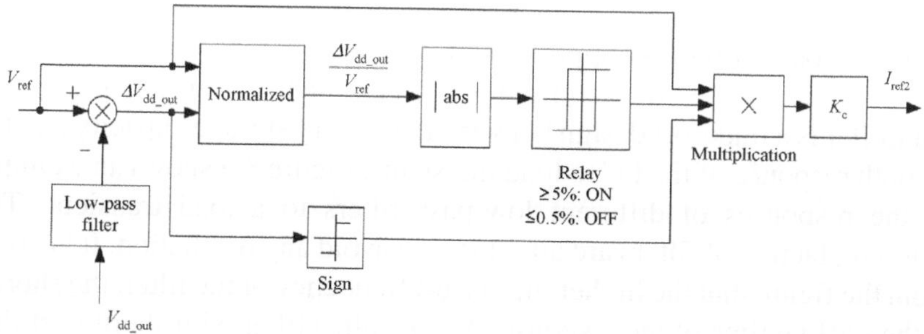

FIGURE 8.4 Schematic diagram of the battery charge/discharge controller.

The battery voltage (V_{dd_out}) is measured through a voltage transducer (VT), and a low-pass filter is used to filter out any high-frequency spikes in the signal. The "normalized" block normalizes the filtered battery voltage deviation (ΔV_{dd_out}) with respect to the reference voltage (V_{ref}). The "labsl" block will give a non-negative output ($|\Delta V_{dd\_out}/V_{ref}|$) regardless of whether ΔV_{dd_out} is positive or negative, that is whether the battery voltage is higher or lower than the reference voltage.

The "relay" block in Fig. 8.4 is a hysteresis block which turns on to give output = 1 when $|\Delta V_{dd\_out}/V_{ref}| > 0.05$ (5%). The output of the relay block will stay on until the input to the block is less than 0.5%, i.e. $|\Delta V_{dd\_out}/V_{ref}| < 0.005$. Otherwise, the output of the relay block will be zero. The battery will be charged (or discharged) at a constant current when the battery voltage is lower than 95% (or higher than 105%) of its nominal value. The charge (discharge) process will stop when the battery voltage is within 0.5% of its nominal value. Therefore, as long as the battery voltage is lower than 95% of its nominal value (V_{ref}), a positive extra reference signal

$$I_{ref2} = \frac{0.05C_b V_{ref}}{1800} \tag{8.2}$$

will be generated for the current controller. This reference value is kept until the voltage is higher than $0.995V_{ref}$. On the contrary, if the battery voltage is higher than $1.05V_{ref}$, a negative extra current reference value $[-(0.05C_b V_{ref})/(1800)]$ is generated and this value will be held until the battery voltage is lower than $1.005V_{ref}$. Otherwise, the output of the battery charge/discharge controller will be zero.

8.3.3 Filter Design

The system load current is fed to a low-pass filter to filter its high-frequency transients. The filtered signal is then used as one input to the current controller which regulates the duty ratio of the dc/dc converter switch. As a result, only smooth current is drawn from the FC, and the transient load current is supplied by the battery.

The choice of the low-pass filter needed for the load transient mitigation controller is a trade-off design between the battery storage capacity and the smooth response of the FC to load transients. Figure 8.5 shows an example of the responses of different low-pass filters to a load transient. The damping factors of filters are all set to 1 to avoid any oscillation. It is noted from the figure that the higher the cut-off frequency of the filter, the shorter is the settling time of the response. As a result, a filter with shorter settling time requires smaller storage battery capacity. But, the response of such a

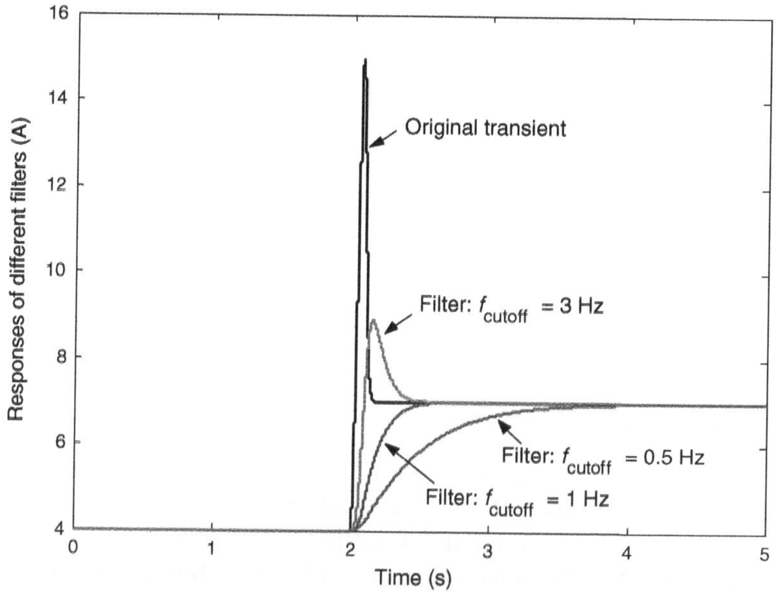

FIGURE 8.5 Example of the responses of different filters to a load transient.

filter may cause undesired overshoot. On the contrary, the lower is the filter cut-off frequency, the longer is the rise time of the filter response. Therefore, the smoother will be the response of the FC. However, in this case, a larger battery capacity is needed. Therefore, a trade-off decision has to be made to choose a filter that is best suited for the application at hand. In practical applications, one can estimate the fastest possible load transient based on the load information at the location where the FC system will be used. Transients with frequencies higher than 1250 Hz can be neglected since they do not affect the FC performance significantly [2]. Then, one decides what overshoot value (current ripple) is tolerable by the FC. An overshoot of less than 10% is considered a good value. Such overshoots will normally not have a significant impact on the FC's healthy operation [2,7]. The filter can therefore be chosen for the fastest load transient with the highest possible cut-off frequency, while its output overshoot will not exceed a prescribed value (10% in this study).

8.4 SIMULATION RESULTS

The PEMFC and SOFC models, discussed in Chapters 3 and 4, and the battery model discussed earlier in this chapter, were used to develop simulation models for the stand-alone FC-battery power generation systems in MATLAB/SIMULINK, using SimPowerSystem block-set. Both PEMFC and SOFC systems are controlled by the load transient mitigation technique proposed in the previous section. In each case, the system performance is investigated under load transients. Power flow is controlled in the FC system, by the load transient mitigation controller, in such a way that the FC stack only provides steady-state power, while the battery will supply transient power to the load. As a result, the FC operating point will move smoothly from its original condition to the new steady-state operating point. Charging and discharging performances of the battery are also investigated.

8.4.1 The Load Transients

Fuel cell systems could be subject to a variety of load transients. Depending on the application, the load transients can be dc or ac. Therefore, the PEMFC and SOFC systems are tested with both ac and dc load transients. Similar load transients, but with different power ratings will be used for the two systems.

8.4.1.1 The dc Load Transients The dc load transient used for performance evaluation of the FC systems is similar to that of a dc motor starting current when a resistance starter box is used to limit the motor starting current. Figure 8.6 shows the load current transient applied to the

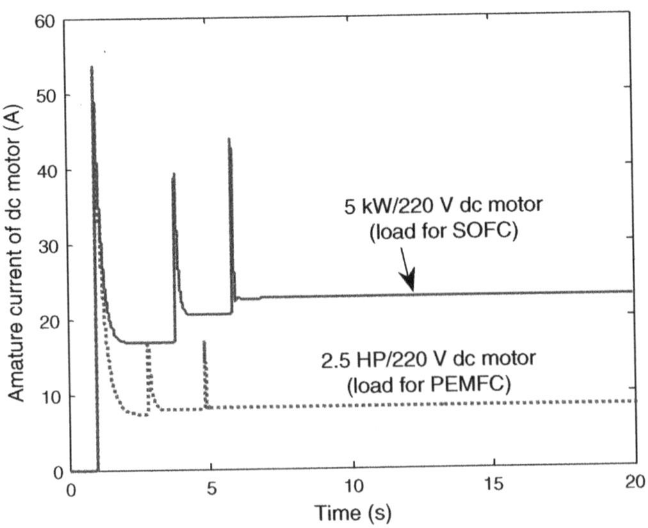

FIGURE 8.6 Load transient of starting a 2.5 HP and a 5 kW 220 V dc motor.

FC systems when a dc motor is started with full voltage in three steps of starting resistance. There is a sharp current transient each time the motor starting resistance is switched (from a higher value to a lower value), as the motor speeds up.

A 2.5 HP/220 V dc motor load is used for the 2-kW PEMFC-battery system and a 5-kW/220 V dc motor is used for the SOFC system, as shown in Fig. 8.6.

8.4.1.2 The ac Load Transients A dynamic ac load model (the response of which is shown in Fig. 8.7) is used to evaluate the performance of the FC systems under ac load current transients. This transient current model is defined as follows:

$$i(t) = I_0 + I_1\,\mathrm{e}^{-\alpha_1 t} - I_2\,\mathrm{e}^{-\alpha_2 t} \tag{8.3}$$

where I_0, I_1, and I_2 are non-negative constants and $\alpha_2 > \alpha_1 > 0$. It is assumed that the load transient starts at $t = 0$ in the above equation, and $i(t)$ is the amplitude of load current at 60 Hz.

The above five constant parameters can be determined by the following equations:

$$\begin{cases} I_0 = i_\infty \\ I_0 + I_1 - I_2 = i_0 \\ \dfrac{\ln(I_2\alpha_2)/(I_1\alpha_1)}{\alpha_2 - \alpha_1} = T_\mathrm{p} \\ I_0 + I_1\,\mathrm{e}^{-\alpha_1 T_\mathrm{p}} - I_2\,\mathrm{e}^{-\alpha_2 T_\mathrm{p}} = i_\mathrm{peak} \\ \dfrac{\ln I_1 - \ln(0.02 I_0)}{\alpha_1} \approx T_\mathrm{S} \end{cases} \tag{8.4}$$

where i_0 and i_∞ are the initial and final values of the transient load current, T_p is the time required for the transient current to reach its peak value (i_peak), and T_S the settling time, which takes the transient to settle to within ±2% of its final value. Therefore, by specifying the initial value (i_0), final value (i_∞), peak value of the transient (i_peak), how fast the transient reaches its peak (T_p), and how long it takes the transient to reach steady state (T_S), we can calculate the values of the five parameters in (8.4). Figure 8.7a shows the ac load transient used for the PEMFC system, starting at $t = 1$ s. The parameters of the load transient are $I_0 = 15.84$ A, $I_1 = 79.2$ A, $I_2 = 95.04$ A, $\alpha_1 = 5$, and $\alpha_2 = 30$.

The parameters of the ac load current transient used for the SOFC system (Fig. 8.7b) are $I_0 = 39.6$ A, $I_1 = 198$ A, $I_2 = 237.6$ A, $\alpha_1 = 5$, and $\alpha_2 = 30$.

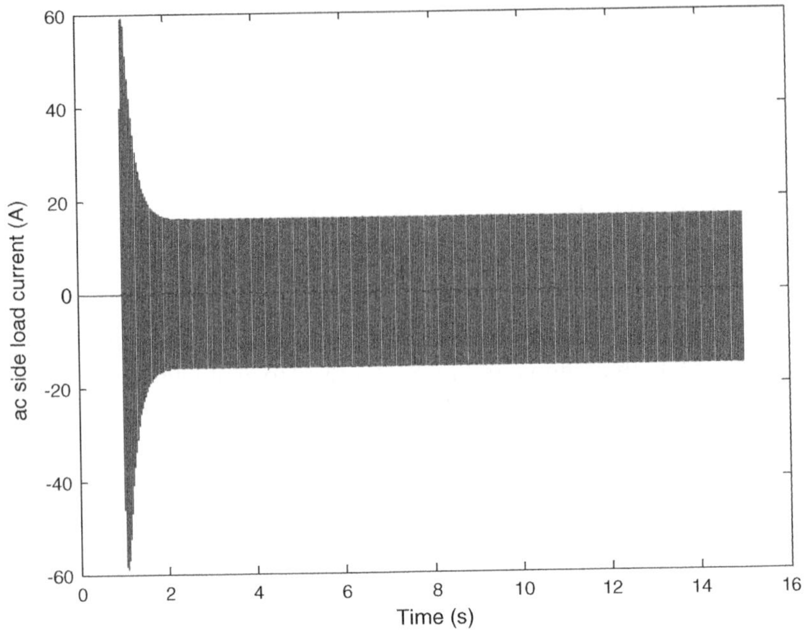

(a)

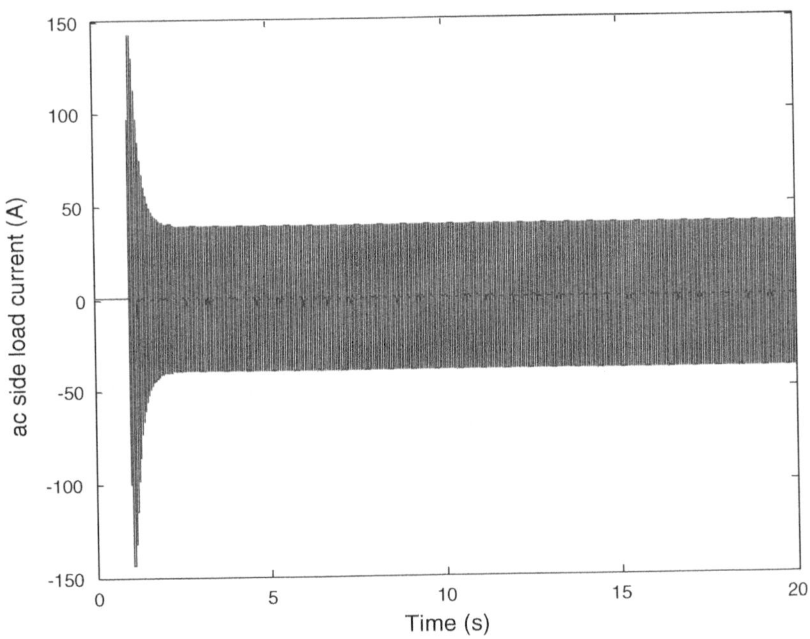

(b)

FIGURE 8.7 The ac load transients used for (a) the PEMFC-battery system and (b) the SOFC-battery system.

8.4.2 Load Transient Mitigation

In this section, simulation results are given for load transient mitigation studies of PEMFC and SOFC systems. Throughout these studies, it is assumed that the system battery is fully charged to its nominal voltage (220 V) before the load transient is applied.

8.4.2.1 PEMFC System The system is first tested under the dc motor load transient shown in Fig. 8.6. The cut-off frequency of the low-pass filter is set at 0.1 Hz and the damping factor is set to 1 to ensure a smooth transition of the FC reference current (I_{ref}) and therefore a smooth change of its operating point from no-load to its load condition. The corresponding current reference signal I_{ref} (see Fig. 8.1) is shown in Fig. 8.8; it shows that the rise time of the filter response to the load transient is less than 10 s, and there is no overshoot. The battery current (i_b) and the dc/dc converter output current (i_{dd_out}) of the PEMFC-battery system responding to the load transient current are also shown in Fig. 8.8. It is noted that i_{dd_out} follows the reference signal (I_{ref}) very well; the two signals almost overlap. It is also noted that the battery picks up the transient load current while the converter output current rises smoothly (as designed) from its previous no-load to its steady-state load condition. The corresponding PEMFC current and voltage responses are shown in Fig. 8.9. The output current and voltage

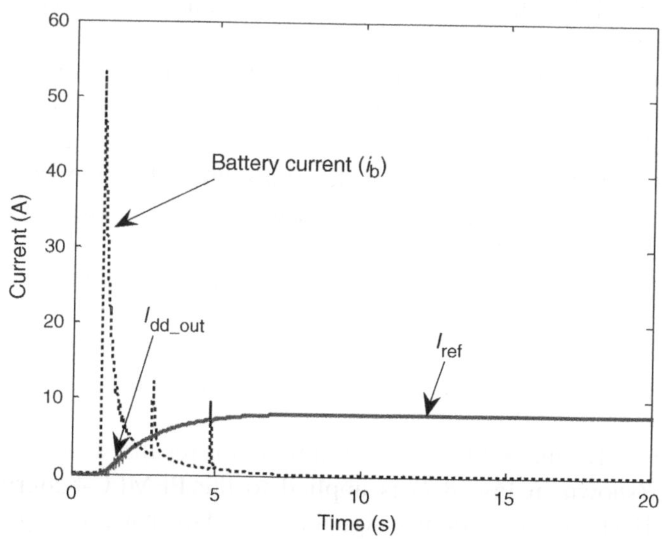

FIGURE 8.8 PEMFC: Reference signal (I_{ref}) for the current controller, battery current (i_b), and converter output current (i_{dd_out}) under the load transient shown in Fig. 8.6.

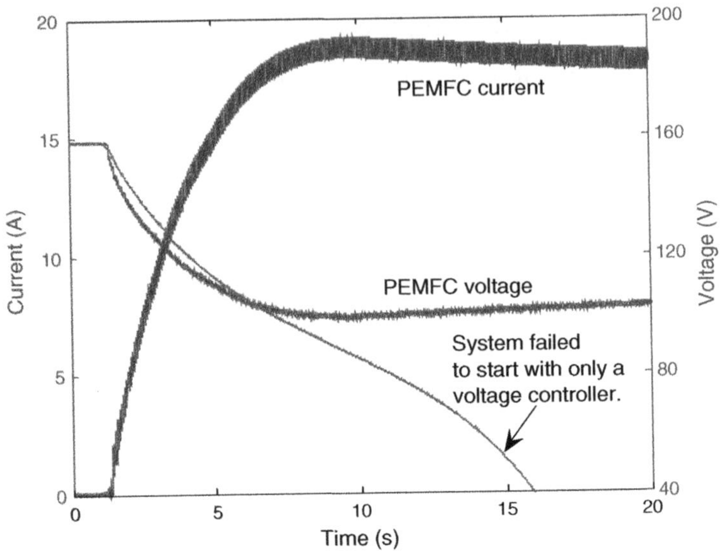

FIGURE 8.9 PEMFC: The output current and voltage responses to the dc load transient shown in Fig. 8.6.

of the PEMFC stack vary smoothly during the load transient moving from their no-load to steady-state (post-loading) operating condition, as desired. It is also noted that the FC current ripple is around 5.5%, which is within the acceptable range (10%).

In order to show the effectiveness of the load transient mitigation controller, the results of a simulation study of the PEMFC system of Fig. 8.1 with a typical voltage controller for the dc/dc converter only is also shown in Fig. 8.9. In this case, the transient mitigation controller is bypassed, and i_{load} and i_{dd_out} in Fig. 8.1 are not used as feedback signals; therefore the low-pass filter and the battery charge/discharge controller are bypassed, L_{dd_out} is taken out, and only the voltage controller is used for the dc/dc converter. As noted from Fig. 8.9, in this case when the dc load transient, shown in Fig. 8.6, is applied to the PEMFC system, the system fails to start the dc motor, therefore justifying the need for the current controller.

Figures 8.10 and 8.11 show the simulation results when the ac load transient (shown in Fig. 8.7) is applied to the PEMFC-battery system. Figure 8.10 shows the reference signal (I_{ref}) and the battery output current, and the corresponding PEMFC output current and voltage are shown in Fig. 8.11. It is noted from these figures that the load transient mitigation

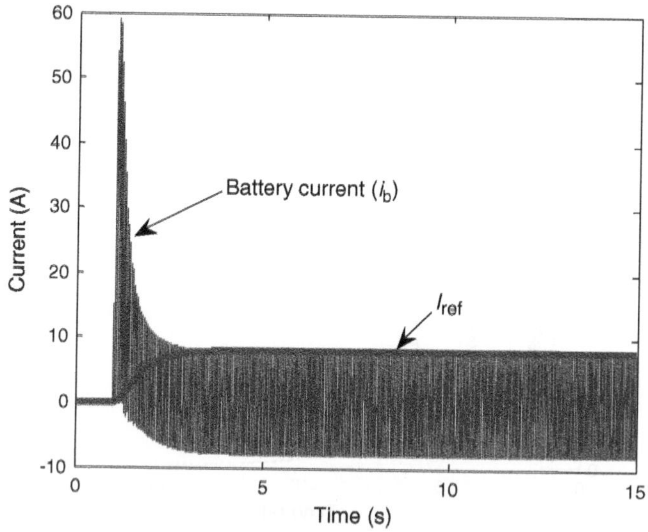

FIGURE 8.10 PEMFC: The control reference signal (I_{ref}) and battery current (i_b) under the ac load transient shown in Fig. 8.7.

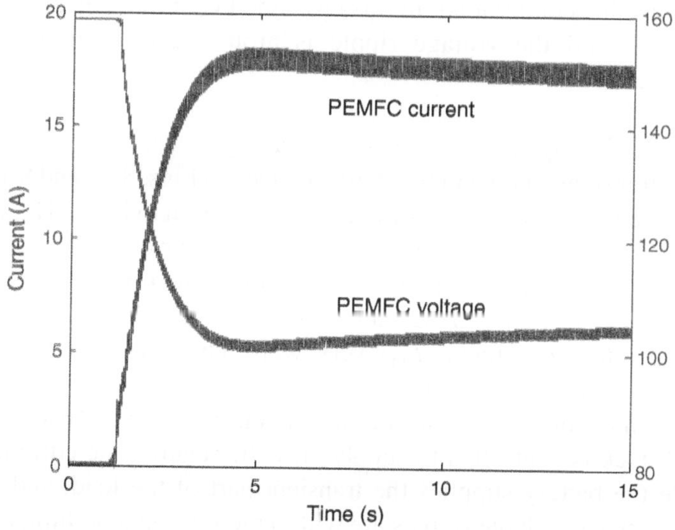

FIGURE 8.11 PEMFC: The output current and voltage responses to the ac load transient shown in Fig. 8.7.

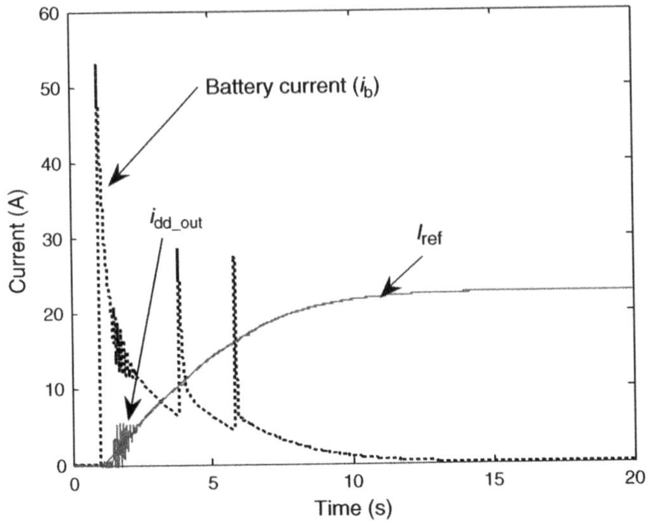

FIGURE 8.12 SOFC: Reference signal (I_{ref}) for the current controller, the battery current (i_b) and the converter output current (i_{dd_out}) under the load transient shown in Fig. 8.6.

technique also works well under ac load transients, and the PEMFC voltage and current vary smoothly from their no-load operating point to their steady-state values under load. In this case, the FC output current ripple is around 5.5%, and the voltage ripple is around 2%, both within the acceptable range.

8.4.2.2 SOFC System The SOFC-battery system responses under the dc load transient (shown in Fig. 8.6) are given in Figs. 8.12 and 8.13. The cut-off frequency of the low-pass filter is again chosen to be 0.1 Hz, and the damping factor is set to 1. Figure 8.12 shows the current reference signal (I_{ref}), the battery current (i_b) and the converter output current (i_{dd_out}) when the SOFC system responds to the load transient current. It is noted that i_{dd_out} follows the reference signal (I_{ref}) very well. The corresponding output current and voltage of the SOFC stack are given in Fig. 8.13. It is noted that the SOFC output voltage and current vary smoothly during the load transient; the SOFC stack is controlled to supply only the steady-state current to the load while the battery supplies the transient part of the load, and the FC output current and voltage ripples are within their acceptable limits.

Figures 8.14 and 8.15 show the simulation results when the ac load transient shown in Fig. 8.7b is applied to the SOFC system. From the reference signal (I_{ref}) and the battery output current (i_b), given in Fig. 8.14,

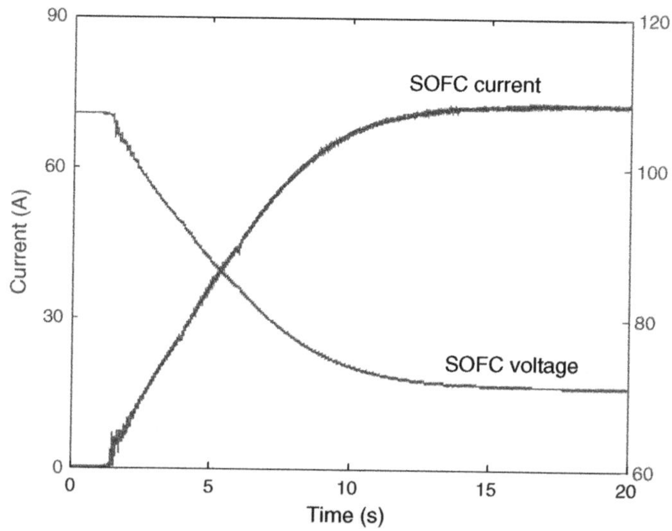

FIGURE 8.13 SOFC: The output current and voltage responses to the dc load transient shown in Fig. 8.6.

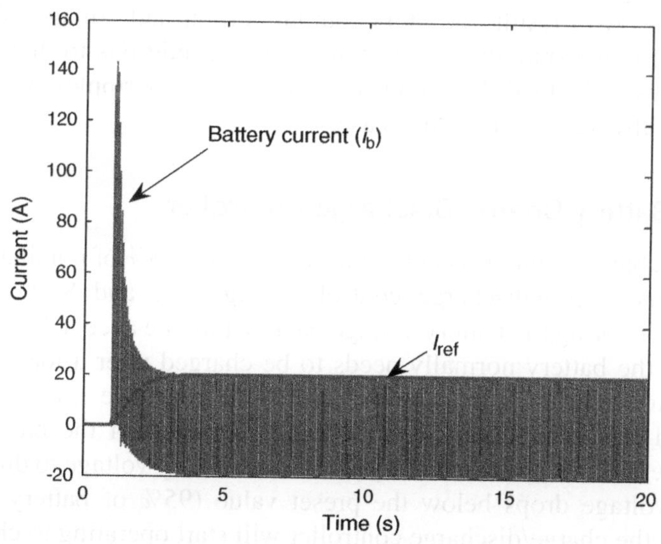

FIGURE 8.14 SOFC: Control reference signal (I_{ref}) and battery current (i_b) under the ac load transient.

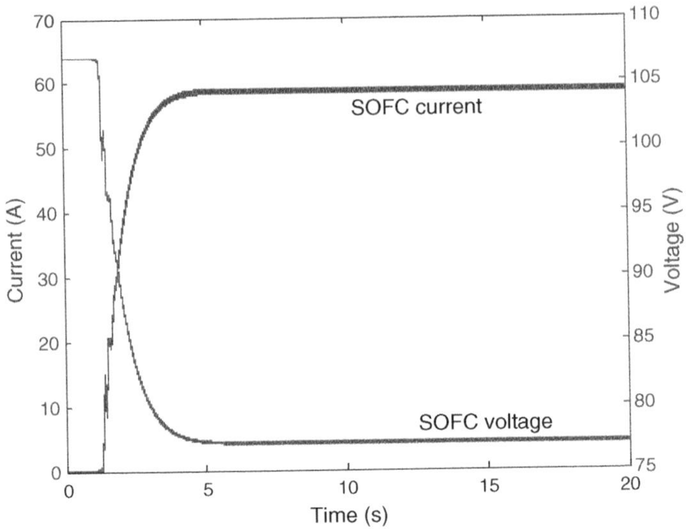

FIGURE 8.15 SOFC: The output current and voltage responses to the ac load transient.

it is clear that the battery responds to the transient part of the load, and the reference signal (for the dc/dc converter) rises smoothly to its new steady-state value. As a result, the SOFC output current and voltage (shown in Fig. 8.15) move smoothly from their no-load conditions to their steady-state values under load. The output current and voltage ripples of the SOFC stack are also very small in this case.

8.4.3 Battery Charge/Discharge Controller

When the battery voltage is out of the acceptable (±5% of nominal) range, the battery charge/discharge controller (Figs. 8.1 and 8.4) will start operating to bring the battery voltage back within the acceptable range. In practice, the battery normally needs to be charged after a load transient applied to the FC-battery system. This is because the transient power delivered from the battery to the load, together with the battery self-discharge characteristic, normally causes the battery voltage to drop. If the battery voltage drops below the preset value (95% of battery nominal voltage), the charge/discharge controller will start operating to charge the battery. Similarly, if the battery voltage goes above its preset limit (105% of battery nominal voltage), the charge/discharge controller will go into discharging mode to bring the battery voltage close to its nominal value.

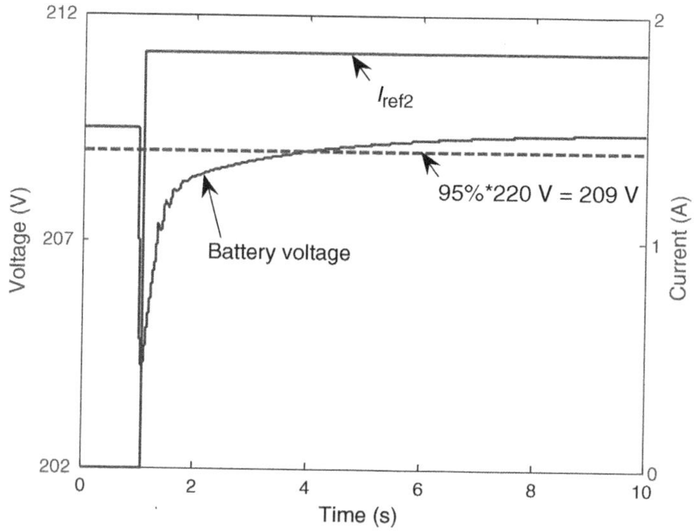

FIGURE 8.16 Battery voltage and extra current reference (I_{ref2}) curves when the battery is being charged.

In this section simulation results are given for battery charging after the dc motor load transient (shown in Fig. 8.6) is applied to the SOFC-battery system.

Figure 8.16 shows the battery voltage and the extra current reference (I_{ref2} in Fig. 8.1) for the FC current controller after the load transient is applied to the FC-battery system. Before the load is applied, the battery voltage is close to its lower limit (i.e., 95% of 220 V = 209 V). During the load transient, the battery voltage drops below 209 V, and as a result, the battery charge/discharge controller starts operating, generating a new value for I_{ref2}. This extra current reference is added to the current reference (I_{ref1}) from the current filter to produce the new current reference signal (I_{ref}) for the current controller (see Fig. 8.1). It is noted that as a result of the new reference current (I_{ref2}), the FC begins charging the battery, and the battery voltage increases to near its nominal value.

Figure 8.17 shows the load transient current, the overall current reference signal (I_{ref}), the filter output reference signal (I_{ref1}), the battery current (i_b), and the corresponding converter output current (i_{dd_out}). It is clear from Figs. 8.16 and 8.17 that during the load transient, the battery picks up the transient load (battery discharges), and the reference currents (I_{ref1} and I_{ref}) slowly increase to bring the FC current to its new steady-state value as well as charging the battery. This smooth transition of FC current is essential for FC improved reliability and durability.

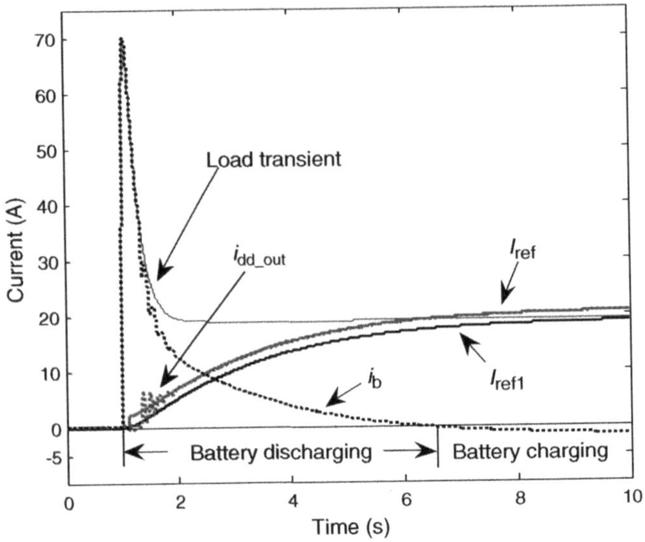

FIGURE 8.17 The load transient, the overall current control reference signal (I_{ref}), the low-pass filter output signal (I_{ref1}) and the corresponding converter output current (i_{dd_out}) when the battery is being charged.

8.5 SUMMARY

In this chapter, a load transient mitigation control strategy is proposed for FC-battery power generation systems. The control strategy consists of a current controller for the dc/dc converter and a battery charge/discharge controller to keep the battery voltage within a desired range. During load transients, FCs are controlled to supply the steady-state power to the load while the battery will supply the load transient power. Simulation results from the studies, carried out for both the PEMFC and SOFC systems under different load transients, show that the FC current can be controlled to move smoothly from one steady-state operating point to another as desired. Meanwhile, the battery picks up the transient load; the battery is properly charged or discharged to keep its voltage within a pre-specified range.

REFERENCES

[1] C. Wang, M.H. Nehrir,and S.R. Shaw, Dynamic models and model validation for PEM fuel cells using electrical circuits, *IEEE Transactions on Energy Conversion*, 20 (2), 442–451, 2005.

[2] R.S. Gemmen, Analysis for the effect of inverter ripple current on fuel cell operating condition, *Transactions of the ASME—Journal of Fluids Engineering*, 125 (3), 576–585, 2003.

[3] J.C. Amphlett, E.H. de Oliveira,R.F. Mann,P.R. Roberge, and A. Rodrigues Dynamic interaction of a proton exchange membrane fuel cell and a lead-acid battery, *Journal of Power Sources*, 65, 173–178, 1997.

[4] D. Candusso, L. Valero and A. Walter, Modelling, control and simulation of a fuel cell based power supply system with energy management, Proceedings, 28th Annual Conference of the IEEE Industrial Electronics Society (IECON 2002), 2, 1294–1299, 2002.

[5] J. Larminie and A. Dicks, *Fuel Cell Systems Explained*, Wiley, Chichester, UK, pp. 362–367, 2001.

[6] F.Z. Peng, H. Li, G. Su, and J.S. Lawler, A new ZVS bidirectional DC–DC converter for fuel cell and battery application, *IEEE Transactions on Power Electronics*, 19 (1), 54–65, 2004.

[7] K. Acharya, S.K. Mazumder, R.K. Burra, R. Williams, and C. Haynes, System-interaction analyses of solid-oxide fuel cell (SOFC) power-conditioning system, Conference Record of the 2003 IEEE Industry Applications Conference, 3, 2026–2032, 2003.

[8] S.K. Mazumder, K. Acharya, C.L. Haynes, R. Williams Jr., M.R. von-Spakovsky, D.J. Nelson, D.F. Rancruel, J. Hartvigsen, and R.S. Gemmen, Solid-oxide-fuel-cell performance and durability: resolution of the effects of power-conditioning systems and application loads, *IEEE Transactions on Power Electronics*, 19 (5), 1263–1278, 2004.

[9] C. Wang and M.H. Nehrir, A physically-based dynamic model for solid oxide fuel cells, *IEEE Transactions on Energy Conversion*, 22 (4), 887–897, 2007.

[10] Z.M. Salameh, M.A. Casacca and W.A. Lynch, A mathematical model for lead-acid batteries, *IEEE Transactions on Energy Conversion*, 7 (1), 93–98, 1992.

[11] H. Louie and K. Strunz, Superconducting magnetic energy storage (SMES) for energy cache control in modular distributed hydrogen-electric energy systems, *IEEE Transactions on Applied Superconductivity*, 17 (2), 2361–2364, 2007.

[12] S.M. Schoenung and W.V. Hasssenzal, Long- vs. Short-Term Energy Storage Technologies Analysis: A life-Cycle Cost Study for the U.S. Department of Energy-Energy Storage Systems Program, *Report SAND2003-2783*, Sandia National Laboratories, Albuquerque, NM, 2003.

[13] C. Wang and M.H. Nehrir, Load transient mitigation for stand-alone fuel cell power generation systems, *IEEE Transactions on Energy Conversion*, 22 (4), 864–872, 2007.

[14] D. Linden (editor), *Handbook of Batteries*, 2nd edn, McGraw-Hill, New York, 1995.

[15] N. Mohan, T.M. Undeland, and W.P. Robbins, *Power Electronics—Converters, Applications and Design*, Wiley, Hoboken, NJ, 2003.

[16] J. Van de Vegte, *Feedback Control Systems*, 3rd edn, Prentice-Hall, NJ, 1994.

[17] D.W. Hart, *Introduction to Power Electronics*, Prentice-Hall, Upper Saddle River, NJ, 1997.

[18] R.D. Middlebrook, Small-signal modeling of pulse-width modulated switched-mode power converters, *Proceedings of the IEEE*, 76 (4), 343–354, 1988.

CHAPTER 9

HYBRID FUEL CELL BASED ENERGY SYSTEM CASE STUDIES

9.1 INTRODUCTION

The steadily increasing energy consumption, the soaring cost and exhaustible nature of fossil fuels, and the intensifying concerns over the global environment have created much interest in alternative energy distributed generation (AEDG) sources, such as green renewable energy sources, microturbine-based (MT) and fuel cell (FC) based power generation systems. Wind and solar photovoltaic (PV) power generation are two of the most promising renewable power generation technologies whose growth has exceeded most optimistic predictions.

Fuel cells are a clean and efficient power generation source. They show great potential to be a key power generation technology of the future because of their numerous merits, such as high efficiency, zero or low emission of pollutant gases, and flexible modular structure. However, none of the above technologies is currently cost efficient or completely dispatchable to operate as an independent power generation source. Wind and solar power are highly dependent on climate, while FC and MT technologies have not matured and their cost is too high at the present time to justify their widespread use. In addition, a fossil-energy-based fuel, such as natural gas, is currently needed for most FC and MT applications.

Modeling and Control of Fuel Cells, By M. Hashem Nehrir and Caisheng Wang

Nevertheless, as explained in Chapter 1, the hydrogen needed for the operation of FCs could be produced from wind- and solar-generated electricity.

The different AEDG sources can complement each other to some extent. In general, multisource hybrid AEDG systems have better potential to provide higher quality and more reliable power than a system comprising a single resource. For this reason, hybrid energy systems have caught worldwide research attention [1–34]. Of course, each hybrid system needs a proper control strategy to manage (prioritize) power generation among the different AEDG sources in the system. Hybrid energy systems can be made of many combinations of different AEDG sources and storage devices, as shown in Fig. 9.1. Wind and solar energy are the most commonly used renewable energy sources in a hybrid system. Fuel cells and MTs also have the potential for use in hybrid energy systems to provide more reliable power and for combined heat and power (CHP) supply. Conventional distributed generation (DG) sources such as gas turbines and diesel generators can also be used with FCs or wind–PV generation systems to improve system performance and reliability.

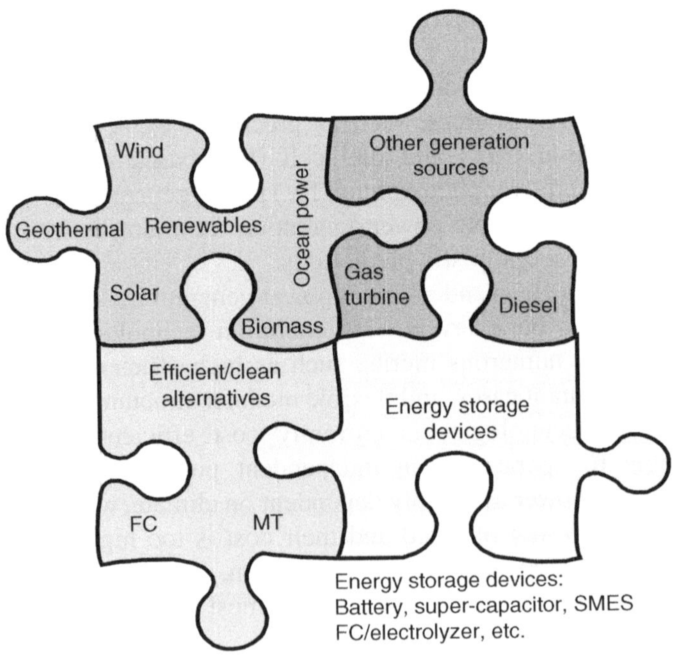

FIGURE 9.1 Different categories of alternative energy sources may be used in a hybrid energy system.

One or a local cluster of AEDG sources and their associated loads, energy storage devices, and distribution system, which can operate as an island (stand-alone) or be interfaced to a utility grid (grid-connected), is referred to as a *microgrid*. Since different AEDG sources have different operating characteristics, it is desired to have a well-defined and standardized procedure for connecting them to a microgrid to give them *plug-and-play* operation capability. Adapted from the concept widely used in computer science and technology, plug-and-play operation here means a device (a DG, an energy storage system, or a controllable load) capable of being added into an existing system (microgrid) without requiring system reconfiguration to perform its designed function, namely, generating power, providing energy storage capacity, or carrying out load control. A suitable system configuration and a proper interfacing circuit (also called power electronic building block or PEBB) may be necessary to achieve the plug-and-play function of a DG system [10].

In this chapter, first some general discussions on hybrid system configurations, system integration issues, and hybrid CHP systems are presented. Then, two case studies of hybrid alternative energy systems with FCs are given.

In the first case study, the PEMFC model developed in Chapter 3 is used to design a stand-alone hybrid wind–PV–FC storage system. Due to intermittent nature of wind and solar energy, stand-alone wind and/or PV energy systems normally require energy storage devices or some other generation sources. The storage device can be a battery bank, supercapacitor bank, superconducting magnetic energy storage (SMES) system, compressed air system, flow battery, FC–electrolyzer (or a regenerative FC) system, etc., or a combination thereof.

In the second case study, the SOFC (solid oxide fuel cell) model developed in Chapter 4 is used to evaluate the over all efficiency of SOFC in CHP operation mode.

9.2 HYBRID ELECTRONICALLY INTERFACED SYSTEMS

There are several ways to integrate different alternative energy sources to form a hybrid system. The methods can be generally classified into two categories: dc-coupled and ac-coupled systems. The latter can further be classified into power frequency ac (PFAC) coupled and high-frequency ac (HFAC) coupled systems. These methods are briefly explained below.

9.2.1 The dc-Coupled Systems

In the dc-coupled configuration, shown in Fig. 9.2, different alternative energy sources are connected to a dc bus through appropriate power electronic (PE) interfacing circuits. The dc energy is converted into 60 Hz (or 50 Hz) ac through a dc/ac inverter, which can be designed and controlled to allow bidirectional power flows. If there are any dc loads, they can be connected to the dc bus directly, or through some dc/dc converters to achieve appropriate dc voltage for the dc loads. The system can supply power to ac loads (connected to the ac bus), or be interfaced to a utility grid. The dc-coupling scheme is simple and no synchronism is needed to integrate different energy sources. But, it also has its own drawbacks. For instance, if the dc/ac inverter in the system is out of service, then the whole system will not be able to supply ac power. To avoid this situation, it is possible to connect several inverters with lower power rating in parallel, in which case synchronization of the output voltage of the different inverters, or synchronization with the grid, if the system is grid-connected, is needed. A proper power-sharing control scheme is also required to achieve a desired load distribution among the different inverters. A comparison of the dc-coupled systems with ac-coupled ones, which will be explained below, is summarized in Table 9.1.

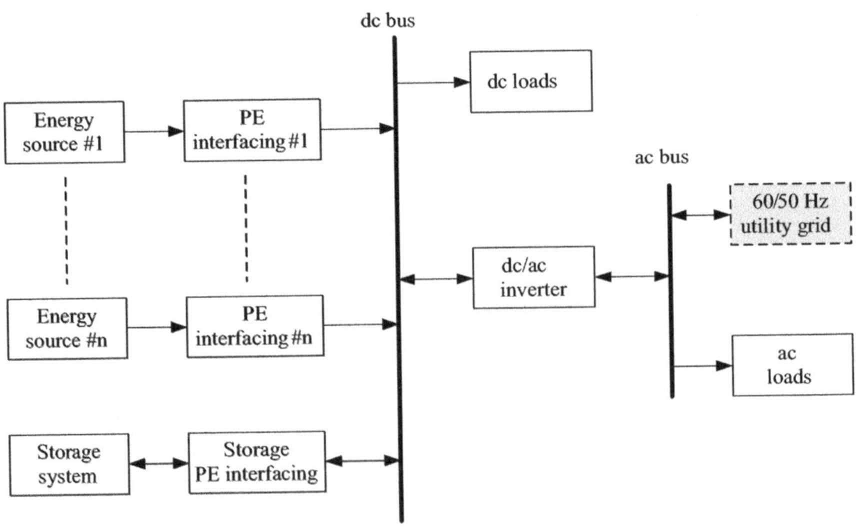

FIGURE 9.2 Schematic of dc-coupled hybrid energy systems.

TABLE 9.1 Comparison of dc- and ac-Coupled Integration Schemes for Hybrid Alternative Energy Systems

Coupling Scheme	Advantage	Disadvantage
dc [1,2,12,29,30]	1. Simple system configuration.	1. Nonstandard dc bus voltage level may cause compatibility problems and/or require high installation and maintenance costs.
	2. Synchronization not needed.	2. If the dc/ac inverter is out of service, the entire system fails to supply ac power. To avoid this situation, multiple inverters with lower power rating may be used.
	3. Single-wired connection possible, if earth is used as a circuit path.	
PFAC [29,31,32]	1. High reliability. If one of the energy sources is out of service, it can be isolated from the system without affecting the others.	1. Synchronization required.
	2. Suitable for grid connection.	2. Need for power factor and harmonic distortion correction.
	3. Standard interfacing and modular structure.	
	4. Easy multivoltage and multiterminal matching.	
HFAC [29,33,34]	1. Higher order harmonics can be easily filtered out.	1. Complex control
	2. Improved efficiency.	2. Higher component and maintenance costs due to high frequency.
	3. Suitable for applications with HFAC loads.	3. Dependence on future advances in power electronics.
	4. Due to high frequency operation, the physical size and weight of transformers and harmonic filters can be smaller (compared to their power frequency counterparts).	4. Concerns about electromagnetic compatibility.

9.2.2 The ac-Coupled Systems

Alternating current (ac) coupling can be divided to two categories: PFAC-coupled and high HFAC-coupled systems. The schematic of PFAC-coupled systems is shown in Fig. 9.3a, where the different energy sources

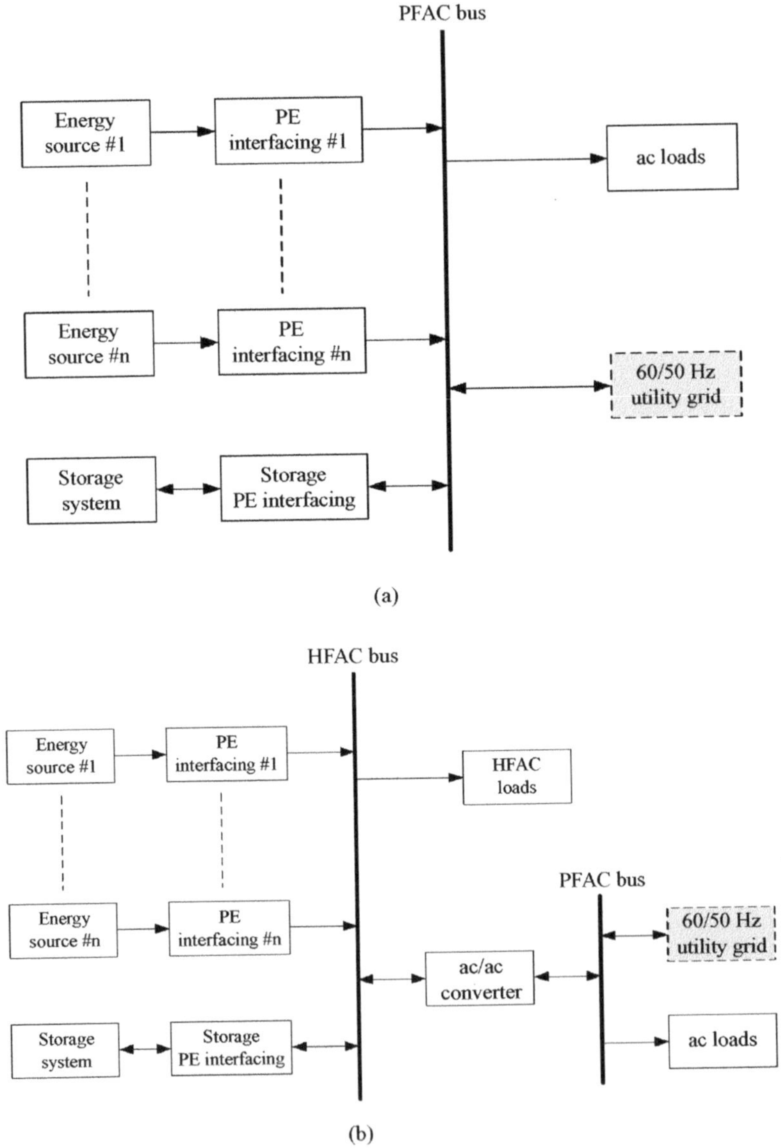

FIGURE 9.3 Schematic of hybrid ac-coupled energy systems:(a) PFAC and (b) HFAC.

are integrated through their own power electronic interfacing circuits to a power frequency ac bus. Coupling inductors may also be needed between the power electronic circuits and the ac bus to achieve desired power flow management.

The schematic of HFAC-coupled systems is shown in Fig. 9.3b. In this scheme, the different energy sources are coupled to a HFAC bus, where HFAC loads are connected. This configuration has been used mostly in applications with HFAC (e.g., 400 Hz) loads, such as in airplanes, vessels, submarines, and in space station applications [29].

The HFAC configuration can also include a PFAC bus (through an ac/ac converter), to which regular ac loads can be connected. In both PFAC and HFAC systems, dc power can be obtained through ac/dc rectification.

Different coupling schemes find their own appropriate applications. Direct current (dc) coupling is the simplest and the oldest type of integration, and to some extent also modular. In particular, the use of power electronic building blocks for interfacing of dc-coupled systems makes them modular. PFAC link is also modular and suitable for grid connection. HFAC coupling is more complicated and more suitable for applications with HFAC loads. Table 9.1 summarizes the advantages and disadvantages of each coupling scheme.

9.2.3 Stand-Alone Versus Grid-Connected Systems

A hybrid alternative energy system can be either stand-alone or grid-connected. For stand-alone applications, the system needs to have sufficient storage capacity to handle the power variations of the alternative energy sources involved. A system of this type can be considered as a stand-alone (island) microgrid [35]. In grid-connected applications, the AEDG sources in a microgrid can supply power both to the local loads and supply/receive power to/from the utility grid. In addition to real power, some DG sources can also be used/adjusted to supply reactive power and provide voltage support to the utility grid. The capacity of the storage device for these systems can be smaller if they are grid-connected since the grid can be used as system backup. However, when connected to a utility grid, important operation and performance requirements, such as voltage, frequency, and harmonic regulations, are imposed on the system [36].

Grid-connected and stand-alone FC power generation systems have been discussed in Chapters 7 and 8, respectively. In this chapter, the design and simulation results for a stand-alone hybrid wind–PV–FC system and CHP operation of a SOFC system are presented.

9.3 FUEL CELLS IN HYBRID COMBINED HEAT AND POWER OPERATION MODE

Combined heat and power (CHP) generation, also called cogeneration, is to simultaneously generate both electricity and useful heat, usually in the form of process heat. Fuel cells are more efficient in this operation mode when their exhaust heat is recovered for space heating, for example, or used to generate more electricity through some other generation devices such as a gas turbine or a MT in combined-cycle operation mode.

High-temperature FCs, such as SOFC and MCFC (molten carbonate fuel cell), are particularly well suited for CHP operation. A block diagram showing the operation of a high-temperature FC power plant in CHP application is given in Fig. 9.4. The FC power plant is the main source to supply electricity to customers through a power electronic system (PES). The high-temperature exhaust gas flows from the FC power plant is used (after passing through a combustor) to operate a MT. The turbine drives a compressor to supply pressurized air to the FC and MT. In addition, a high-speed permanent magnet synchronous generator (PMSG), for example, can be driven by the turbine to generate additional electricity to be supplied to the customers through its power electronic interfacing unit. The exhaust flows of the MT are fed through a recuperator to heat the air. The residual thermal energy in the exhaust gas flows of the recuperator can also be utilized for space heating, as shown in Fig. 9.4.

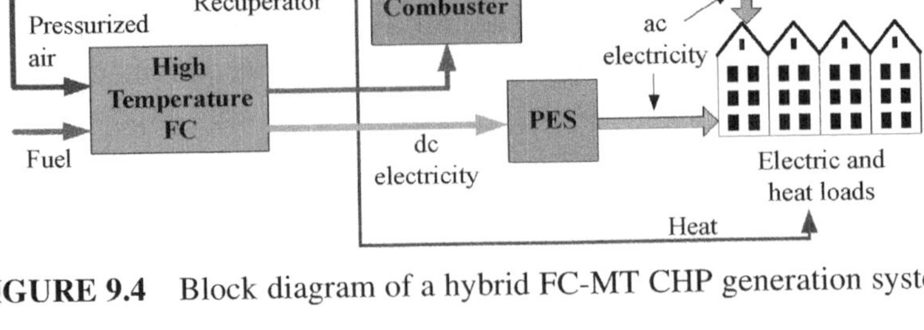

FIGURE 9.4 Block diagram of a hybrid FC-MT CHP generation system.

9.4 CASE STUDY I: A HYBRID STAND-ALONE WIND–PV–FC SYSTEM

In this case study, a multisource hybrid alternative distributed generation system consisting of wind, PV, fuel cell, and electrolyzer is discussed. Wind and PV are the primary power sources of the system to take full advantage of the available renewable energy. The FC–electrolyzer combination is used as a backup and long-term storage system. For stand-alone applications, a battery is also used in the system for short-term energy storage to supply fast transient and ripple power. The different energy sources in the system are integrated through an ac link bus. A power management strategy is designed for the system to coordinate power flows among the different energy sources. The details of the system configuration, power management strategy, unit sizing, maximum operation of the power generation units involved, and simulation results are given in this section.

9.4.1 System Configuration

Figure 9.5 shows the system configuration for an ac-coupled hybrid wind–PV–FC system. In the system, the renewable wind and solar power are the primary power source, while the FC–electrolyzer combination is used as a backup and storage system. This system can be considered as a "green" power generation system because the main energy sources and storage system are all environmental friendly; the system can operate in stand-alone or grid-connected mode. When there is excess wind and/or solar generation available, the electrolyzer turns on to begin producing hydrogen, which is delivered to the hydrogen storage tanks. If the H_2 storage tanks become full, the excess power will have to be diverted to other dump loads, not shown in Fig. 9.5. When there is a deficit in power generation, the fuel cell stack will begin to produce energy using hydrogen from the reservoir tanks, or in case they are empty, from the backup H_2 tanks. The battery bank is used only for supplying transient power to fast load transients, ripples and spikes in stand-alone applications. In grid-connected applications, the battery bank can be taken out from the system; the utility grid will take care of transient power. The different energy sources are connected to a 60 Hz ac bus through appropriate power electronic interfacing circuits. The system can be easily expended, that is, future energy sources can be integrated into the system as desired, as shown in Fig. 9.5.

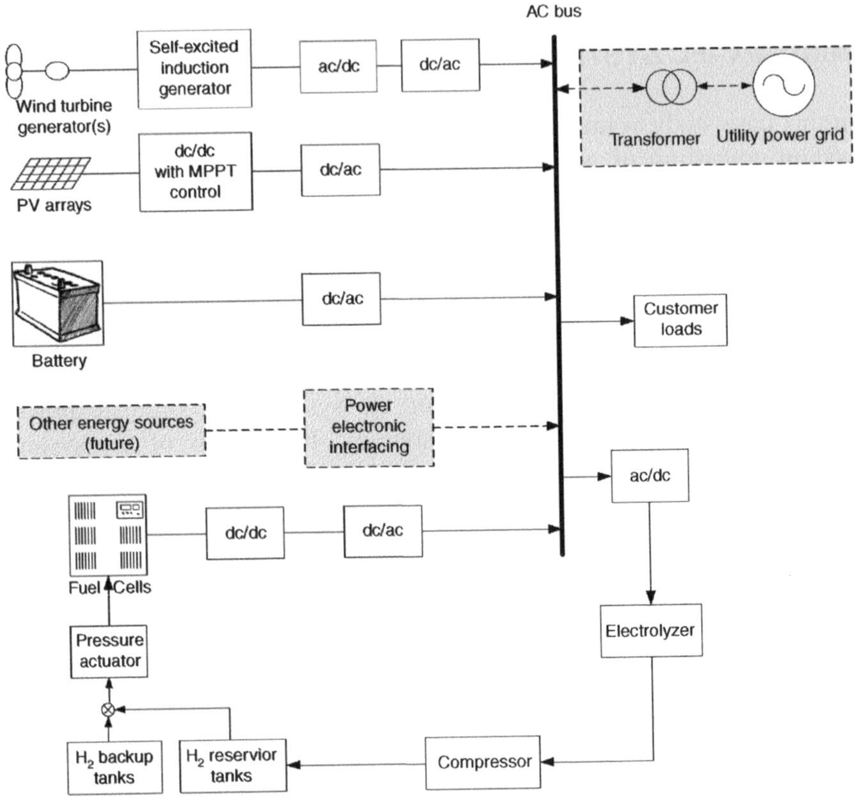

FIGURE 9.5 Schematic diagram of a wind–PV–FC hybrid energy system.

Rechargeable batteries are the most common type of devices used for energy storage. However, they have their drawbacks; the specific energy (Wh/kg) of batteries so far is not as high as desired for high-energy-density applications, such as electric vehicles and outer space applications. Table 9.2

TABLE 9.2 Specific Energy Range of Battery-Based and Fuel Cell-Based Storage Systems [29,43,52]

Type	Specific Energy (Wh/kg)
Lead–acid	30–40
Lithium ion	130–140
Nickel–cadmium	20–35
Nickel–metal hydride	50–70
Zinc–air	150–200
FC–electrolyzer	~350
URFC	>500

gives a comparative summary of the specific power range of batteries and other storage devices. In addition, the life cycle of batteries is not as long as desired. For example, the typical cyclic service life for lead–acid batteries is 500 deep cycles. If a system has regular deep charge–discharge cycles, this type of battery will require replacement in less than 2 years.

In Fig. 9.5, the combination of the fuel cell stack, electrolyzer, and hydrogen tanks can be used for energy storage and backup generation. This combination can achieve a specific energy of 300 Wh/kg or higher [49]. Moreover, such storage systems also have longer cyclic service life. A more elegant solution is to utilize regenerative fuel cell (RFC) stacks, which can act as both fuel cell and electrolyzer. Compared to the conventional fuel cell/electrolyzer combination, a RFC can be much smaller and more compact. RFC is also called the unitized regenerative fuel cell (URFC), where the electrolyzer and fuel cell are combined into one unit and only one of the two modes can be operated at a time. A schematic diagram for URFC is shown in Fig. 9.6.

A URFC energy storage system can achieve a specific energy density greater than 500 Wh/kg [50,51], which is much higher than any commercially available battery system listed in Table 9.2. Although the theoretical efficiency of a URFC can reach 80%, the practical round-trip efficiency is generally below 45%, which is lower than the round-trip efficiency of a battery, typically around 80%. Nevertheless, due to its extremely high specific energy density, URFCs show great potential of being used as energy storage systems, particularly for the applications where energy storage system mass is a constraint.

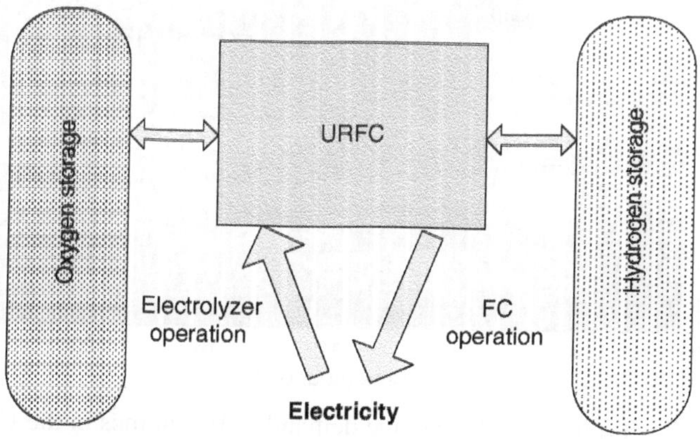

FIGURE 9.6 Schematic diagram of a URFC.

9.4.2 System Unit Sizing

The unit sizing procedure discussed in this section is for a stand-alone hybrid residential electricity supply system in the U.S. Pacific Northwest region with the system structure shown in Fig. 9.5. The unit-sizing procedure can be applied to other applications (with appropriate modifications).

The hybrid system is designed to supply power for five homes. A typical hourly average load demand profile for each home in the region, reported in [37], is used in the simulation study. The total load demand of the five homes is shown in Fig. 9.7. A 50-kW wind turbine is assumed to be available for the hybrid system. The following unit sizing procedure is used to determine the size of PV arrays, fuel cells, the electrolyzer and the battery.

Throughout this section, the capacity factor of each renewable energy system, k_{cf}, defined by Equation 9.1 is used to estimate the availability of each renewable energy source:

$$k_{cf} = \frac{\text{Actual average output power over a period of time } T}{\text{Nominal output power of the renewable energy system}} \tag{9.1}$$

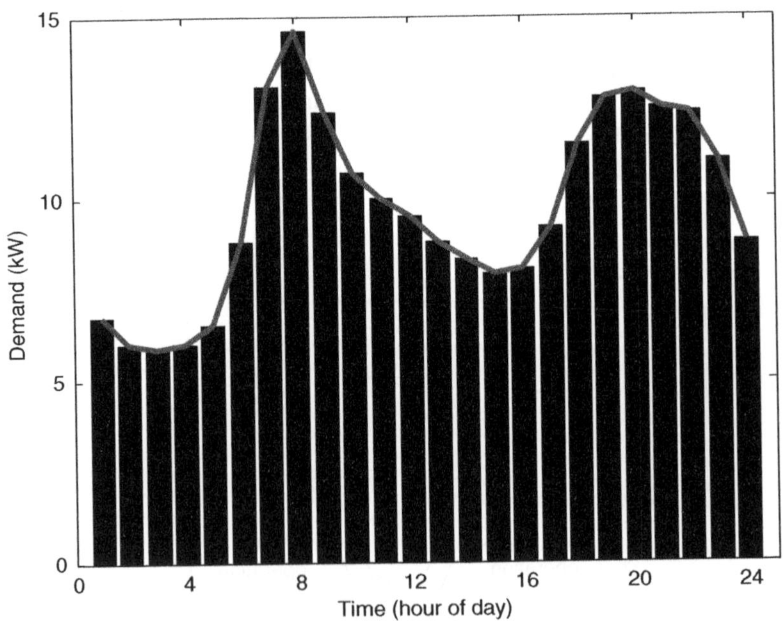

FIGURE 9.7 Typical hourly average demand of five homes in the U.S. Pacific Northwest area.

In the above equation T is normally taken as 1 year. For the wind and solar data reported in [3,24], the capacity factor of the wind turbine (k_{cf_wtg}) and the PV (k_{cf_PV}) array used in the proposed hybrid system for the southwestern part of Montana are taken as 13% and 10%, respectively.

The purpose of unit sizing is to minimize the difference between the generated power ($\bar{P}_{gen}$) from the renewable energy source and the demand ($\bar{P}_{dem}$) over a period of time T (taken as 1 year). Therefore

$$\Delta P = \bar{P}_{gen} - \bar{P}_{dem} = k_{cf\_wtg} \times P_{wtg,rated} + k_{cf\_PV} \times P_{PV,rated} - \bar{P}_{dem} \quad (9.2)$$

where $P_{wtg,rated}$ is the power rating of the wind turbine generator and $P_{PV,rated}$ is the power rating of the PV array.

To balance generation and demand, the rated power for the PV arrays is

$$P_{PV,rated} = (\bar{P}_{dem} - k_{cf\_wtg} \times P_{wtg,rated})/k_{cf\_PV} \quad (9.3)$$

From Fig. 9.7, the average load demand over 24 h is 9.76 kW. Then, according to Equation 9.3, the size of the PV array is calculated to be 32.6 kW. A 33-kW array is used, for which the information is given in Table 9.3.

The FC–electrolyzer combination is taken as the storage for the system. The fuel cell needs to supply the peak load demand when there is no wind and solar power. Therefore from Fig. 9.7, the size of the fuel cell stack needed is 14.6 kW. To leave a safe margin, an 18-kW fuel cell array is used. This array is made using the 500-W PEMFC stack model developed in Chapter 3.

The electrolyzer should be able to handle the excess power from the wind and solar power source. The maximum possible excess power is

$$P_{wtg,max} + P_{PV,max} - P_{dem,min} = 50 + 33 - 6.4 = 76.6\,\text{kW} \quad (9.4)$$

The possibility is very small that both the wind and solar power reach their maximum points at the same time, while the load demand is at its lowest value. According to the data reported in [30], the excess available power normally is less than half of the maximum possible value. Therefore, a 50-kW electrolyzer, that is, over 60% of the maximum available power, given in Equation 9.4 is used.

The battery capacity can be determined based on the information available about the nature of transient power at the load site. In this study, a 10-kWh battery bank is used. The details of the system component parameters are listed in Table 9.3.

TABLE 9.3 Component Parameters of the Proposed Hybrid Energy System

Wind energy system—Wind turbine	
Rated power	50 kW
Cut in speed (cutout speed)	3 m/s (25 m/s)
Rated speed	14 m/s
Blade diameter	15 m
Gear box ratio	7.5
Wind energy system—induction generator	
Rated power	50 kW
Rated voltage	670 V
Rated frequency	60 Hz
PV array	
Module unit	153 cells, 173 W @1 kW/m^2, 25°C
Module number	$16 \times 12 = 192$
Total array power rating	$192 \times 173 \approx 33$ kW
Fuel cell array	
PEMFC stack	500 W
PEMFC array	$6 \times 6 = 36$
PEMFC array power rating	$36 \times 0.5 = 18$ kW
Or, SOFC stack	5 kW
SOFC array	$2 \times 2 = 4$
SOFC array power rating	4×5 kW $= 20$ kW
Electrolyzer	
Rated power	50 kW
Number of cells	40 (in series)
Operating voltage	60–80 V
Battery	
Capacity	10 kWh
Rated voltage	400 V

9.4.3 System Component Characteristics

To develop an overall power management strategy for the system and to investigate the system performance, dynamic models for the main components in the proposed hybrid system have been developed using MATLAB/SIMULINK™ [38]. The models are for the following: wind energy conversion system (WECS), PV, FC, and electrolyzer. In this section, the characteristics of the above system components are discussed.

9.4.3.1 The Wind Energy Conversion System Model The power extracted from wind (P_{wind}) is

$$P_{\text{wind}} = \frac{1}{2}\rho A v^3 C_p(\lambda, \theta)\ (\text{W}) \tag{9.5}$$

where ρ is the air density (kg/m^3), A the area swept by the wind turbine rotor blades (m^2) and v the wind velocity (m/s). C_p is called the power coefficient or the rotor efficiency and is a function of tip speed ratio (*TSR* or λ) and pitch angle (θ) [39,40].

A variable-speed pitch-regulated wind turbine is considered in this study. Figure 9.8 shows the groups of C_p–λ curves of the wind turbine used in this study at different pitch angles [40]. It is noted from the figure that the value of C_p can be changed by changing the pitch angle (θ). In other words, the output power of the wind turbine can be regulated by controlling the pitch angle.

A self-excited induction generator (SEIG) model [38,41,42] was developed and used as a part of the WECS model. The ratings of the SEIG are given in Table 9.3.

Figure 9.9 shows the output power of the WECS versus wind speed. The output power is kept constant through pitch angle control when wind speed is higher than the rated wind speed of the wind turbine. When wind speed is higher than the cutout speed, the system is taken out of operation.

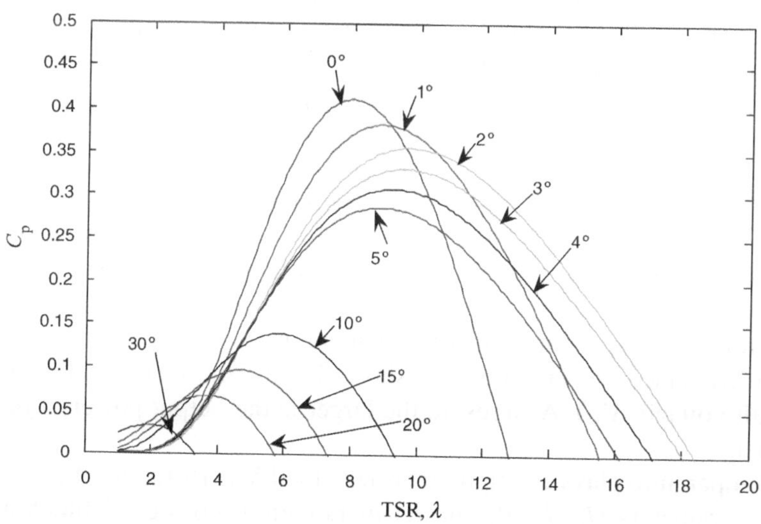

FIGURE 9.8 C_p–λ characteristics of the wind turbine at different pitch angles.

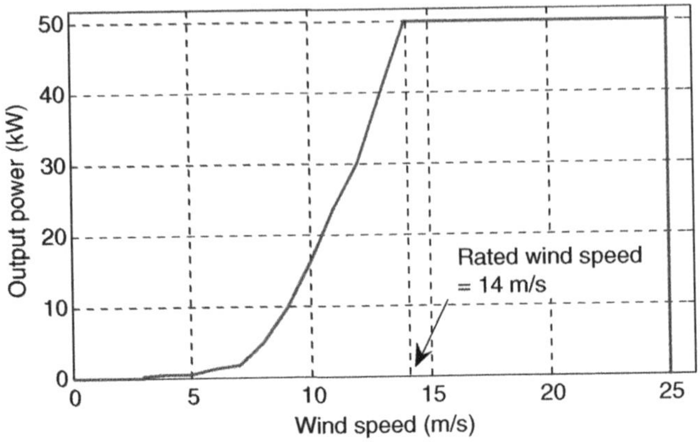

FIGURE 9.9 Wind turbine output power characteristic.

9.4.3.2 The Photovoltaic Array Model The PV effect is a physical process through which solar energy is converted directly into electrical energy. The physics of a solar PV cell is similar to the classical *p–n* junction diode [43]. The relationship between the output voltage V and the load current I of a PV cell or module can be expressed as [18,43]:

$$I = I_L - I_0\left[\exp\left(\frac{V + IR_s}{\alpha}\right) - 1\right] \tag{9.6}$$

where I_L is the current of the PV cell (A) due to solar illumination, I_0 the saturation current (A), I the load current (A), V the PV output voltage (V), R_s the series resistance of the PV cell (Ω), and α the thermal voltage timing completion factor of the cell (V).

$$\alpha = A \times N_s \times \frac{kT}{q} \tag{9.7}$$

where A is the completion factor, N_s the number of cells in series, k the Boltzmann's constant (1.38065×10^{-23} J/K), q the magnitude of charge of an electron (1.6022×10^{-19} C), and T the cell temperature (K).

The I–V characteristic curves of the PV model used in this study under different irradiances (at 25°C) are given in Fig. 9.10 [38]. The higher is the solar irradiance, the larger are the short-circuit current (I_{sc}) and the open-circuit voltage (V_{oc}). As a result, the larger is the output power of the PV model.

Temperature plays an important role in the PV performance because the four parameters (I_L, I_0, R_s and α) in Equation 9.6 are all functions of temperature. The effect of the temperature on the PV model performance is

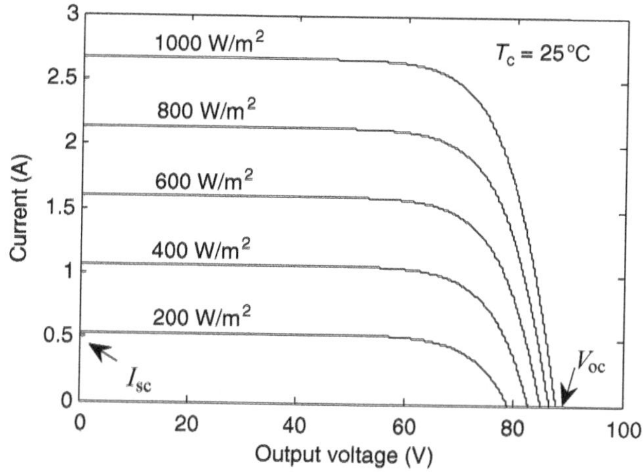

FIGURE 9.10 *I–V* characteristic curves of the PV model at different irradiances.

illustrated in Fig. 9.11. The lower the temperature, the larger is the PV open-circuit voltage and the higher the maximum available power.

9.4.3.3 The Fuel Cell and Electrolyzer Models The fuel cell stack models developed in Chapters 3 and 4, and the electrolyzer model developed in Chapter 5 can be used in this study. The PEMFC model and the electrolyzer model are used in this case study, and the SOFC model is used in the next case study.

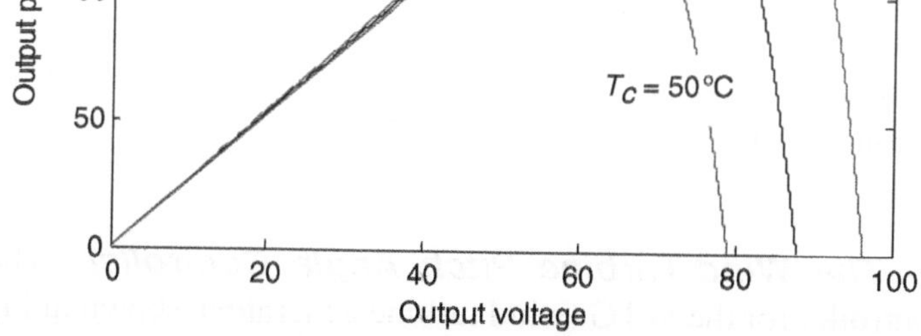

FIGURE 9.11 *P–V* characteristic curves of the PV model at different operating temperatures.

9.4.4 System Control

9.4.4.1 *The Overall Power Management Strategy* A power management strategy is needed to manage power flows among the different energy sources of a multisource energy system. Figure 9.12 shows the block diagram for the overall power management strategy for the proposed hybrid wind–PV–FC–electrolyzer system.

The WECS is controlled by a pitch angle controller to provide maximum possible power at different wind speeds. The PV power generation unit is controlled by a maximum power point tracking (MPPT) controller to provide maximum available PV power as solar irradiance varies. The net power (P_{net}) is the difference between the power generated by wind and PV sources and the load demand, which will determine the operation of the FC–electrolyzer system.

$$P_{net} = P_{wind} + P_{PV} - P_{load} - P_{sc} \tag{9.8}$$

In Equation 9.8, P_{sc} is the self-consumed power, that is, the power consumed by the auxiliary system components such as for the cooling systems, control units, and the gas compressor. In this case study, only the power consumed by the compressor (P_{comp}) is considered.

The governing control strategy is that when there is excess wind–PV-generated power (i.e., $P_{net} > 0$), it is supplied to the electrolyzer (P_{elec}) to generate hydrogen that is delivered to the hydrogen storage tanks through a gas compressor. Therefore, the power balance equation given in Equation 9.8 can be written as

$$P_{wind} + P_{PV} = P_{load} + P_{elec} + P_{comp}, \quad P_{net} > 0 \tag{9.9}$$

When there is a deficit in power generation (i.e., $P_{net} < 0$), the FC stack begins to produce power (P_{FC}) for the load, using hydrogen from the storage tanks. The power balance equation for this situation can be written as

$$P_{wind} + P_{PV} + P_{FC} = P_{load}, \quad P_{net} < 0 \tag{9.10}$$

The power controller has been discussed in detail in Chapter 7 (see Section 7.3.2).

9.4.4.2 *The Wind-Turbine Pitch Angle Controller* The pitch angle controller for the WTG (wind turbine generator), shown in Fig. 9.12, employs two proportional-integral (PI) controllers, as shown in Fig. 9.13. These controllers control the wind flow around the wind turbine blade by

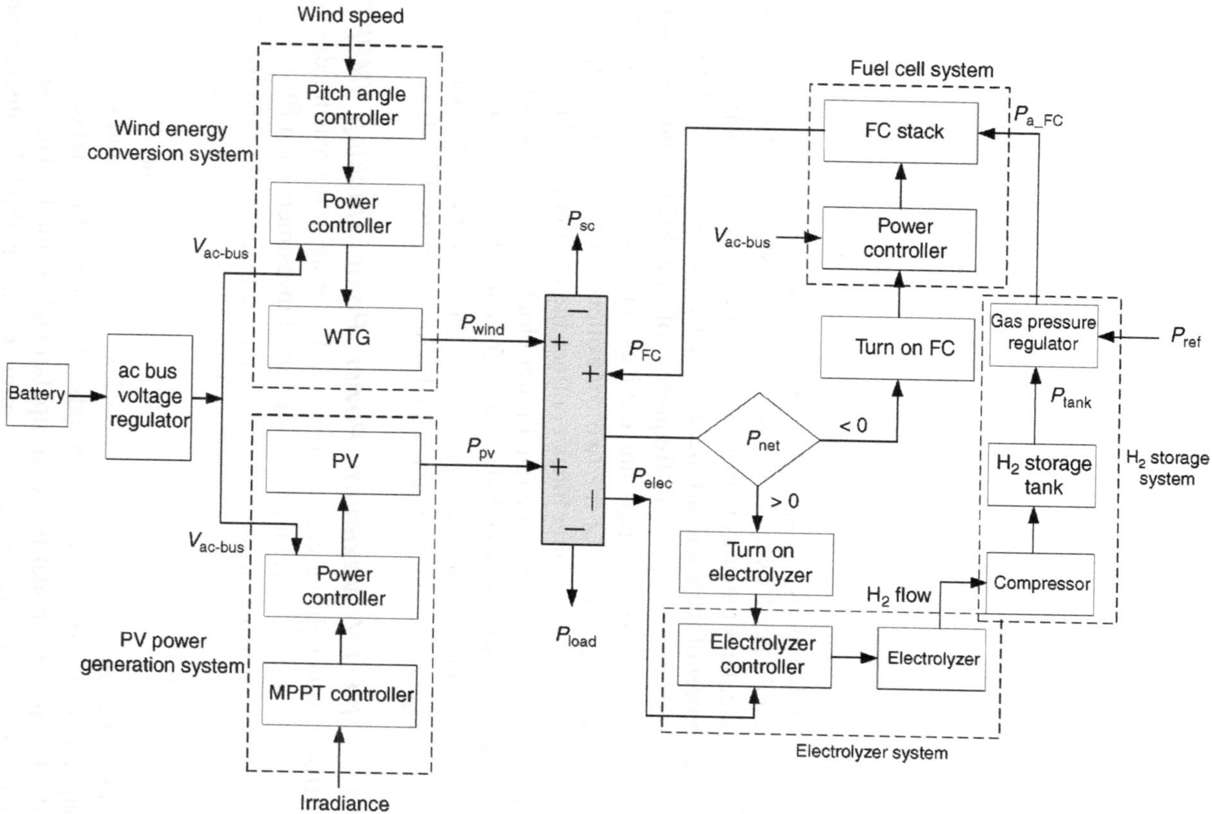

FIGURE 9.12 Block diagram of the overall power management strategy for the hybrid wind–PV–electrolyzer system.

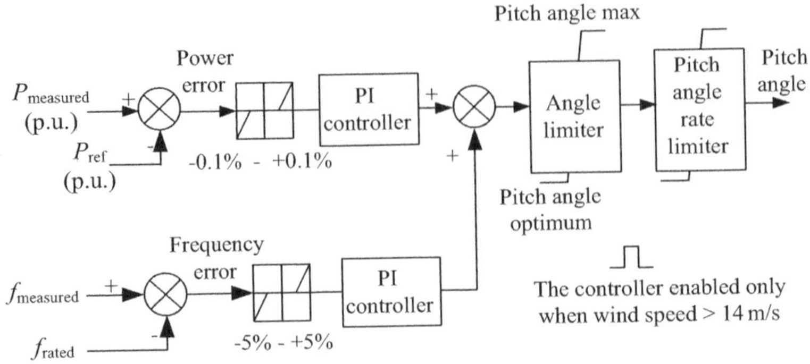

FIGURE 9.13 Pitch angle controller.

changing the blade pitch angle, thereby controlling the toque exerted on the turbine shaft. If the wind speed is less than the rated wind speed of the wind turbine, the pitch angle is kept constant at an optimum value (2° in this study). If wind speed exceeds the rated wind speed of the WTG, the controller calculates the power error (ΔP) between the reference power and the output power of the wind turbine, and the frequency error (Δf) between the measured stator electrical frequency of the SEIG and its rated frequency. These error signals are used as input to the PI controllers, and the sum of the output of the two controllers determines the required pitch angle. The angle limiter block confines the minimum and maximum pitch angle, and the pitch angle rate limiter block limits the rate of change of pitch angle. Most modern wind turbines consist of huge rotor blades, and there is a limit on how fast their pitch angle can be varied. The maximum rate of change of the pitch angle is usually in the order of 3–10°/s.

9.4.4.3 The PV Maximum Power Point Tracking (MPPT) Control Although the capital cost of a PV array is decreasing due to recent technological developments, the initial investment for a PV system is still high compared to conventional electricity generation techniques [48]. Therefore, it is a very natural desire to draw the maximum power possible from a PV array.

Many MPPT techniques have been proposed in the literature, a summary of which is given in [44]. The important techniques can be grouped into (1) look-up table methods, (2) perturbation and observation methods, and (3) model-based computational methods. The two major techniques in the "model-based computational methods" category are voltage-based MPPT (VMPPT) and current-based MPPT (CMPPT) techniques. As obvious from their name, in VMPPT, the maximum power point is determined based on adjusting the load on the PV array until the maximum voltage point is obtained, where as in CMPPT, information about the maximum short-circuit current of PV array is used for determining the maximum power point [45]. In this design, a CMPPT method is used to obtain the maximum power for the PV array of the hybrid wind–PV–FC system under study.

The main idea behind CMPPT is that the current at the maximum power point (I_{mp}) has a strong linear relationship with the short-circuit current I_{sc}. I_{sc} can either be measured on-line under different operating conditions or computed from a validated model. Figure 9.14 shows the curve of I_{mp} versus I_{sc} at a constant temperature. It is noted that I_{mp} has a very good linear dependence on I_{sc}, which can be expressed as

$$I_{\text{mp}} = k_{\text{cmppt}} \times I_{\text{sc}} \tag{9.11}$$

where k_{cmppt} is the proportionality constant used for control of CMPPT.

The control scheme for the CMPPT used in the PV system (as a part of the hybrid system under study) is shown Fig. 9.15. The value of I_{mp} calculated from Equation 9.11 is used as the reference signal to control the buck dc/dc converter in the figure so that the output current of the PV system matches I_{mp}. Also through a PI controller, the current error signal ($I_{\text{PV,err}}$) is also used to generate a power angle signal (θ), which is used to generate the dq reference values for the inverter as

$$\begin{aligned} V_{d,\text{pu}} &= \cos(\theta) \\ V_{q,\text{pu}} &= \sin(\theta) \end{aligned} \tag{9.12}$$

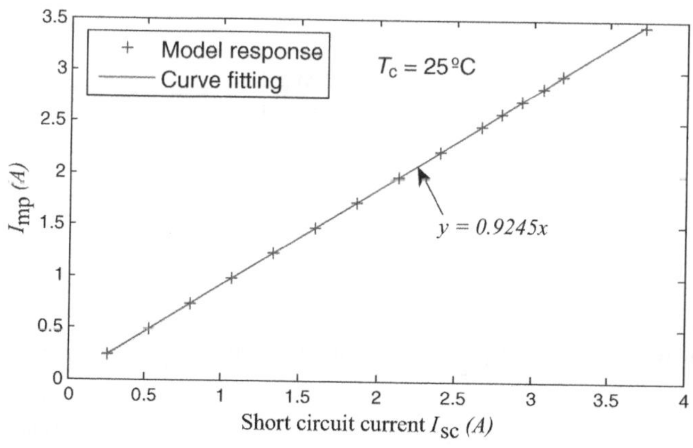

FIGURE 9.14 Linear relationship between I_{sc} and I_{mp} under different solar irradiances.

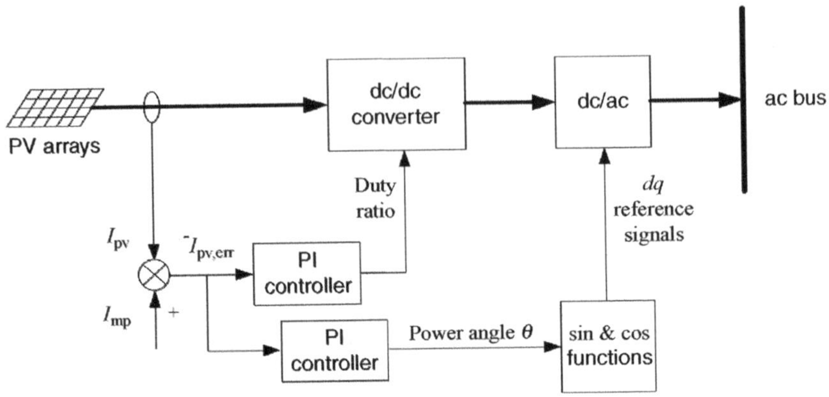

FIGURE 9.15 CMPPT control scheme.

The *dq* reference signals are then used to control the power flow of the inverter, as discussed in Chapter 7, so that the maximum available power is delivered to the ac bus in Fig. 9.15.

9.4.4.4 The ac Bus Voltage Regulator Figure 9.16 shows the block diagram of the ac bus voltage regulator. The ac bus voltage is measured and transformed into *dq* values ($V_{d,q}$) through the "*abc*/*dq* transformation" block, as explained in Chapters 6 and 7. $V_{d,q}$ are then compared with the reference *d,q* voltage values ($V_{d,q(\mathrm{ref})}$), and the voltage error signals are fed into a PI voltage controller. The controller output is

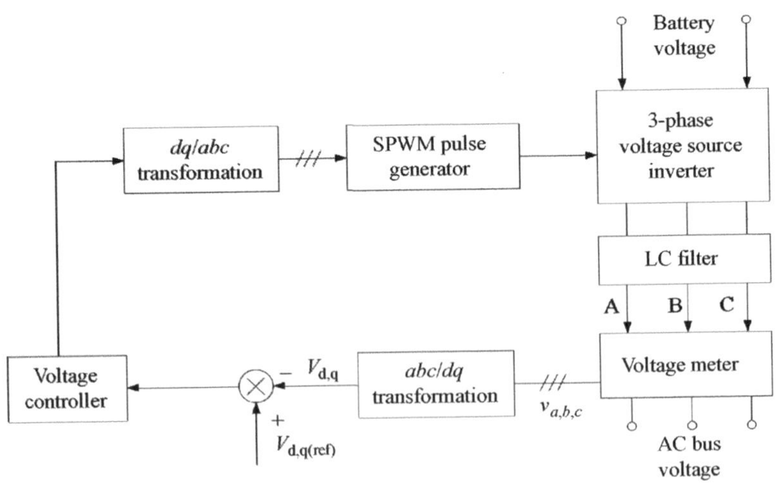

FIGURE 9.16 Block diagram of the ac voltage regulator.

transferred back into the control signals in *abc* coordinates through the "*dq*/*abc* transformation" block. These control signals are used to generate proper SPWM pulses for the inverter switches, which shape the inverter output voltage (the ac bus voltage).

9.4.4.5 The Electrolyzer Controller The block diagram of the electrolyzer controller is shown in Fig. 9.17. As discussed in Chapter 5, an electrolyzer can be considered as a voltage-sensitive nonlinear dc load. For the given electrolyzer within its rating range, the higher the dc voltage applied to it, the larger is the load current and the more H_2 is generated. The function of the electrolyzer controller is to control the controllable ac/dc rectifier to obtain a proper output dc voltage ($V_{dc,elec}$) so that the excess available wind–PV-generated power can be fully utilized by the electrolyzer to generate H_2. For example, when the excess available power is increased, the controller controls the rectifier to give a higher dc voltage ($V_{dc,elec}$) to the electrolyzer. As a result, the power consumed by the electrolyzer (P_{elec}) becomes greater, matching the excess available power.

9.4.5 Simulation Results

A simulation system test-bed, developed for the wind–PV–FC–electrolyzer energy system in MATLAB/SIMULINK, is used to evaluate its performance under different environmental conditions (solar and wind availability, and temperature). As discussed in the section on unit sizing,

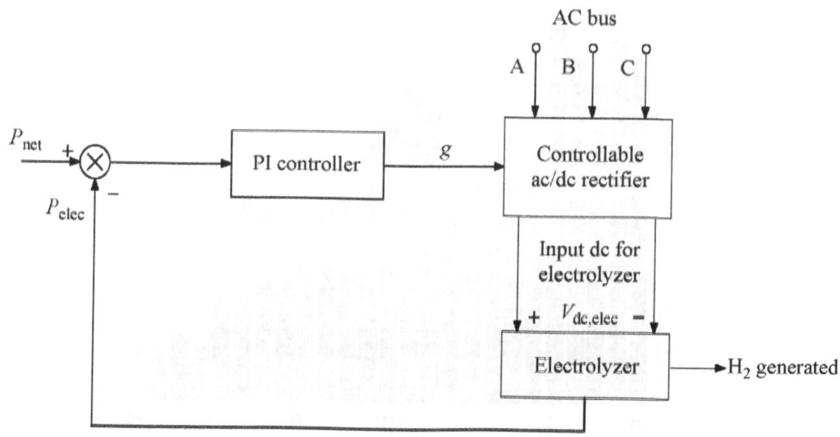

FIGURE 9.17 Block diagram of the electrolyzer controller.

the system is designed to supply electric power demand of five houses in the US Pacific Northwest region. The hourly average residential load demand profile and actual weather data, obtained from the online records of a weather station affiliated with the Pacific Northwest Cooperative Agricultural Weather Network (AgriMet) at Deer Lodge, Montana, [46], were used in the simulation study. The weather data collected on February 1, 2006 were used for a winter scenario study. Simulation studies can be carried out for other scenarios as long as the weather and load demand data are available [38].

The wind speed data, collected at the height of 2 m is corrected to the turbine hub height (assumed to be 40 m) using the following expression [24,47]:

$$W_{s1} = W_{s0}\left(\frac{H_1}{H_0}\right)^{\alpha} \tag{9.13}$$

where W_{s1} (m/s) is wind speed at the hub height of H_1 (m), W_{s0} (m/s) the wind speed at the height of H_0 (m), and α the wind speed correction exponent, taken as 0.13 in this study, as suggested in Refs. [24] and [47].

Figure 9.18 shows the corrected hourly wind speed profile over 24 h. The hourly solar irradiance and air temperature data are shown in Figs. 9.19 and 9.20, respectively.

The output power from the wind energy conversion unit of the hybrid energy system over a 24-h simulation period is shown in Fig. 9.21. When

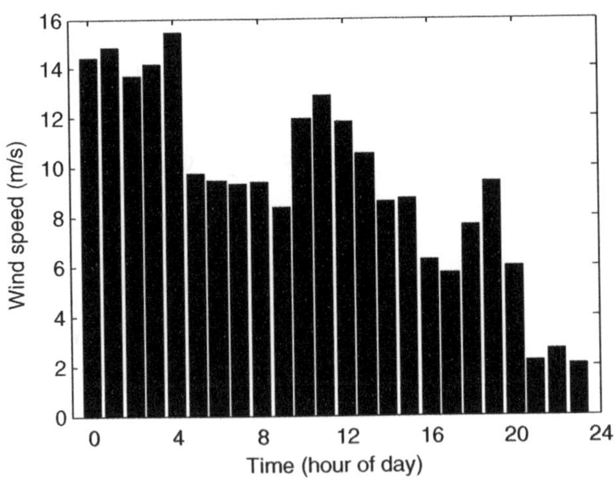

FIGURE 9.18 Wind speed data for the winter scenario simulation study.

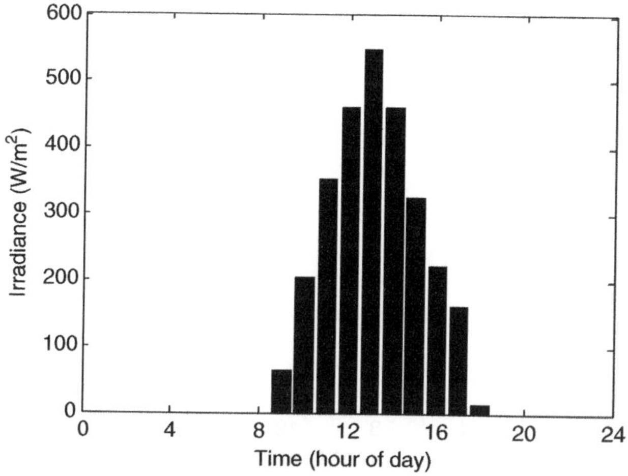

FIGURE 9.19 Solar irradiance data for the winter scenario simulation study.

the wind speed is less than the wind turbine cut-in speed (3 m/s), there is no wind power generated, and when the wind speed is over 14 m/s, the output power is limited to 50 kW by the pitch angle controller (see Figs. 9.9 and 9.13). For wind speeds between 3 and 14 m/s, the wind-generated power is approximately proportional to the cube of wind speed.

The output power of the PV array over the 24-h simulation period is shown in Fig. 9.22. The PV array output power is controlled by the CMPPT

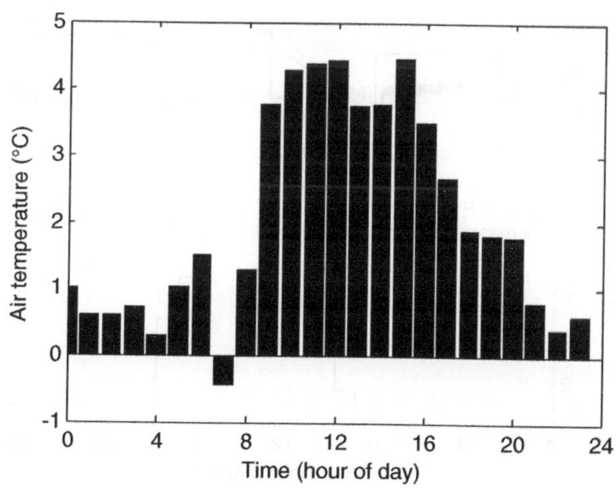

FIGURE 9.20 Air temperature data for the winter scenario simulation study.

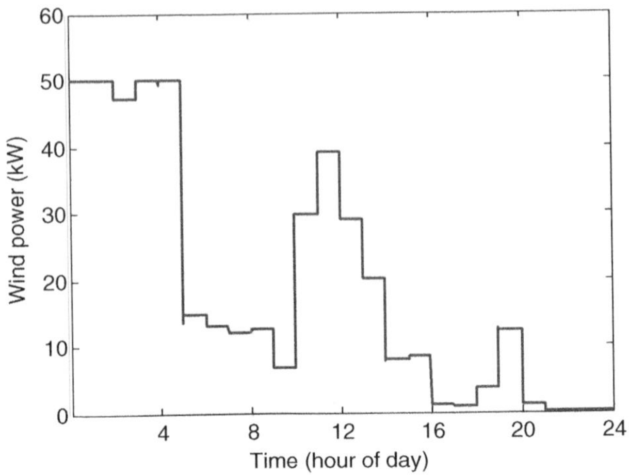

FIGURE 9.21 Wind power for the winter scenario study.

controller (Fig. 9.15) to provide maximum output power under different solar irradiances.

As mentioned before, temperature plays an important role in the PV module's performance (see Fig. 9.11). Figure 9.23 shows the PV temperature response over the simulation period. Two main parameters for determining the temperature of the PV module are solar irradiance (Fig. 9.19) and surrounding air temperature (Fig. 9.20); the higher the temperature, the lower is the maximum available power from the PV array.

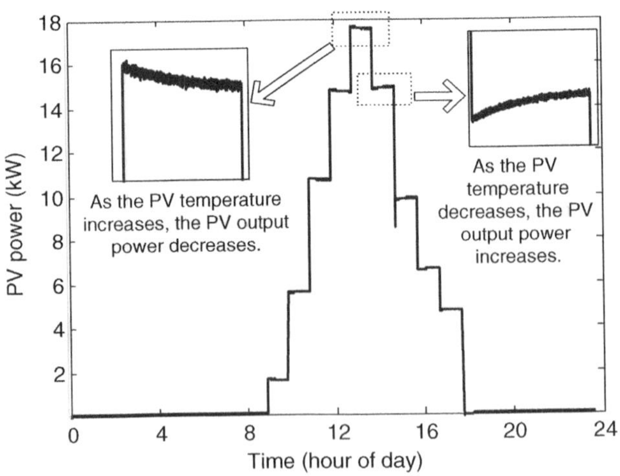

FIGURE 9.22 PV power for the winter scenario.

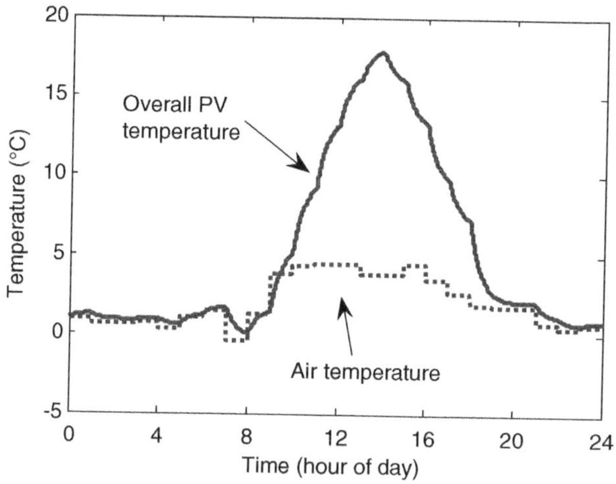

FIGURE 9.23 The PV temperature response over the simulation period for the winter scenario.

Figures 9.22 and 9.23 also show the effect of temperature on PV performance.

When $P_{net} > 0$ (see Eqs. 9.8 and 9.9, there is excess power available for H_2 generation, which is supplied to the electrolyzer, as shown in Fig. 9.24. Figure 9.25 shows the H_2 generation rate over the simulation period. The corresponding dc voltage applied to the electrolyzer and the

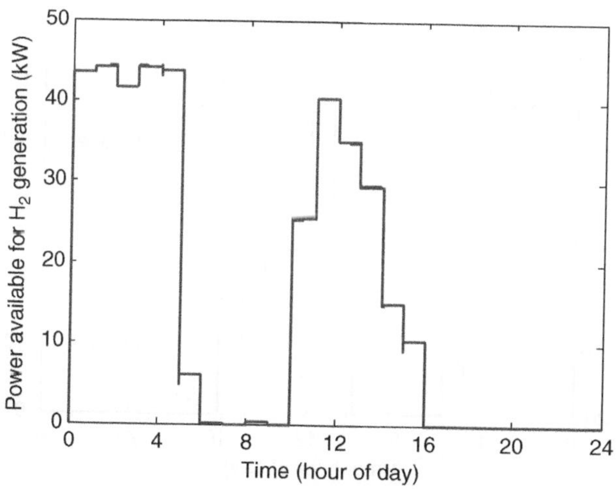

FIGURE 9.24 Power available for H_2 generation in the winter scenario.

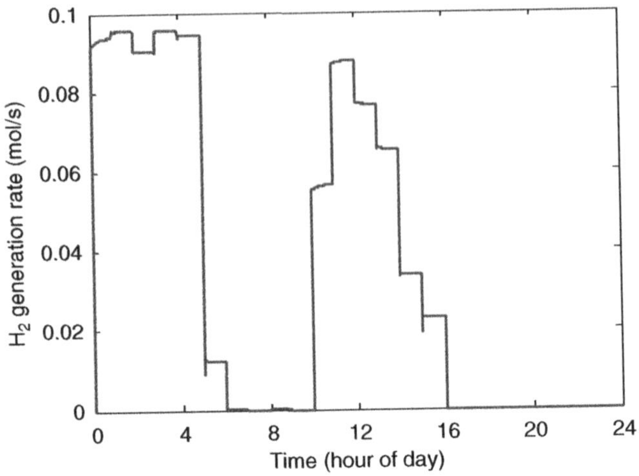

FIGURE 9.25 H_2 generation rate for the winter scenario study.

electrolyzer current are shown in Fig. 9.26. The more is the excess available power P_{net}, the higher is the dc input voltage to the electrolyzer, and as a result, the more is the generated H_2.

When $P_{net} < 0$, the sum of wind and PV generated power is not sufficient to supply the load demand. Under this condition, the fuel cell turns on to supply the power shortage. Figure 9.27 shows the power delivered by the FC stack.

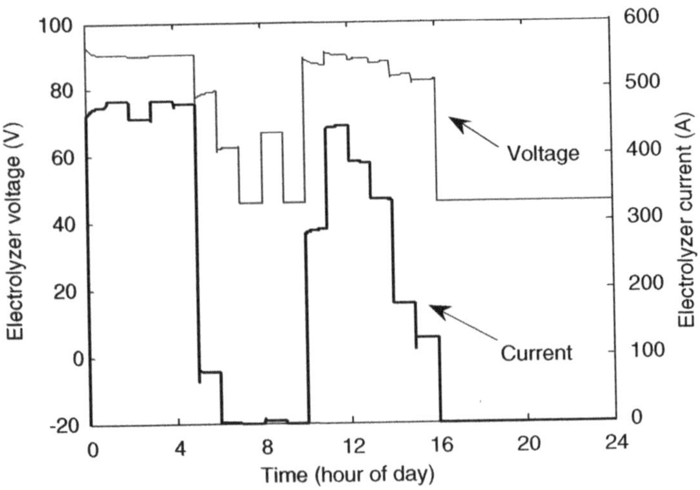

FIGURE 9.26 Electrolyzer voltage and current for the winter scenario study.

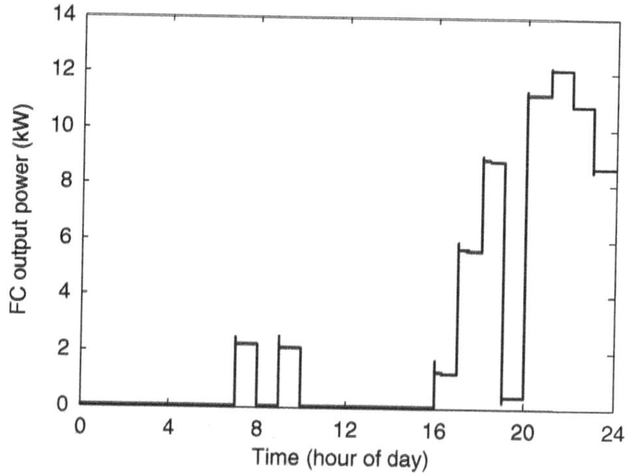

FIGURE 9.27 Power supplied by the FC stack in the winter scenario.

9.5 CASE STUDY II: SOFC EFFICIENCY EVALUATION IN HYBRID OPERATION MODE

This case study is an extract from Ref. [53], with permission from the American Society of Mechanical Engineers (ASME). The contributing authors are C.M. Colson, M.H. Nehrir (Electrical and Computer Engineering Department, Montana State University—Bozeman); M.D. Deibert (Chemical and Biological Engineering Department, Montana State University. Bozeman); M.R. Amin (Mechanical and Industrial Engineering Department, Montana State University—Bozeman); and C. Wang (Engineering Technology Division, Wayne State University).

As discussed in Chapter 2, SOFCs show great potential for hybrid operations. Hybrid SOFC applications encompass combined cycle (CC) systems and combined heat and power (CHP) systems, the focus of this case study. In both categories of hybrid systems, high-temperature SOFC exhaust gases are utilized for further energy extraction. Conventional CC systems are most commonly seen in gas turbine (GT) applications, where the unutilized turbine exhaust heat is recovered in a heat recovery steam generator (HRSG). This steam then drives a secondary turbine generator for additional electricity production. On the other hand, CHP systems, also known as cogeneration systems, take advantage of SOFC exhaust gas streams for uses such as residential or commercial heating, steam generation, or potentially for further electricity production, therefore resulting in higher efficiencies. In this case study, first, a discussion of SOFC efficiency is presented based on the first and second laws of thermodynamic presented in Chapter 2. Then, the heating values of SOFC gaseous fuel and the electrical efficiency and heat transfer potential of SOFC will be examined without the support of heat recovery subsystems. Finally, the SOFC overall efficiency in the CHP operation mode will be developed. The SOFC model developed in Chapter 4 is used for efficiency evaluation.

9.5.1 Thermodynamic Laws and SOFC Efficiency

The first law of thermodynamics for an open system, such as a fuel cell, relates the difference between the enthalpy flow in/out of the system, to the rates of heat flow and work flow to/from the system. The first law states [57]

$$dQ - dW + \sum_{\text{input streams}} \bar{n}_i \hat{H}_i = \sum_{\text{output streams}} \bar{n}_i \hat{H}_i \tag{9.14}$$

where dQ, dW, and $\hat{H}_i$ represent the rate of system heat flow (J/s), rate of system work flow (J/s), and specific enthalpy (J/mol) of the ith input or output stream, respectively. The $\bar{n}_i$ represents molar flow rate (mol/s). Neglecting the kinetic and potential energy content of each energy stream, as discussed in Chapter 2, the system enthalpy can be written as ($H = U + PV$, where H, U, V, P are specific enthalpy, specific internal energy, and specific volume, respectively and pressure is a nonspecific quantity). The system differential enthalpy can be written as

$$dH = dQ - dW \tag{9.15}$$

where dH, dQ, and dW are the rates of enthalpy, heat, and work flow in an open system. Within the fuel cell system, dW is in the form of output electrical work flow.

It must be noted that, while the first law describes the conservation of energy, heat and work are fundamentally different than enthalpy in that the former represent modes of energy transfer to and from the system. Work and heat quantities have the same units, but are not simply comparable to enthalpy that represents properties of the system's input and output energy streams.

According to the second law of thermodynamics, the amount of reversible (ideal) heat transfer for a system is expressed by

$$dQ_{\text{ideal}} = T\,dS \tag{9.16}$$

where dQ_{ideal} is the reversible rate of energy transferred as heat (J/s), T the temperature (K), and dS the entropy change (J/K s).

Referring to Chapter 2, change in Gibbs energy for a constant temperature process can be written as

$$dG = dH - T\,dS \tag{9.17}$$

where dG is the Gibbs energy change, which represents the maximum amount of system energy that may be converted to useful work flow (output electrical energy). In the case of SOFCs, Gibbs energy is the maximum amount of electrical energy that could theoretically be extracted from the fuel–oxidizer reaction.

According to the first law of thermodynamics, SOFC efficiency is described by the ratio of useful work (output electrical energy) and the difference between the system's stream energy flows (enthalpies). First law efficiency, or sometimes referred to as thermal efficiency, is defined as

$$\eta = \frac{\text{useful energy}}{\text{total enthalpy change}} \tag{9.18}$$

The second law of thermodynamics expresses the efficiency as the ratio of the total work flow from the system to the maximum work flow, which is the difference between the Gibbs energy flow of the feed (fuel and oxidizer) and product streams.

The maximum amount of energy that may be converted to work is exactly equal to the Gibb's energy change of a system. Therefore, the representation of ideal efficiency for a reversible isothermal reaction occurring within a fuel cell is [54]:

$$\eta_{\text{ideal,electrical}} = \frac{-\text{Gibb's energy change}}{-\text{enthalpy change of reaction}} \tag{9.19}$$

Work potential, also known as exergy, is conserved in perfectly reversible systems, but can be destroyed by irreversible processes [55]. Exergy analysis highlights the fundamental difference between a heat engine and the fuel cell: a heat engine transfers heat to useful mechanical work, while the fuel cell as an electrochemical device transfers chemical energy to electrical energy. The combustion process within the heat engine is largely irreversible, and therefore destroys work potential. The electrochemical conversion process that occurs within the fuel cell differs from a pure combustion process in that less entropy is generated in the former process.

It can be concluded that fuel cell ideal electrical efficiency, as defined by Equation 9.19, which relates the change in Gibb's energy to the change in the enthalpy reaction, only describes fuel cell ideal electrical efficiency.

However, fuel cell operation in CHP applications takes advantage of more than just the electrical power produced by the fuel cells. The ideal electrical efficiency defined by Equation 9.19 does not describe the entirety of the thermodynamic system and may not be appropriate for describing CHP applications.

SOFC electrical efficiency under different loading conditions can be obtained as follows:

$$\eta_{\text{electrical}} = \frac{VI}{-\Delta H_{\text{rxn}}} \tag{9.20}$$

where V, I are SOFC output voltage and current, and ΔH_{rxn} is the enthalpy flow (J/s). It is noted that enthalpy flow represents the total enthalpy flow of the inlet fuel species applied to the SOFC anode. The usage of enthalpy flow is explained later in this section.

A critical aspect of SOFC operation that affects the electrical efficiency, as defined by Equation 9.20, is a property of operation called fuel utilization (FU). Because an SOFC is a direct energy conversion device rather than a combustion device, not all available fuel is consumed during operations. In fact, under constant fuel flow operating mode, the amount of fuel that is consumed by the SOFC is directly related to its load demands, as shown in Fig. 9.28. As a result, to find the true ideal electrical efficiency of the SOFC, it is only appropriate to calculate the enthalpy flow of the

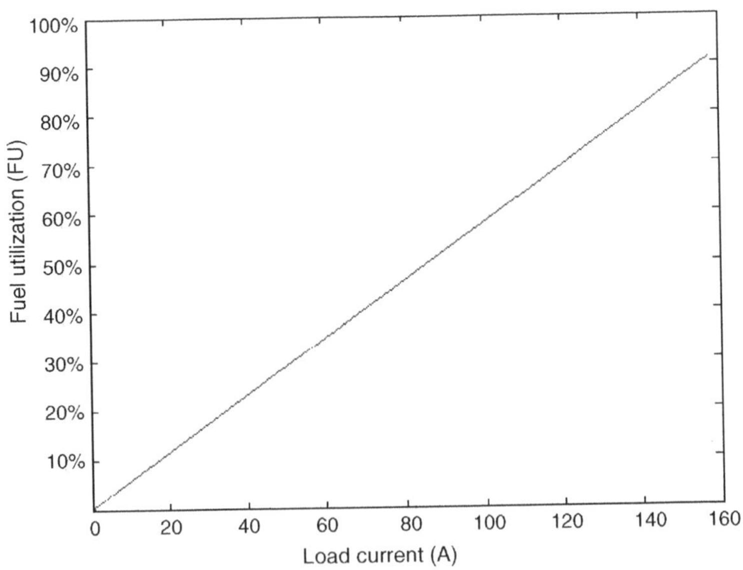

FIGURE 9.28 Simulated SOFC stack fuel utilization versus load current.

actual quantity of fuel that is converted by the SOFC, rather than the total quantity of fuel mass transported to the SOFC anode inlet. Equation 9.20, therefore, is modified to include the FU percentage, as follows:

$$\eta_{\text{electrical}} = \frac{VI}{(-\Delta H_{\text{rxn}})(\text{FU})} \tag{9.21}$$

where FU is fuel utilization (%). In the SOFC model described in Chapter 4, a constant fuel flow is applied, but FU varies dynamically.

Figure 9.29 shows the simulated electrical efficiency of the SOFC model versus load current for various operating temperatures. The data shows that, at higher currents less energy is converted to useful work in the form of electrical power, however, higher temperature operation has a beneficial effect on SOFC efficiency. In the ideal case, the available amount of Gibb's energy from the hydrogen–oxidizer reaction decreases slightly as temperature increases, but realistic simulations show that higher temperatures provide more favorable thermochemical parameters, which overcome the decrease in Gibb's energy. This phenomenon is

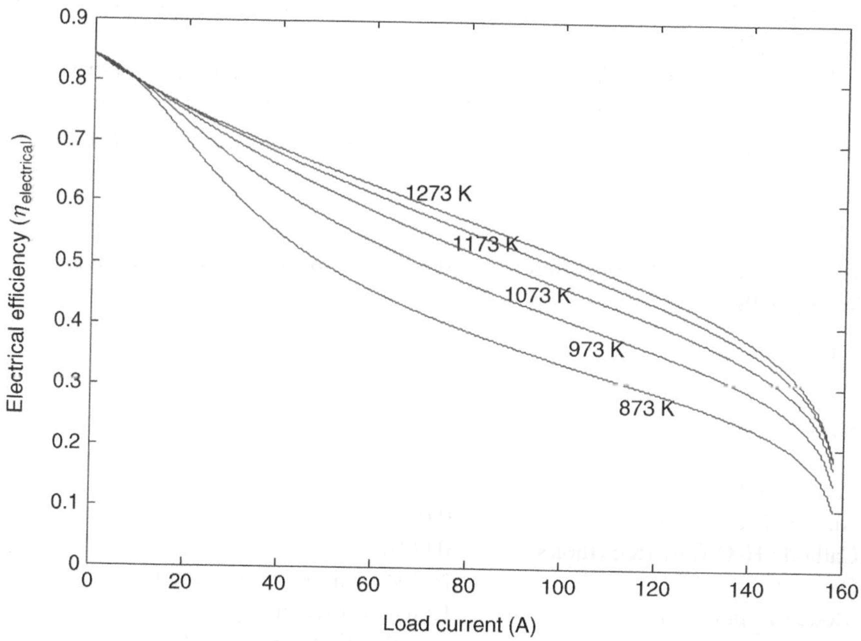

FIGURE 9.29 Simulated SOFC stack electrical efficiency versus load current at different operating temperatures.

characterized by the SOFC experiencing improved reaction and mass transport rates, as well as a slightly lower cell resistance, as operating temperature goes up [56]. However, as load current increases, SOFC electrical efficiency tends downward and operating temperature tends upward. This phenomenon results from the thermochemical response, resulting in ohmic, concentration, and charging effect voltage losses, described in Chapter 4. As operating temperature rises in response to higher load (within the SOFC rating), and although the electrical efficiency goes down, the higher temperature can be more favorably exploited in CHP operation mode. It is noted that due to the fuel utilization effect within the SOFC, electrical efficiency does not appear to go to zero at low load currents. In fact, at zero current, FU is equally zero, which results in an undefined zero-over-zero efficiency value. We want to emphasize that the efficiency curve shown in Fig. 9.29 is merely based on Equation 9.21, that is, the fuel cell only (without support systems). As discussed later and shown in Figs. 9.31 and 9.32, a practical fuel cell system always requires some initial support energy to start the system. In other words, even if there is no electrical power output, a practical fuel cell still consumes a base amount of fuel to operate. Under this case, of course, the efficiency will be zero if the load current is zero, as shown in Fig. 9.32. The simulation parameters used for the SOFC model to obtain the curves of Fig. 9.29 are listed in Table 9.4.

When examining a SOFC in CHP operation, both the Gibb's energy and exhaust heat must be combined because the fuel cell will generate electrical energy, and heat transfer is harnessed from the fuel cell exhaust

TABLE 9.4 SOFC Parameters Used for the Simulation Results Given in Fig. 9.28

Parameters	Value
Number of cells in stack	96
Stack temperature (K)	873–1273
Anode input pressure (atm)	1.1
Cathode input pressure (atm)	1.1
Anode H_2 flow rate (mol/s)	0.0096
Cathode H_2O flow rate (mol/s)	0.0001
Oxidant	Not stoichiometrically limited
Mode of operation	Constant fuel supply, utilization varies due to electrochemical and thermal conditions

gas for conversion to useful work. For SOFC, work potential due to heat transfer is defined as [55]:

$$dW_{\text{exhaust}} = \left(1 - \frac{T_{\text{sink}}}{T_{\text{fuel cell}}}\right) dQ_{\text{exhaust}} \tag{9.22}$$

where dW_{exhaust} is the rate at which work is performed by the heat engine (J/s), T_{sink} the temperature at which heat engine rejects heat to the environment (K), T_{fuelcell} the fuel cell exhaust temperature (K), and dQ_{exhaust} the heat flow of exhaust (J/s).

Therefore, for any defined duration of time, a complete measure of the fuel cell system work potential can be obtained by combining Equation 9.22 with the Gibbs energy change. As a result, the ideal efficiency expression for a high temperature fuel cell can be written by modifying Equation 9.22 to include W_{exhaust}:

$$\eta_{\text{ideal}} = \frac{-\Delta G + [1 - (T_{\text{sink}})/(T_{\text{fuel cell}})]Q_{\text{exhaust}}}{(-\Delta H)(\text{FU})} \tag{9.23}$$

where η_{ideal} is the ideal SOFC efficiency that incorporates ideal electrical and ideal heat recovery efficiency components in a defined period of time, ΔG the change in Gibbs energy (J), and ΔH the change in enthalpy for the system (J).

9.5.2 Hydrogen Fuel Heating Values

The fuel–oxidizer chemical reactions that take place within a SOFC were given in Chapters 2 and 4. Standardized enthalpies of formation can be used to determine the enthalpy of reaction (also referred to as enthalpy of combustion) for the fuel–oxygen reaction. For the reaction within a fuel cell, the standard enthalpy of reaction is determined by [57]:

$$\hat{H}^0_{\text{rxn}} = \sum_{\text{products}} v_i \Delta \hat{H}^0_i - \sum_{\text{reactants}} v_i \Delta \hat{H}^0_i \tag{9.24}$$

where $\hat{H}^0_{\text{rxn}}$ is the standard enthalpy change of reaction (J/mol), or heat of reaction, v_i the stoichiometric coefficient of each product or reactant species, and $\hat{H}^0_i$ the standard heat of formation for each reaction species (J/mol). The standard enthalpy of reaction for the hydrogen–oxidizer reaction that occurs within a fuel cell depends on the resultant state of water, being in liquid state (−285.84 kJ/mol) or in gaseous state (−241.83 kJ/mol) [57].

For modeling an open system, such as a fuel cell, calculating the enthalpy flow is useful for evaluating the energy balance described in Equation 9.18. The enthalpy flow is defined as [58]:

$$\Delta H_{\text{rxn}} = \sum_{\text{products}} \bar{n}_i \hat{H}_i - \sum_{\text{reactants}} \bar{n}_i \hat{H}_i \tag{9.25}$$

where ΔH_{rxn} is the enthalpy flow for an open system (J/s), $\bar{n}_i$ the molar flow rate of products or reactants (mol/s), and $\hat{H}_i$ the specific enthalpy for each reaction species (J/mol).

The enthalpy of reaction for a fuel–oxidizer reaction is referred to as the heating value. When pure hydrogen is used as fuel and reacted with oxygen, only water is produced. The final state of product water determines the appropriate heating value for use when describing the energy content of the fuel. Higher heating value (HHV) refers to the energy content of a fuel if the reactants are returned to standard temperature, typically 25°C (298 K). Lower heating value (LHV) refers to the energy content of a fuel if the reactants are returned to a temperature (appropriate for the system pressure) that yields water in the gaseous state, near its condensation point [59]. LHV represents a smaller quantity of energy than HHV because the standard enthalpy change of condensation (latent heat of condensation) quantity that would be released from the product water in changing state from gas to liquid is not included.

There is significant debate over which heating value is appropriate to use when assessing fuel cells. It may not be appropriate to utilize HHV when calculating the efficiency of SOFC due to its high operating temperature. With operating temperatures in the range of 600–1000°C between its inlet and outlet ports, water produced at the SOFC anode will not condense to the liquid state until undergoing heat transfer, downstream of the fuel cell. Similarly, the LHV of the hydrogen fuel may not be appropriate for SOFC applications because SOFCs do not operate at temperatures in the vicinity of the condensation point of water.

What contributes to the debate over higher and lower heating values is that the effect of using one or the other may be counterintuitive. HHV represents a larger magnitude of energy contained within the same amount of fuel. It would seem that a conservative estimate of the energy evolved from a fuel–oxygen reaction would therefore be more in-line with the smaller quantity of energy, that is, LHV. However, when it comes to efficiency calculations, because the heating value is in the denominator, see Equation 9.19, the smaller the amount of energy contained within a fuel yields a higher efficiency for the given amount of reaction output.

It should always be noted that the enthalpy change in Equation 9.19 is a function of temperature, which just cannot be simply represented by a fixed value.

9.5.3 SOFC Electrical Efficiency

The physically based dynamic SOFC model developed in Chapter 4 facilitates modeling of multiple physical parameters as they occur within a SOFC. With voltage, current, and temperature data, it is possible to model the SOFC electrical efficiency and also utilize the gas phase heat capacity equations to model SOFC efficiency in CHP operation mode in real time under varying operating conditions.

Most literature evaluates fuel cell performance merely on the production of electricity and neglects any thermal recovery downstream in a CHP configuration [60,61]. Figure 9.30 gives a comparison of SOFC electrical efficiencies versus load current obtained from the SOFC dynamic model (with parameters given in Table 9.4) for hydrogen HHV (−285.83 kJ/mol

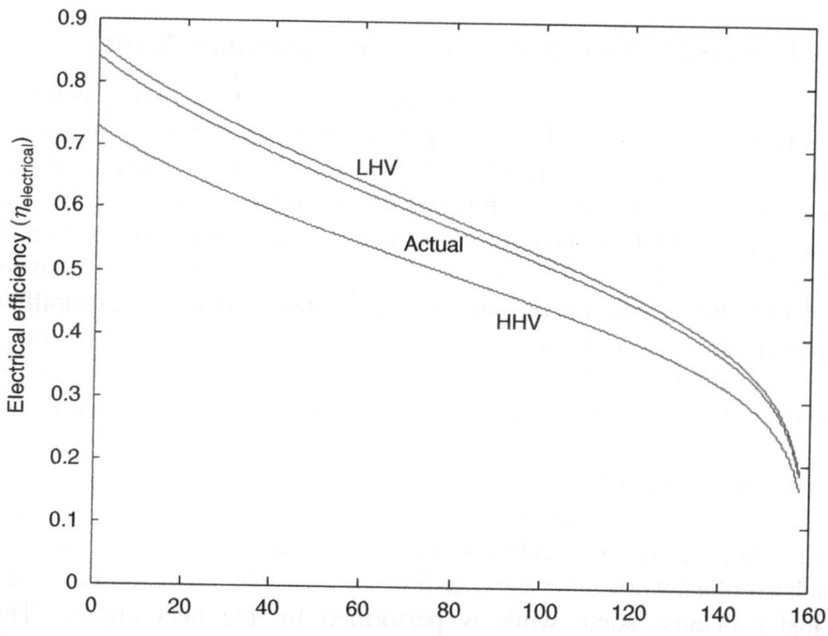

FIGURE 9.30 Comparisons between SOFC stack model electrical efficiencies utilizing hydrogen HHV and HLV, and the calculated enthalpy flow for actual SOFC operating parameters at 1273 K.

of $H_2 = -141.79$ MJ/kg of H_2) and LHV (-241.83 kJ/mol of $H_2 = -119.96$ MJ/kg of H_2), as obtained based on Equation (9.20). The hydrogen HHV and LHV, multiplied by the molar flow rate of consumed hydrogen (as obtained from the model) was used to obtain the denominator in Equation 9.20. These results are compared to the actual enthalpy flow calculated from SOFC simulation data and used in Equation 9.20. The purpose of showing Fig. 9.30 is to demonstrate the utility of determining actual enthalpy flow for modeling purposes. This does not imply that the widely accepted use of LHV and HHV for general fuel-to-work approximations are invalid, conversely, this clearly illustrates how they bracket the efficiency potential of a SOFC.

It is clear from Fig. 9.30 that SOFC electrical efficiency is not congruent with either HHV or LHV, but as expected, is somewhere between the two. This demonstrates that at temperatures other than those corresponding exactly to the hydrogen HHV and LHV, it is not accurate to use either value when calculating efficiency. The actual electrical operating efficiency of a SOFC can therefore be determined dynamically at its actual operating temperatures.

9.5.4 SOFC Efficiency in Hybrid CHP Operation Mode

In this section, we will incorporate the SOFC as a subsystem within a simple CHP system, including required front-end support subsystems. CHP has the capability to produce electricity and exhaust streams of high temperature gas that can be further utilized for residential/commercial heating and cooling or be used for further electricity production in turbine applications [62]. Energy that is not extracted by the heat recovery system is then rejected to the environment. Figure 9.31 shows a simplified representation of the energy conversion flow that occurs within a SOFC–CHP system. Fuel energy and an external source of energy are inputs to the system. Electrical power and work from heat transfer are the useful outputs of the system. The overall system efficiency may be raised by utilizing some SOFC exhaust heat directly for support processes.

A simple reversible heat engine model for a thermal recovery system is proposed for analysis of hybrid systems. The heat engine model works on the basis of a hot reservoir that transfers heat to a cold reservoir. As the heat transfer occurs, some work is performed by the heat engine. This combination of heat transfer and work production can be utilized for useful purposes such as generating electricity, preheating SOFC inlet gasses, providing process steam, or producing residential hot water, etc. [63].

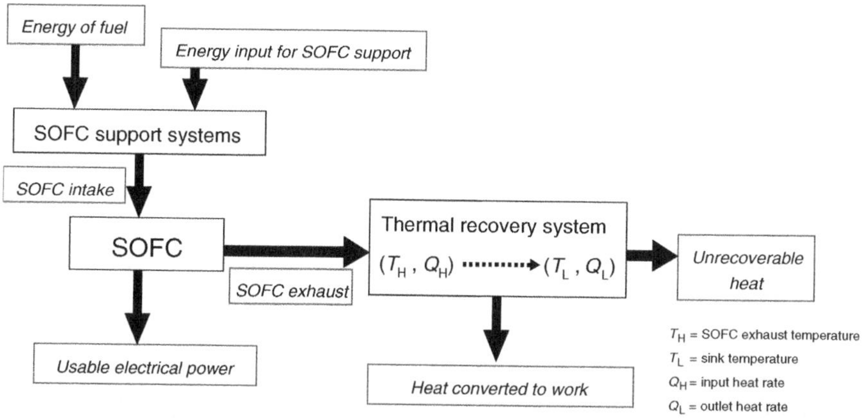

FIGURE 9.31 Simplified efficiency flow model for SOFC–CHP system.

There exist numerous cogeneration processes that utilize varying means of work extraction from a hot exhaust stream. To create a generalized model of a heat recovery process for the SOFC–CHP system, a number of simplifications are made. The simulated heat recovery system takes in hot SOFC exhaust and transfers its heat to a cold reservoir. As the heat transfer is taking place, the heat recovery system converts an arbitrary 38% of the heat energy present to mechanical work. This is based on a conventional HRSG-turbine system efficiency of 30–40% [64]. In an ideal Carnot heat engine, the rate of system work flow (dW) is equivalent to the difference between the input enthalpy flow (dQ_{hot}) and the output enthalpy flow (dQ_{cold}). Ideal Carnot heat engine efficiency is the ratio of system work flow over the net enthalpy flow, as shown by a manipulation of terms in Equation 9.22 and also in Equation 9.23. The maximum work flow that may be extracted from a heat engine is bounded by the Carnot limit. Therefore, for this simple nonideal heat engine model, only 38% of the maximum work flow available between a cold reservoir reference at 390 K and the dynamically varying hot SOFC exhaust gas temperature is converted to shaft work. In a static example where hot SOFC exhaust is at 1273 K, the 38% model heat engine efficiency conservatively compares to an ideal Carnot efficiency of 69.3%. The initial heat content of each SOFC exhaust product is calculated based on its concentration within the mass flow rate of exhaust. Because specific heat capacity is compound specific and a function of temperature, the gas phase heat capacity equations from Ref. [65] are utilized in Equation 9.26 to determine the specific heats of each exhaust product ($c_{p,i}$), as the SOFC operating temperature varies under changing load conditions. The total heat flow of the SOFC exhaust gasses are computed according to the sum:

$$dQ_{\text{exhaust}} = \sum_i \left\{ \int_{T_{\text{sink}}}^{T_{\text{fuel cell exhaust}}} [\bar{n}_i c_{p,i}(T)\, dT] \right\} \qquad (9.26)$$

where dQ_{exhaust} is the exhaust heat flow (J/s), $\bar{n}_i$ the molar flow rate of each exhaust product (mol/s), $T_{\text{fuel cell exhaust}}$ the temperature of SOFC exhaust product (K), T_{sink} the cold sink temperature (K), and $c_{p,i}$ the specific heat capacity (J/mol K) of exhaust species, i.

In order to simulate the support subsystems required to sustain an operating SOFC, a lumped model is used. Initial reactants are assumed to be at ambient temperature (298 K). The compression and preheating processes are assumed to be approximately 70% efficient (based on the first law of thermodynamics). Improved support subsystem models are needed for more accurate efficiency evaluation of SOFC–CHP operations.

Using the support and heat recovery system models, simulations were conducted to determine how much additional work could be extracted from the fuel cell. Figure 9.32 shows a comparison of the electrical efficiency of a SOFC with that of a SOFC–CHP system that both incorporate the energy

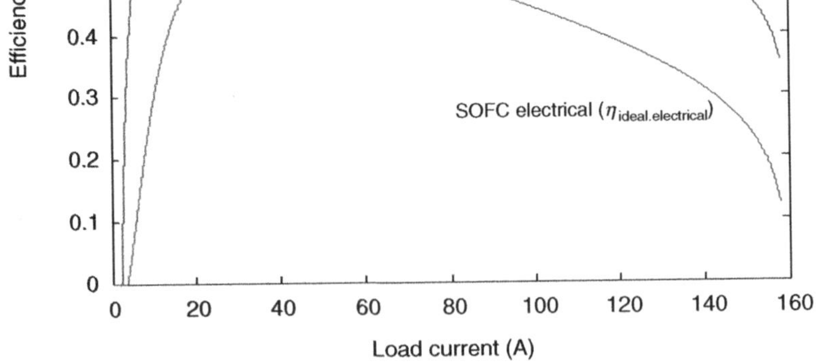

FIGURE 9.32 Comparison of SOFC stand-alone electrical efficiency and SOFC–CHP system (both including input species preconditioning support) efficiency. Parameters given in Table 9.4 were used in the simulation.

lost in conditioning the inlet gasses for SOFC operation. The results show the inefficiency of operations at low electrical load. The principle of SOFC operation is that energy contained within the hydrogen fuel is partially converted to electrical power. A large quantity of input energy from an external source is still required to support SOFC operations at light electrical loading. This energy far exceeds the electrical power output and heat recovery benefits at low electrical load. However, as the SOFC load increases, the output electrical power rises to a level that, when combined with the gains from heat recovery, far exceed the energy expended in supporting the SOFC operation.

The purpose of Fig. 9.32 is to contrast SOFC stand-alone system to that of a SOFC–CHP system. In literature regarding fuel cell capabilities, it can be confusing to the reader when presented with electrical efficiency performance of a fuel cell which does not typically take into account the energy consumed in supporting its operations. In Fig. 9.32, however, the SOFC electrical efficiency is shown as would be consistent in an actual stand-alone system requiring inlet species support. The figure highlights the difference between SOFC electrical and SOFC–CHP system efficiencies, especially at low load. While the SOFC does convert fuel energy to electrical energy efficiently, the electrical energy produced is significantly offset by the large amount of energy consumed to sustain SOFC operations at high temperatures. SOFC–CHP hybrid operations capitalize on the energy that would otherwise be lost to the environment and make SOFC operations more viable for converting fuel to useful work. Apparent in Fig. 9.32 is how beneficial CHP processes can be to the overall energy conversion capability of the SOFC system.

The electrical efficiency obtained from the 5-kW SOFC model (developed in Chapter 4) at full load (100 A) is approximately 50%, whereas the CHP efficiency is approximately 70%. Simulation results given in Fig. 9.32 show that coupling a high temperature fuel cell, such as SOFCs, with a CHP heat recovery system can maximize the energy extracted from an inlet fuel and can have significant efficiency benefits.

9.6 SUMMARY

In this chapter, a general overview on hybrid system configuration and system integration was given. A case study for a hybrid wind–PV–FC energy system was carried out; the system configuration and control strategy for power management were given and discussed, and the results of simulation studies were presented using actual weather data and practical load profiles. The results show that a hybrid system can achieve a desired performance with appropriate control and power management scheme.

In another case study, the efficiency of a hybrid SOFC system in CHP operation mode was evaluated. Simulation results show that coupling a high-temperature fuel cell, such as SOFCs, with a CHP heat recovery system can maximize the energy extracted from an inlet fuel and can have significant efficiency benefits.

REFERENCES

[1] K. Agbossou, M. Kolhe, J. Hamelin, and T.K. Bose, Performance of a stand-alone renewable energy system based on energy storage as hydrogen, *IEEE Transactions on Energy Conversion*, 19 (3), 633–640, 2004.

[2] K. Agbossou, R. Chahine, J. Hamelin, F. Laurencelle, A. Anourar, J.-M. St-Arnaud, and T.K. Bose, Renewable energy systems based on hydrogen for remote applications, *Journal of Power Sources*, 96, 168–172, 2001.

[3] D.B. Nelson, M.H. Nehrir, and C. Wang, Unit sizing and cost analysis of stand-alone hybrid wind/PV/fuel cell power generation systems, *Renewable Energy*, 31 (10), 2006. 1641–1656,

[4] R. Lasseter, Dynamic models for micro-turbines and fuel cells, *Proceedings, 2001 PES Summer Meeting*, Vol. 2, 761–766, Vancouver, Canada, 2001.

[5] Y. Zhu and K. Tomsovic, Development of models for analyzing the load-following performance of microturbines and fuel cells, *Journal of Electric Power Systems Research*, 62, 1–11, 2002.

[6] S.R. Guda, C. Wang, and M.H. Nehrir, Modeling of microturbine power generation systems, *Electric Power Components and Systems*, 34 (9), 2006.

[7] K. Sedghisigarchi and A. Feliachi, Dynamic and transient analysis of power distribution systems with fuel cells—Part I: Fuel-cell dynamic model, *IEEE Transactions on Energy Conversion*, 19 (2), 423–428, 2004.

[8] K. Sedghisigarchi and A. Feliachi, Dynamic and transient analysis of power distribution systems with fuel cells—Part II: control and stability enhancement, *IEEE Transactions on Energy Conversion*, 19 (2), 429–434, 2004.

[9] S.H. Chan, H.K. Ho, and Y. Tian, Multi-level modeling of SOFC-gas turbine hybrid system, *International Journal of Hydrogen Energy*, 28 (8), 889–900, 2003.

[10] Henry Louie and Kai Strunz, Superconducting magnetic energy storage (SMES) for energy cache control in modular distributed hydrogen-electric energy systems, *IEEE Transactions on Applied Superconductivity*, 17 (2), 2361–2364, 2007.

[11] K. Strunz and E.K. Brock, Stochastic energy source access management: infrastructure-integrative modular plant for sustainable hydrogen-electric co-generation, *International Journal of Hydrogen Energy*, 31 (9), 1129–1141, 2006.

[12] H. Dehbonei, Power conditioning for distributed renewable energy generation, Ph.D. Dissertation, Curtin University of Technology, Perth, Western Australia, 2003.

[13] P.A. Lehman, C.E. Chamberlin, G. Pauletto, and M.A. Rocheleau, Operating experience with a photovoltaic-hydrogen energy system, *Proceedings, Hydrogen'94: The 10th World Hydrogen Energy Conference*, Cocoa Beach, *FL*, June 1994.

[14] A. Arkin and J.J. Duffy, Modeling of PV, electrolyzer, and gas storage in a stand-alone solar-fuel cell system, *Proceedings of the 2001 National Solar Energy Conference*, Annual Meeting, American Solar Energy Society, Washington, DC, 2001.

[15] L.A. Torres, F.J. Rodriguez, and P.J. Sebastian, Simulation of a solar-hydrogen-fuel cell system: results for different locations in Mexico, *International Journal of Hydrogen Energy*, 23 (11), 1005–1010, 1998.

[16] S.R. Vosen and J.O. Keller, Hybrid energy storage systems for stand-alone electric power systems: optimization of system performance and cost through control strategies, *International Journal of Hydrogen Energy*, 24 (12), 1139–1156, 1999.

[17] Th.F. El-Shatter, M.N. Eskandar, and M.T. El-Hagry, Hybrid PV/fuel cell system design and simulation, *Renewable Energy*, 27 (3), 479–485, 2002.

[18] Ø. Ulleberg and S.O. Mørner, TRNSYS simulation models for solar-hydrogen systems, *Solar Energy*, 59 (4–6), 271–279, 1997.

[19] J.C. Amphlett, E.H. de Oliveira, R.F. Mann, P.R. Roberge, and A. Rodrigues. Dynamic interaction of a proton exchange membrane fuel cell and a lead-acid battery, *Journal of Power Sources*, 65, 173–178, 1997.

[20] D. Candusso, L. Valero, and A. Walter, Modelling, control and simulation of a fuel cell based power supply system with energy management, *Proceedings of the 28th Annual Conference of the IEEE Industrial Electronics Society (IECON 2002)*, Vol. 2, 1294–1299, Sevilla, Spain, 2002.

[21] M.T. Iqbal, Modeling and control of a wind fuel cell hybrid energy system, *Renewable Energy*, 28 (2), 223–237, 2003.

[22] H. Sharma, S. Islam, and T. Pryor, Dynamic modeling and simulation of a hybrid wind diesel remote area power system, *International Journal of Renewable Energy Engineering*, 2 (1), 2000.

[23] R. Chedid, H. Akiki, and S. Rahman, A decision support technique for the design of hybrid solar-wind power systems, *IEEE Transactions on Energy Conversion*, 13 (1), 76–83, 1998.

[24] W.D. Kellogg, M.H. Nehrir, G. Venkataramanan, and V. Gerez, Generation unit sizing and cost analysis for stand-alone wind, photovoltaic, and hybrid wind/PV systems, *IEEE Transactions on Energy Conversion*, 13 (1), 70–75, 1998.

[25] F. Giraud and Z.M. Salameh, Steady-state performance of a grid-connected rooftop hybrid wind-photovoltaic power system with battery storage, *IEEE Transactions on Energy Conversion*, 16 (1), 1–7, 2001.

[26] E.S. Abdin, A.M. Osheiba, and M.M. Khater, Modeling and optimal controllers design for a stand-alone photovoltaic-diesel generating unit, *IEEE Transactions on Energy Conversion*, 14 (3), 560–565, 1999.

[27] F. Bonanno, A. Consoli, A. Raciti, B. Morgana, and U. Nocera, Transient analysis of integrated diesel-wind-photovoltaic generation systems, *IEEE Transactions on Energy Conversion*, 14 (2), 232–238, 1999.

[28] T. Monai, I. Takano, H. Nishikawa, and Sawada, Response characteristics and operating methods of new type dispersed power supply system using photovoltaic fuel cell and SMES, *2002 IEEE PES Summer Meeting*, Vol. 2, 874–879, Chicago, Illinois, 2002.

[29] F.A. Farret and M.G. Simões, *Integration of Alternative Sources of Energy*, John Wiley & Sons, Inc., New York, 2006.

[30] Øystein Ulleberg, Stand-alone power systems for the future: Optimal design, operation and control of solar-hydrogen energy systems, Ph. D. Dissertation, Norwegian University of Science and Technology, Trondheim, 1998.

[31] P. Strauss and A. Engler, ac coupled PV hybrid systems and microgrids state-of-the-art and future trends, *Proceedings of the 3rd World Conference on Photovoltaic Energy Conversion*, Vol. 3 (12–16), 2129–2134, Osaka, Japan, May 2003.

[32] G. Hegde, P. Pullammanappallil, and C. Nayar, Modular ac coupled hybrid power systems for the emerging GHG mitigation products market, *Proceedings of the Conference on Convergent Technologies for Asia-Pacific Region*, Vol. 3 (15–17), 971–975, Bangalore, India, Oct. 2003.

[33] S.K. Sul, I. Alan, and T.A. Lipo, Performance testing of a high frequency link converter for space station power distribution system, *Proceedings of the 24th Intersociety: Energy Conversion Engineering Conference (IECEC-89)*, 1 (6–11), 617–623, Washington, D.C., Aug.1989.

[34] H.J. Cha and P.N. Enjeti, A three-phase ac/ac high-frequency link matrix converter for VSCF applications, *Proceedings of the IEEE 34th Annual Power Electronics Specialist Conference 2003 (PESC'03)*, 4 (15–19), 1971–1976, Acapulco, Mexico, June 2003.

[35] N. Hatziargyriou, H. Asano, R. Iravani, and C. Marnay, Microgrids, *IEEE Power and Energy Magazine*, 5 (4), 2007.

[36] IEEE Std 1547, *IEEE Standard for Interconnecting Distributed Resources with Electric Power Systems*, 2003.

[37] J. Cahill, K. Ritland, and W. Kelly, *Description of Electric Energy Use in Single Family Residences in the Pacific Northwest 1986–1992, Office of Energy Resources, Bonneville Power Administration,* Portland, OR, December 1992.

[38] C. Wang, Modeling and control of hybrid wind/photovoltaic/fuel cell distributed generation systems, Ph.D. Dissertation, Montana State University, Bozeman, MT, 2006.

[39] J.G. Slootweg, Wind power: modeling and impact on power system dynamics, Ph.D. Dissertation, Department of Electrical Engineering, Delft University of Technology, Delft, Netherlands, 2003.

[40] S. Heier, *Grid Integration of Wind energy Conversion Systems*, Wiley, New York, 1998.

[41] P.C. Krause, O. Wasynczuk, and S.D. Sudhoff, *Analysis of Electric Machinery*, IEEE Press, New York, 1995.

[42] E. Muljadi, C.P. Butterfield, H. Romanowitz, and R. Yinger, Self-excitation and harmonics in wind power generation, *Transactions of the ASME Journal of Solar Energy Engineering*, 127 (4), 581–857, 2005.

[43] M.R. Patel, *Wind and Solar Power Systems*, 2nd edn, CRC Press LLC, Boca Raton, FL, 2006.

[44] M.A.S. Masoum, H. Dehbonei, and E.F. Fuchs, Theoretical and experimental analyses of photovoltaic systems with voltage and current-based maximum power-point tracking, *IEEE Transactions on Energy Conversion*, 17 (4), 514–522, 2002.

[45] J.H.R. Enslin, M.S. Wolf, D.B. Snyman, and W. Swiegers, Integrated photovoltaic maximum power point tracking converter, *IEEE Transactions on Industrial Electronics*, 44 (6), 1997.

[46] AgriMet Historical Dayfile Data Access, Available at http://www.usbr.gov/pn/agrimet/webaghrread.html.

[47] P. Gipe, *Wind Power: Renewable Energy for Home, Farm, and Business*, Chelsea Green Publishing Company, White River Junction, VT, 2004.

[48] IEA Newsletter PV Power, Issue 26, Available at http://www.iea-pvps.org/pvpower/index.htm, June 2007.

[49] F. Mitlitsky, N.J. Colella, B. Myers, and C.J. Anderson, Regenerative fuel cells for high altitude long endurance solar powered aircraft, *Proceedings of the 28th Intersociety Energy Conversion Engineering Conference, Atlanta, GA*, 8–13, Aug. 1993.

[50] F. Mitlitsky, B. Myers, and H.A. Weisberg, Regenerative Fuel Cell Systems R&D, Lawrence Livermore National Laboratory, *Department of Energy Hydrogen Program Review*, 1998.

[51] F. Barbir, T. Molter, and L. Dalton, Regenerative fuel cells for energy storage: efficiency and weight trade-offs, *IEEE Aerospace and Electronic Systems Magazine*, 20 (3), 35–40, 2005.

[52] D. Linden, *Handbook of Batteries*, 3rd edn, McGraw-Hill, New York, 2002.

[53] C.M. Colson, M.H. Nehrir, M.C. Deibert, and M.R. Amin, Efficiency evaluation of solid oxide fuel cells in combined-cycle operations, *ASME Journal of Fuel Cell Science and Technology*, in press, to be published in 2009.

[54] A. Rao, J. Maclay, and S. Samuelsen, Efficiency of electrochemical systems, *Journal of Power Sources*, 134, 181–184, 2004.

[55] C. Haynes, Clarifying reversible efficiency misconceptions of high temperature fuel cells in relation to reversible heat engines, *Journal of Power Sources*, 92, 199–203, 2001.

[56] A.J. Appleby and F.R. Foulkes, *Fuel Cell Handbook*, Van Nostrand Reinhold, New York, 1989.

[57] R. Felder and R. Rousseau, *Elementary Principles of Chemical Processes*, John Wiley & Sons, New York, 1978.

[58] R. O'Hayre, S.-W. Cha, W. Colella, and F.B. Prinz, *Fuel Cell Fundamentals*, Wiley, New York, 2006.

[59] C.C. Lee and S. Dar Lin, *Handbook of Environmental Engineering Calculations*, McGraw-Hill, New York, 2.46–2.48, 2000.

[60] W. Jamsak, S. Assabumrungrat, P.L. Douglas, N. Laosiripojana, and S. Charojrochkul, Theoretical performance analysis of ethanol-fuelled solid oxide fuel cells with different electrolytes, *Chemical Engineering Journal*, 119, 11–18, 2006.

[61] F.A. Coutelieris, S. Douvartzides, and P. Tsiakaras, The importance of the fuel choice on the efficiency of a solid oxide fuel cell system, *Journal of Power Sources*, 123, 200–205, 2003.

[62] M. Bischoff, Large stationary fuel cell systems: status and dynamic requirements, *Journal of Power Sources*, 154, 461–466, 2006.

[63] F. Calise, M. Dentice d'Accadia, A. Palombo, and L. Vanoli, Simulation and exergy analysis of a hybrid solid oxide fuel cell (SOFC)-gas turbine system, *Energy*, 31, 3278–3299, 2006.

[64] M.P. Boyce, *Gas Turbine Engineering Handbook*, 3rd edn, Elsevier, Amsterdam, 86–91, 2006.

[65] P.J. Linstrom and W.G. Mallard, (editors), NIST Chemistry WebBook—Gas Phase Thermochemistry Data, NIST Standard Reference Database Number 69 June 2005, National Institute of Standards and Technology, Available at http://webbook. nist.gov, 2005.

CHAPTER 10

PRESENT CHALLENGES AND FUTURE OF FUEL CELLS

10.1 INTRODUCTION

Sustainable energy is an increasingly significant challenge for humanity. Revolutionary innovations in renewable energy generation and conversion technologies, and consumption efficiencies are required to promote global socioeconomic and sociopolitical stability. As mentioned previously, fuel cell systems are anticipated to play a significant role in future energy conversion, with applications ranging from centralized and distributed power generation to transportation and portable electronics. Highly efficient, and able to operate on both hydrogen and hydrocarbon feeds, fuel cells will serve well in the transition to a hydrogen economy. Widespread deployment of fuel cell systems is currently impeded by several challenges, which can be categorized into three distinct, yet interrelated areas: (1) cost, (2) fuel and fuel infrastructure, and (3) materials and manufacturing. Many challenges (e.g., cost) are common among all fuel cell system types, whereas others (e.g., fuel and fuel infrastructure) are specific to fuel cell type, and their end-use application.

This chapter presents basic fuel cell system operations (expanding upon Chapters 1 and 2), current challenges for deployment (separated in three areas) and discussion of ongoing research and development (R&D) efforts.

Modeling and Control of Fuel Cells, By M. Hashem Nehrir and Caisheng Wang

10.2 FUEL CELL SYSTEM OPERATIONS

Figure 1.4 (of Chapter 1) displays a basic schematic of fuel cell system operations, including fuel processing, fuel cell stack, power conditioner, and possible energy storage device. Each of these processes presents unique technical challenges depending upon the fuel cell system type and end-use application. The different components of the fuel cell system, shown in Fig. 1.4, and the challenges involved in their operations/processes are discussed in detail below, with emphasis on power conditioning and DG applications, which are the objectives of this book.

10.2.1 Fuel Processor

Figure 10.1 presents an expanded view of a typical fuel processor (with the mass flow from left to right) for a fuel cell stack, including individual unit operations, which are specific to type of fuel, fuel cell, and end-use application.

As described in Chapter 1, hydrogen is the most common fuel cell fuel. Hydrogen can be derived from renewable (e.g., biogas, wind, or solar-driven electrolysis) or nonrenewable (e.g., fossil fuel-oil/gas/coal) sources. End-use application, fuel cell type, and fuel availability dictate fuel source selection. DG applications will likely utilize existing infrastructure (e.g., natural gas) for close proximity to load centers, or renewable sources in off-grid applications. Different fuel cell types require varying fuel specifications, and therefore fuel processing operations are designed according to fuel source and fuel cell type. Of the available fuel cell types, PEMFC and SOFC show the greatest potential for DG applications (see Chapter 1). The following presents general and specific fuel processing considerations for PEMFC and SOFC in DG applications.

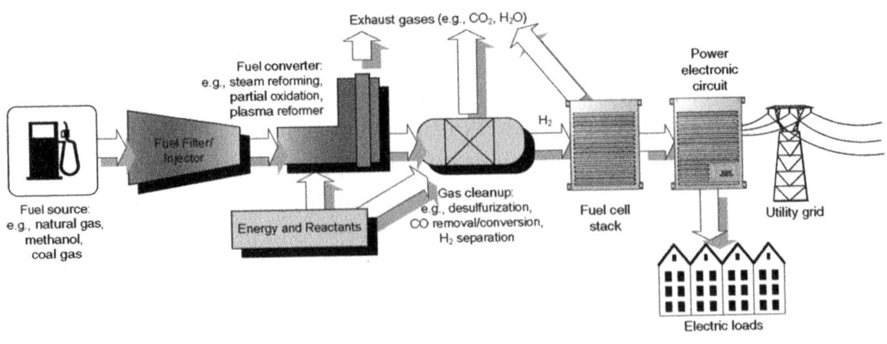

FIGURE 10.1 Fuel processing schematic for fuel cell systems.

In general, designs of fuel processors for fuel cells need to consider the following: operating pressure and temperature amenable with fuel cell type and application; flexibility to meet the application-specific load following and/or start-up requirements; and size (footprint) and/or mass restrictions. Numerous fuel processing technologies are available to convert hydrocarbon fuel into hydrogen rich gas streams, such as steam reforming, partial oxidation and plasma reforming to name a few. Each technology has its relative advantages and disadvantages, some of which were described in Chapter 1. Additional references [1,2,20] provide detailed discussions on fuel processing technologies for fuel cells.

PEMFCs often require pure hydrogen for reliable and durable operation, which complicates fuel processing (especially with sulfur-containing hydrocarbon fuel sources). Due to the low-temperature operation and use of precious metal catalysts, PEMFCs have poor tolerance to carbon monoxide (CO) and sulfur-containing gases (e.g., H_2S). Therefore, depending upon the fuel source, significant efforts must be directed toward the "gas cleanup" operation within a fuel processor, as shown in Fig. 10.1. This often involves hydrogen separation (e.g., using a separation membrane, which is often made of palladium) following the reformation of typical fuels (e.g., natural gas—see Chapter 1 for details). Hydrogen separation and/or other gas cleanup operations can also introduce a significant energy expense (so-called parasitic load), and thereby lower overall system efficiency in converting the fuel's chemical energy into useable electrical energy (in addition to increased capital costs).

Compared with PEMFCs, SOFCs offer fuel flexibility (e.g., both H_2 and CO are fuel for SOFCs), including the ability to internally-reform some fuels, for example, methane (CH_4). This is due primarily to the elevated operating temperatures (e.g., >600°C), which also permits use of nonprecious electrocatalysts, for example, nickel (Ni) versus platinum (Pt). Like PEMFCs, SOFCs are generally intolerant of sulfur-containing gas impurities, for example, sulfur hydride (H_2S); however, their high operating temperatures allows a wider selection of electrocatalysts (e.g., various ceramic oxides), some of which demonstrate sulfur tolerance.

10.2.2 Fuel Cell Stack

At the heart of any fuel cell system is the fuel cell stack. An expanded schematic of a typical planar SOFC stack assembly is presented in Fig. 10.2. This general design is also applicable to planar PEMFC stacks. Tubular and other designs are presented in Chapter 2 and elsewhere [1–3].

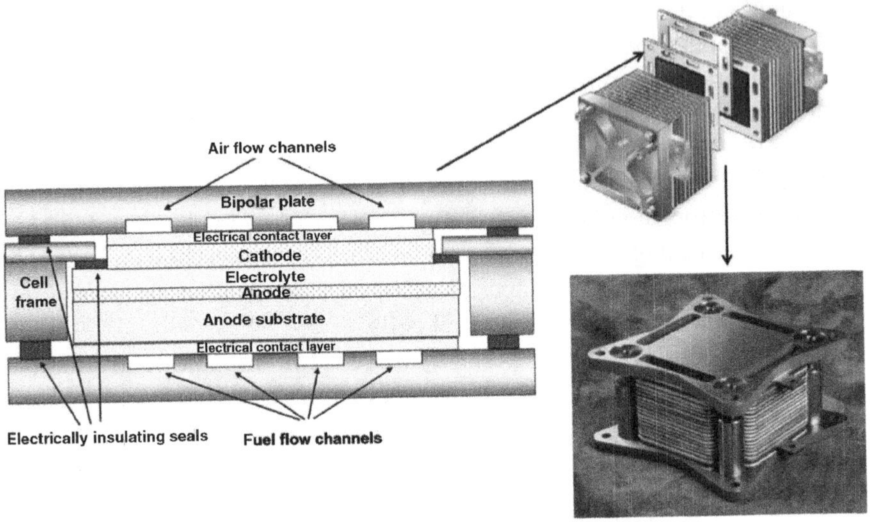

FIGURE 10.2 Typical planar SOFC stack assembly [21,22]. Photo courtesy of Versa Power Systems.

The basic planar fuel cell stack includes

- individual cells, comprised of anode, cathode and electrolyte, typically termed membrane-electrode assembly (MEA) or positive electrode, electrolyte, negative electrode (PEN);
- interconnects, also known as bipolar plates;
- electrically insulated seals, either rigid or compliant (viscous);
- structural components, for example, cell frame; and,
- reactant and exhaust gas (e.g., air and fuel) flow channels.

Stack designs vary in geometry, cell number, gas manifold and thermal management. The primary function of the fuel cell stack is to convert the fuel into useable energy (e.g., electric and thermal). Because of the lower temperature operation, PEMFC systems often produce only electrical energy, whereas SOFC systems produce both electric and thermal energy, commonly referred to as combined heat and power (CHP). Exceptions to this include PEMFC systems which produce useable hot water (e.g., $\sim$80 °C). In comparison with conventional combustion systems, fuel utilization within a fuel cell system is rarely 100%. Unused fuel is often recycled, or used in downstream and/or balance of plant (BOP) processes, such as preheating the fuel and oxidant via combustor and heat exchanger.

To facilitate varying power applications, fuel cell stacks often adopt a modular design such that they can be connected in series or parallel to accommodate end-use load requirements. As detailed later in this chapter, fuel cell stack performance and durability is often limited by deleterious materials interactions, which is the subject of numerous materials science investigations.

10.2.3 Power Conditioner System

As discussed in the previous chapters, fuel cell systems need power electronic interfacing circuits. Though there are varieties of power conditioning systems (PCSs) for fuel cells, a generic fuel cell power electronic interfacing system is shown in Fig. 10.3 [4–7]. The fuel cell output voltage is first boosted up to a higher voltage level through a dc–dc converter. For residential applications, the dc bus voltage (V_{dc}) may vary from 350 to 480 V depending on the actual applications and the circuit topologies. A higher dc bus voltage can be used for larger systems to compensate the larger voltage drops across the inverter and filter. The inverter with a filter will supply three-phases or single-phase ac power to the load. The system can be either operated as stand-alone or grid-connected. If it is grid-connected, a step-up transformer may be needed as well to further boost up the ac voltage to match the grid value. The energy storage device (battery or super-capacitor with appropriate interfacing circuits) can be either

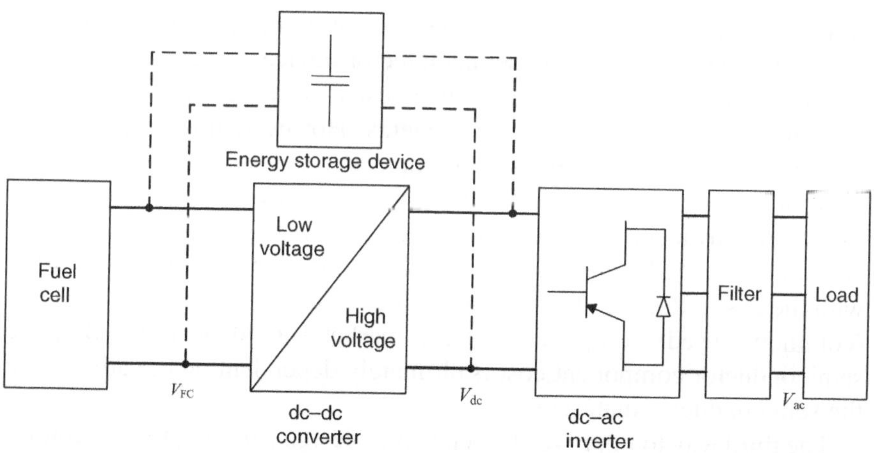

FIGURE 10.3 Block diagram of a generic power conditioning system for fuel cells.

connected at the low voltage side or the high voltage side of the dc–dc converter.

In Fig. 10.3, there are mainly just two types of circuits: dc–dc converter and dc–ac inverter. Sometimes, they are built into one device. Inverters at megawatt level or even larger have been successfully developed with high efficiency. The efficiency of the state-of-art dc–ac inverters has reached over 95%. But, for dc–dc converters, there are more challenges. First of all is the efficiency; the efficiency of a typical dc–dc converter available in the market is normally around 90% or less. The second challenge is the converter capacity; the majority of available dc–dc converters have a power rating less than 10 kW. Therefore, the focus in this section is given to the dc–dc converters for fuel cell systems.

There is still room for improvement in the performance of dc–ac inverters for fuel cells, including improving inverter efficiency and reducing their cost. The power electronic research and development project under the DOE's Solid-State Energy Conversion Alliance (SECA) program is aimed to improve the inverter's efficiency to 99% (like a transformer) and reduce its cost as well [8].

For a practical converter design, there are a number of requirements to be met, for example, voltage, current and power ratings, total mass and volume, and so on. Sometimes, a compromise may have to be made between satisfying one requirement and some others. Nevertheless, there are always two ultimate objectives for a dc–dc converter design: efficiency improvement and cost reduction. The possible means to achieve these two objectives are reviewed below.

Major loss components of the state-of-the-art dc–dc converter are device conduction and switching losses. The cost of a converter is also mainly determined by the switching semiconductor devices. Therefore, the first way to improve efficiency and reduce cost is to design a more efficient topology with less switching components. However, this may result in more peripheral devices and more complicated control scheme.

Another fundamental way to improve the converter efficiency is to use more advanced semiconductor devices with less conducting and switching losses. However, this efficiency objective may have a conflict with the cost reduction. For the cost reduction purpose, the most available (commonly-used) components in the market should be used. But, the semiconductor component cost is ultimately dependent on the advance in the semiconductor materials.

The third way to improve efficiency is to design a more effective control scheme. This can also contribute to the cost reduction. The switching losses can be much reduced by applying soft-switching techniques, namely, zero voltage switching (ZVS), or zero current switching (ZCS). The introduction of a digital signal processor (DSP)-based controller makes this goal much easier to be achieved.

Another difficulty for large-scale fuel cell power generation systems is the limited power rating of dc–dc converters. One way to address this issue is to connect smaller dc–dc converters in parallel to achieve large power capacity. In this case, current sharing among converters becomes crucial because uneven current distribution may cause inductor saturation, uneven thermal stresses, degraded converter performance and even failure of converters. The current-sharing techniques for paralleled power converters are well documented [9–14] in the past years. Among various current-sharing schemes, droop method and active current sharing are the most widely used techniques. The droop method is an inactive approach, which relies on the internal and/or externally added resistances of the paralleled modules to maintain a relatively uniform current distribution among them. The droop method normally does not require any communication among modules, but the system output voltage regulation is poor. Compared with the droop method, the active current-sharing technique can achieve a near-perfect current distribution and a better output voltage regulation. Active current-sharing schemes can be further categorized into "democratic" scheme and "master-slave" scheme [9–14]. In a "democratic" scheme, load currents are shared in proportion to the modules' ratings. While for a "master-slave" approach, the converter with the highest relative output will become the "master" and the slave converters are then controlled by a current value, which is slightly lower than that of the master converter. An active current control is usually implemented with additional load sharing circuitry based on output current sensing for each converter. The additional circuit can be added as an inner loop, outer loop or implemented with an external dedicated current-sharing controller.

The dc–dc load sharing control is not only important for obtaining a larger power capacity, but also important for achieving a scalable structure for a fuel cell-based system. In such cases, dc–dc converters do not need to be designed particularly for a certain system, which can reduce system cost to some extent because of mass production. Therefore, for large fuel cell system applications, load sharing capability of the dc–dc converters in the system needs particular consideration.

Although the above discussions are based on dc–dc converters, most of them are also valid for dc–ac inverters. Through the discussions, it is noted that converter design and development is a multiobjective optimization problem. Therefore, another research area in the future can be the multiobjective optimization for the power conditioning systems for fuel cells.

10.2.4 Balance of Plant (BOP) Systems

Although not presented in Fig. 1.4, processes before, after and between the fuel cell system operations are considered BOP. BOP processes include input pretreatment (e.g., fuel and water vaporization); gas delivery (e.g., fuel pumps and air blowers); exhaust manifolds (from both fuel processor and fuel cell); thermal management (e.g., heat exchangers); and, diagnostics and controls. These are critical systems, which have significant impact on fuel cell system operation and performance. Many BOP systems (e.g., pumps, blowers, diagnostics, and controls) require electric energy, which decreases overall efficiency of the fuel cell system. Therefore, it is imperative to design and integrate efficient, reliable and durable BOP systems within the fuel cell system. Current research and development efforts are directed toward efficient BOP design through modeling coupled with experimentation [15–17].

10.3 PRESENT CHALLENGES AND OPPORTUNITIES

As mentioned in the introduction, cost, fuel production and fuel infrastructure, and materials and manufacturing are the three major challenges impeding the rapid deployment of fuel cell systems. These challenges are discussed below in context on ongoing research and development efforts to overcome them.

10.3.1 Cost

Underpinning the major impediments against fuel cell commercialization is cost. For fuel cell systems, many complex economic factors are at play. These include costs associated with materials and manufacturing, fuel, fuel processing and infrastructure and, as a result, the cost of the generated electricity. Competition from conventional energy sources is also increasingly difficult because of continual improvements in efficiency in addition to reliable operation on inexpensive fuels (e.g., coal). It is generally accepted that current fuel cell technology is cost prohibitive in its ability to compete economically with other technologies for most applications. In various niche applications where efficiency, emissions, and noise are paramount (e.g., space or military applications), and/or with substantial subsidies, fuel cell systems are both economically and technologically attractive, and have demonstrated adequate performance. Detailed economic analyses have been performed to estimate cost contributions of different FC system components and manufacturing processes, with varied conclusions. The reader is referred to available government-contracted reports and other documents for further discussions on fuel cell system cost and cost reduction, for example [18,19]. Particular attention should be paid toward raw material costs for FC system components, as these often dominate overall system costs. It is anticipated, however, that mass production and economies of scale will significantly lower system costs.

10.3.2 Fuel and Fuel Infrastructure

Most fuel cells (PEMFCs, AFCs, PAFCs) require *almost* pure H_2 for reliable operation. Chapter 1 presented different sources of hydrogen including reforming of fossil-fuels (e.g., coal, natural gas) and water electrolysis using nuclear and/or renewable energy. Each of these methods has relative advantages and disadvantages for hydrogen production, as discussed in Chapter 1 and in relevant references, for example [1–3]. Subsequent to producing hydrogen, another significant challenge comes in the form of hydrogen storage and delivery. Depending upon the production location, hydrogen may be used through adjacent fuel cell systems immediately, or it must be stored and transported to its end-use location. Promising hydrogen storage technologies include compression (e.g., in carbon composite tanks), metal hydrides (e.g., $NaBH_4$, $LiBH_4$, $LiAlH_4$, or MgH_2) and cryogenic storage, that is, liquid H_2. Choice of hydrogen storage technology is highly dependent upon end-use application, and cost and energy penalties must be carefully considered. Hydrogen transport is made difficult because of hydrogen's high permeability in most materials, which renders many existing gas pipelines unsuitable. Numerous pipeline coating technologies are being researched for this application, however currently, there is no one technological winner regarding hydrogen transport or storage.

Also, as new policies mandate significantly reduced carbon emissions, more efficient energy conversion devices, such as fuel cells, will gain popularity. SOFCs in particular are currently being engineered to operate in "clean coal" power plants, in which coal is gasified, producing hydrogen rich "synthesis gas" to be used in fuel cells. Carbon capture and sequestration (CCS) technologies are also under design for these advanced SOFC-containing energy systems. Coal gas can also be used as fuel for internal combustion engines (ICEs), gas turbines, as feedstock for petrochemicals, as well as liquid transportation fuels.

Technological solutions for hydrogen production, storage and delivery will substantially accelerate progress toward energy sustainability (and the hydrogen economy), regardless of whether or not fuel cell systems become economically feasible. This is because conventional energy conversion devices, for example, ICEs, can be operated directly on hydrogen to produce useable energy and environmentally benign bi-product, water. Even though fuel-to-electricity conversion efficiency favors electrochemical devices, such as fuel cells, ICEs have demonstrated low-cost and durable operation, and may be preferred for fuel-to-mechanical energy conversion, such as in transportation applications.

10.3.3 Materials and Manufacturing

Materials and manufacturing issues impeding wide-spread deployment are also specific to the fuel cell type. For example, as mentioned earlier, low temperature (<200 °C) PEMFC systems often require precious metals for electrocatalysts (e.g., platinum), which challenges many cost goals. The amount of electrocatalyst used in the electrodes is referred to as catalyst loading, which is typically given in units of mg/cm^2. The current typical value of catalyst loading used is $\sim$1.0 mg/cm^2 [20]. The U.S. Department of Energy (DOE) has set a catalyst loading goal of <0.4 mg/cm^2 (total anode + cathode) to become competitive in automotive markets [20]. Other challenges with PEMFC materials include the need for more (ionically) conductive electrolytes, more active electrocatalysts, improved cell interconnection, and improved water and thermal management [20]. PEMFC research and development efforts are being directed toward these areas and also increasing operating temperatures to improve the electrocatalytic activity of less expensive electrode catalysts. In addition, new cell designs aimed at reducing catalyst loading with improved interconnection materials promise to reduce PEMFC costs to commercially-viable thresholds.

On the contrary, high operating temperatures (e.g., >600 °C) of SOFC systems (which negates precious metal electrocatalysts) often leads to undesired interfacial gas and solid-state reactions among component materials. Additional challenges for SOFC system materials include the need for improved seals, increased tolerance to fuel impurities, more active cathode materials and corrosion resistant interconnects. Because SOFC stack designs vary significantly, so too do the materials and manufacturing challenges. SOFC systems can generally be categorized by different operating temperature regimes: high temperature (>800°C), intermediate temperature (600–800°C), and low temperature (<600°C). Some SOFC systems are also being designed around proton-conducting electrolytes, which may offer increased efficiency, but require pure hydrogen fuel. Proton-conducting SOFC systems are at a relatively early-stage in development compared to the conventional oxygen ion-conducting SOFCs. High temperature SOFC systems have demonstrated long term stability (years) and acceptable performance operated on desulfurized natural gas. Unfortunately, the costs associated with the exotic oxide cell materials, the gas reformer/cleanup and BOP make high temperature SOFC systems economically prohibitive. Intermediate-temperature SOFC systems are promising cost-wise due to thinner electrolytes and inexpensive interconnection, however, deleterious materials interactions at interfaces challenge their durability. Low-temperature SOFC systems often have flexibility in inexpensive materials and exhibit less interfacial reactions, but suffer on account of decreased electrocatalytic activity of common electrodes. Research and development efforts are focused on each of these issues for the various SOFC systems to realize commercial viability.

For both PEMFC and SOFC systems, consistent, high-throughput production processes are needed to reduce overall costs. Development of fuel cell production systems is underway at many industrial centers worldwide. Major North American-based industrial developers for PEMFC systems include Ballard Power Systems, United Technologies, Plug Power, ReliOn. Likewise, major North American-based industrial developers for SOFC systems include Acumentrics, Siemens, Cummins Power, General Electric, Delphi Automotive Systems and Fuel Cell Energy/Versa Power Systems. European SOFC developers include (among many others); Rolls-Royce Fuel Cell Systems (RRFCS) and Ceres Power in the United Kingdom, Hexis in Switzerland, and ZTEK in Germany. Other industrial fuel cell developers are located throughout Asia, Australia, and elsewhere. Estimates on future global fuel cell markets vary greatly, with many in the tens of billions (USD) by 2010. Future market factors, such as carbon emission regulations, are expected to drive and reward successful industrial fuel cell developers.

10.4 U.S. FUEL CELL R&D PROGRAMS

As discussed in Chapter 1, the U.S. DOE is the primary federal sponsor of fuel cell-related research and development. Within the DOE, the office of fossil fuels sponsors development of SOFC systems and other fossil fuel-conversion technologies. The SOFC-related activities of this office are conducted or managed through the DOE's National Energy Technology Laboratory (NETL). DOE has also entered into an interagency agreement with the National Institute of standards (NIST) in the Department of Commerce to have NIST lead an effort to evaluate various advanced technology options for high-megawatt PCSs and to identify technologies requiring development to meet the DOE's desired cost and performance goals. These DOE and NIST activities are explained in Section 10.4.1.

Also within DOE, the office of renewable energy and energy efficiency sponsors PEMFC and other hydrogen-conversion, storage and transmission system development. A summary of the activities of this office was given in Chapter 1.

Other significant federal sponsors of fuel cell research and development include the Department of Defense (DoD) and the National Aeronautic and Space Administration (NASA). The National Science Foundation (NSF) funds fundamental research focused on key limitations to fuel cell performance, in addition to hydrogen storage and transmission. Fuel cell development is an extremely dynamic field, with public and private companies competing to deliver products to commercial markets, in addition to several various military applications. Generally, fuel cell applications can be divided into several categories: stationary, transportation, portable, and specialty. It is difficult to predict which application will enjoy the most commercial success. Specialty applications, such as military, space, mining, remote sensors, and others will likely realize early success, as many are less economically sensitive, and more performance-driven. Experience within these applications will presumably expand opportunities for other, more common applications (i.e., stationary, transportation, and portable).

10.4.1 DOE's SOFC-Related Programs

The U.S. DOE NETL, in partnership with private industry, educational institutions and other national laboratories through the Solid-State Energy Conversion Alliance. The SECA industry teams include six SOFC manufacturers: Acumentrics, Cummins Power Generation, Delphi Automotive Systems, FuelCell Energy, General Electric, and Siemens Power Generation [23]. SECA is leading the research, development and demonstration of low-cost, environmentally-friendly, fuel-flexible SOFCs and has three major focus areas: Cost Reduction, Coal-Based Systems Manufacturing, and Core Technology R&D. The SECA Cost Reduction goal is to reduce the manufactured cost of SOFC energy systems to \$400/kW by 2010. Concurrently, Coal-Based Systems is pursuing the scaling, aggregation and integration of the technology for use in large integrated gasification fuel cell (IGFC) systems for central power generation applications in order to efficiently and cleanly utilize the existing U.S. large coal reserves. The performance of IGFC systems will be consistent with the Office of Fossil Energy's Advanced Power production goals, which includes efficiencies of more than 45%, using coal higher heating value to electricity, with 90% carbon capture [24]. Megawatt level proof-of-concept systems will be available for testing through the FutureGen program. This program is a partnership between the U.S. government and industry to pursue an innovative project focused on the design, construction and operation of a state of the art power plant that is intended to eliminate environmental concerns associated with coal utilization. In support of these goals, SECA Industry Teams are establishing the requisite manufacturing base, and the SECA Core Technology Program is providing vital R&D on technical hurdles [25,26].

FutureGen is the world's first "zero-emission" coal plant which will lead the way to a new generation of zero-emission coal-based power plants [27]. FutureGen will demonstrate the potential of commercial scale Integrated Gasification Combined Cycle (IGCC) technologies and advanced coal technologies now at the R&D stage to build a 275 MW integrated plant that will capture and sequester 90% of the potential carbon dioxide emissions. The DOE announced its reconstructed approach for FutureGen in January 2008 with more focus given to CO_2 capture and storage (CCS) technology. Neverthless, coal gasification remains as one of the fundamental components in the clean coal technology for future electricity generation. High temperature SOFCs, capable of running on synthesis gas, right fit in developing combined cycle systems with gas turbines to generate electricity in a cleaner and more efficient way. SECA SOFCs are one of the R&D technologies that will be tested within the FutureGen Program. Plans are being developed to test three 5 MW high efficiency SOFCs at FutureGen, which will operate on hydrogen and on coal-derived synthesis gas with integrated carbon capture.

Megawatt level power conditioning systems are required to convert the low-voltage power produced by fuel cell modules in central station scale plants to the very much higher voltage levels required for delivery of power to the utility grid. This requirement is of particular interest to the DOE's FutureGen and SECA Programs; activities in this area are lead by NIST [25,26,28]. SECA has set a cost goal of \$40/kW to a maximum of \$100/kW for SOFC-based power plant PCS, which is generally recognized as a goal that cannot be met with today's technology [28,29]. To address this challenge, DOE and NIST have entered into an interagency agreement to have NIST lead an effort to evaluate various advanced technology options for high-megawatt PCSs and to identify technologies requiring development to meet the overall PCS cost and efficiency (>90%) goals for the SECA central station fuel cell power plants.

Various PCS approaches that focus on the use of advanced technologies for low-, medium-, and high-voltage architectures are being considered by NIST. The advanced component technologies being considered include advanced power semiconductor devices made with silicon carbide (SiC) material, advanced nanocrystalline magnetic materials for filter inductors and transformers, advanced capacitor technologies, advanced power electronic component cooling systems, and modular power electronic package and interconnect approaches. The details of specific technologies deemed to have the potential to reduce PCS cost and the common needs for high-megawatt PCSs across the fuel cell industry and federal government agencies are addressed in [29].

10.5 FUTURE OF FUEL CELLS: A SUMMARY AND AUTHOR OPINIONS

There are several technical and economic challenges to wide-spread deployment of fuel cell systems. These challenges are closely related to existing and future regulatory policies on energy efficiency and emissions (especially carbon dioxide). Three major challenges to fuel cell commercialization (cost, fuel and fuel infrastructure, and, materials and manufacturing) were outlined. Revolutionary progress in these areas must be made in the future to realize widespread fuel cell use. In particular, the availability of fuel cell fuel (hydrogen) will enable demonstration and rapid development of fuel cell systems. Fuel cells are particularly well-suited for integration with alternative energy systems such as wind and solar, in which they can be used to aviate intermittency issues inherent in these systems. Reversible fuel cells can be used as electrolyzers, producing hydrogen (and oxygen) from water using electricity generated from other sources, which can then be stored and used subsequently via fuel cells to produce electricity during peak power demand periods.

It is expected that in future energy system designs, fuel cells will serve as sub-systems of integrated hybrid power generation systems. Other parts of the hybrid power generation system may include primary energy sources, such as coal, natural gas, and hyrdoelectric, or alternative energy power generation sources, such as solar, wind, geothermal, microturbine, and biomass. In addition, energy storage (in particular, hydrogen storage) and power electronic subsystems will be important parts of the integrated hybrid power generation systems.

Since anthropogenic climate change is related to carbon dioxide emissions, it is the author's opinion that every feasible measure to reduce emissions should be implemented immediately. As this urgency becomes realized globally, international policy-makers will develop strategies including a portfolio of solutions and mandates for energy conservation and efficiency (both at source and end-use). These mandates will further drive development and commercialization of fuel cell systems for numerous applications. Therefore, it is anticipated that fuel cells and other high efficiency energy conversion devices will continue to experience increases in attention and development funding. While some applications are inherently more difficult than others (e.g., FC for primary transportation power versus FC for small stationary power), all applications should be developed such that early adopters can give feedback on critical performance data to facilitate future development. In general, the future of energy conversion devices will undoubtedly capitalize on the many advantages of fuel cell systems, which will become common place in many applications, and hopefully, sooner-than-later.

REFERENCES

[1] J. Larminie and A. Dicks, *Fuel Cell Systems Explained*, 2nd edn, Wiley, Hoboken, NJ, 2003.

[2] N. Sammes, *Fuel Cell Technology: Reaching Toward Commercialization*, Springer, New York, 2006.

[3] S. Singhal, and K. Kendall, *Solid Oxide Fuel Cells: Science and Technology*, Elsevier, Amsterdam, 2003.

[4] J. Wang, F.Z. Peng, J. Anderson, A. Joseph, and R. Buffenbarger, Low cost fuel cell converter system for residential power generation, *IEEE Transactions on Power Electronics*, 19 (5), 1315–1322, 2004.

[5] F.Z. Peng, H. Li, G.-J. Su, and J.S. Lawler, A new ZVS bidirectional DC–DC converter for fuel cell and battery application, *IEEE Transactions on Power Electronics*, 19 (1), 54–65, 2004.

[6] R. Sharma and H. Gao, Low cost high efficiency DC–DC converter for fuel cell powered auxiliary power unit of a heavy vehicle, *IEEE Transactions on Power Electronics*, 21 (3), 587–591, 2006.

[7] C. Liu, A. Johnson, and J.-S. Lai, A novel three-phase high-power soft-switched DC/DC converter for low-voltage fuel cell applications, *IEEE Transactions on Industry Applications*, 41 (6), 1691–1697, 2005.

[8] J. Lai, A low-cost soft-switched DC/DC converter for solid oxide fuel cells, SECA FY 2006 Report, pp. 207–210, 2006.

[9] S. Luo, Z. Ye, R.-L. Lin, and F.C. Lee, A classification and evaluation of paralleling methods of power supply modules, *Proceedings of IEEE Power Electronics Specialist Conference*, 2 (pt. 2), 901–908, 1999.

[10] V.J. Thottuvelil and G.C. Verghese, Analysis and control design of paralleled DC/DC converters with current sharing, *IEEE Transactions on Power Electronics*, 13 (4), 635–644, 1998.

[11] J.-J. Shieh, Peak-current-mode based single-wire current-share multi-module paralleling DC power supplies, *IEEE Transactions on Circuits and Systems I*, 50 (12), 1564–1568, 2003.

[12] M. Ponjavic and R. Djuric, Current sharing for synchronized DC/DC operating in discontinuous condition mode, *IEE Proceedings on Electric Power Applications*, 152 (1), 119–127, 2005.

[13] Y. Panov and M.M. Jovanoiv, Stability and dynamic performance of current-sharing control for paralleled voltage regulator modules, *IEEE Transactions on Power Electronics*, 17 (2), 172–179, 2002.

[14] P. Li and B. Lehman, A design method for paralleling current mode controlled DC–DC converters, *IEEE Transactions on Power Electronics*, 19 (3), 748–756, 2004.

[15] A. Litka, Hybrid ceramic/metallic recuperator for SOFC generator, *Office of Fossil Energy Annual Fuel Cell Report*, 205–207, 2007.

[16] H. Ghezel-Ayagh, Advanced control modules for hybrid fuel cell/gas turbine power plants, *Office of Fossil Energy Annual Fuel Cell Report*, 208–210, 2007.

[17] M.C. Johnson, Hot anode recirculation blower for SOFC systems, *Office of Fossil Energy Annual Fuel Cell Report*, 211–212, 2007.

[18] "Grid independent, residential fuel-cell conceptual design and cost estimate," NETL Final Report, TIAX, LLC (Subcontract #736222-300005), October 2002.

[19] D.B. Nelson, M.H. Nehrir, and J. Gerez, Economic evaluation of grid-connected fuel cell systems, *IEEE Transactions on Energy Conversion*, 20 (2), 2005.

[20] *Fuel Cell Handbook*, 7th edn, EG&G Services, Inc., Science Applications International Corporation, DoE, Office of Fossil Energy, National Energy Technology Laboratory, 2004.

[21] L.R. Pederson, P. Singh, and X.-D. Zhou, Application of vacuum deposition methods to solid oxide fuel cells, *Vacuum*, 80 (10), 1066–1083, 2006.

[22] Robotic assembly of fuel cells could hasten hydrogen economy, *ScienceDaily*, November 9, 2005.

[23] W. Surdoval, DOE's SECA and FutureGen Programs: Progress and Plans, Proceedings, 2008 IEEE PES General Meeting, July 20–24, Pittsburgh, PA.

[24] Federal Register, Vol. 70, No. 233, pp. 72634–72635.

[25] National Energy Technology Laboratory, SECA 8th Annual Workshop, DOE/NETL 2007/1299 (CD), 2007.

[26] National Energy Technology Laboratory, 2007 Office of Fossil Energy Fuel Cell Program Annual Report, DOE/NETL-2007/1288 (CD), 2007.

[27] U.S. Department of Energy, Office of Fossil Energy, FutureGen Integrated Hydrogen, Electric Power Production and Carbon Sequestration Research Initiative, Report to Congress 2004.

[28] A.R. Hefner Jr. Advanced power conditioning system technologies for high-megawatt fuel cell power plants, Proceedings, 2008 IEEE PES General Meeting, July 20–24, Pittsburgh, PA.

[29] Proceedings, The High Megawatt Converters Workshop, January 24, 2007, NIST Headquarters, Gaithersburg, MD.

APPENDIX A

INSTRUCTION FOR RUNNING THE PEMFC AND SOFC MODELS AND THEIR DISTRIBUTED GENERATION APPLICATION MODELS

The ftp site associated* with this book contains the following files:

- A physically based dynamic model for a 500-W PEM Fuel Cell (PEMFC) stack developed in Chapter 3,
- A physically based dynamic model for a 5-kW Solid Oxide Fuel Cell (SOFC) stack, developed in Chapter 4,
- A PEMFC distributed generation (PEMFCDG) system model, discussed in Chapter 7. The PEMFCDG system model is based on the 500-W PEMFC stack model.
- A SOFC distributed generation (SOFCDG) system model, discussed in Chapter 7. The SOFCDG system model is based on the 5-kW SOFC stack model.

The following sections give a brief instruction for running the above models. Basic knowledge of MATLAB/Simulink or PSpice is required to run the models. The reader is referred to the *MATLAB User's Guide* and *PSpice User's Guide* for help on how to run simulations using MATLAB/Simulink and PSpice.

*The ftp site address is:ftp://ftp.wiley.com/sci_tech_med/fuel_cells

I. PEMFC MODEL

The dynamic PEMFC model is an autonomous module operated under constant channel pressure with no control on the input fuel flow into the fuel cell (FC). The FC stack will adjust the input fuel flow according to its load current to keep the channel pressure constant. In addition to the model developed in MATLAB/Simulink, an equivalent electrical circuit model for the same PEMFC stack has also been developed using PSPICE. Both models have been validated through experimental data obtained on a 500-W Avista (now Relion) PEMFC stack. The models can be used in studies related to PEMFC performance evaluation, controller design, FC vehicle studies, and for developing models of high power PEMFC power plants, to name a few.

A. PEMFC MATLAB/SIMULINK Model

- *Steps to Run the Model*
  1. Look for the folder "PEMFC_Model\PEMFC_Simulink" and open the file of the model named "PEMFC500W.mdl" in MATLAB/SIMULINK. The model was built using MATLAB Version 7.0.4.365 (R14) Service Pack 2.
  2. Set proper input quantities for the model. The FC (load) current can either be set explicitly, or be measured from an actual load and then be fed back to the model.
  3. Choose a proper solver and set the simulation time. The recommended solver for this model is "ode23tb." Select the solver by clicking "Simulation" in the menu and then select "Simulation parameters."
  4. Click "Start simulation" to run the model.
- *Model Input Quantities*

  Fuel cell load current, I (A). Maximum load current is 25 A.

  Anode channel pressure, Panode (atm).

  Cathode channel pressure at, Pcathode (atm).

  Room temperature, Troom (K).

  Initial temperature of FC, Tinitial (K).
- *Model Output Quantities*

  Output terminal voltage, V (V).

  FC temperature, Tout (K).

- *Example Files*
  1. Before running the MATLAB/SIMULINK model, three example load current files need to be loaded into Workspace. To do this, one can just run the file "firstload.m" under the same directory. It will load the three example files into the MATLAB Workspace.
  2. Any load current data can be used for simulation studies. However, the load current is limited to 25 A in the model. Figure A.1 shows the SIMULINK diagram for an example, where a resistor *R* is connected to the PEMFC model and the load current is fed back to the model as input.
  3. Proper simulation time needs to be set before simulation. Simulation time is determined by the length of simulation study desired and/or the length of the actual input data. Simulation time for the following three examples are:

     InputI_stdy_ideal: ideal fuel cell load current in steady state, simulation time: 4900 s.

     InputI_stdy: real fuel cell load current in steady state, simulation time: 3900 s.

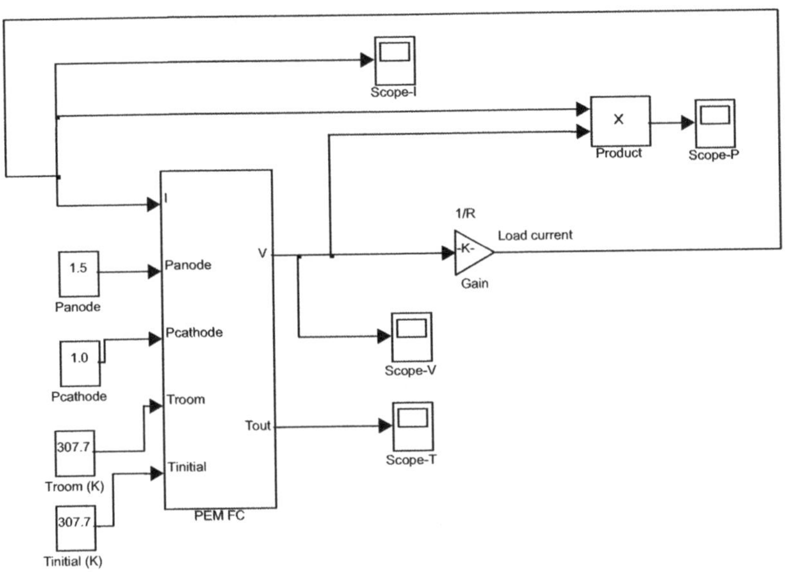

FIGURE A.1 Example for using the PEMFC model in MATLAB/Simulink.

InputI-dynshrt4: real fuel cell load current in transient state, simulation time: 2040 s.

B. PEMFC PSPICE MODEL

- *Steps to Run the Model*

  1. Look for the folder of "PEMFC_Model\PEMFC_PSpice\Version 10.1.0.p001." Open the model file "ETModel.opj" in Capture CIS (Version 10.0.0.p001).
  2. Look for a block named "PEMFC500." Connect the model in your simulation system properly. The FC model block can be used as a regular PSpice component. An example showing how to run the model is given below.

- *Input Terminals*

  Room temperature: Troom (degrees Kelvin presented in volts).

  Ground (see Fig. A.2).

- *Output Terminals*

  Output terminal, Vout.

  FC temperature, Tout (degrees Kelvin presented in volts).

- *Example*

  Figure A.2 shows an example system consisting of the above PSPICE PEMFC model and a super-capacitor (the series combination of a large capacitor and a small resistor) connected across the PEMFC model output terminals. The load current is defined in a file named "iload_c," which is saved under of the model file folder. Load current

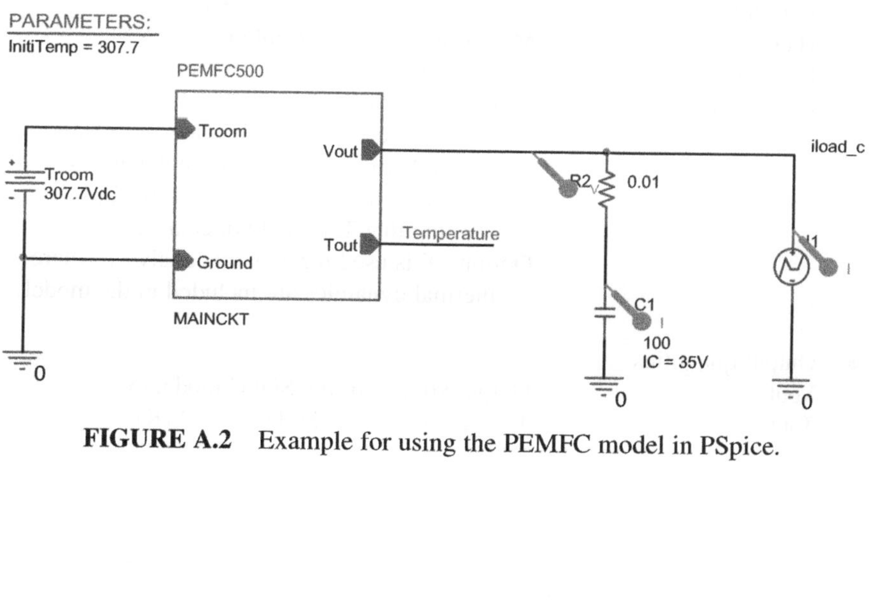

FIGURE A.2 Example for using the PEMFC model in PSpice.

for a desired simulation study can be defined as needed, as long as the current does not exceed 25 A. Make sure the load definition file is clearly specified in the system. Otherwise, one may get an error for finding the load file. It is not necessary to model the load only as a current source, which is pre-defined by a file in the example. For example, the load could be just as simple as a resistor.

II. SOFC MODEL

The SOFC model has 9 input quantities and two main outputs, as listed in the Table A.1. The "Scope" block in MATLAB/Simulink can be used to measure any quantities of interest. Some scope blocks have already been used in the model to measure the following quantities: Load current, FC output power, FC output voltage, FC temperature, activation voltage drop, ohimc voltage drop, concentration voltage drop, fuel utilization, and effective partial pressure of H_2, H_2O, and O_2.

TABLE A.1 SOFC Model Input and Output Quantities

♦ Input quantities	
I	Fuel cell load current (A). Maximum current is 160 A
Pa	Anode channel pressure (atm)
MH2	Mole flow rate of H_2 input at the anode (mol/s)
MH2O	Mole flow rate of H_2O input at the anode (mol/s)
Pc	Cathode channel pressure (atm)
Tairinlet	Air temperature at air inlet (K)
Mair	Mole flow rate of air (mol/s)
Tfuelinlet	Fuel temperature at fuel inlet (K)
TSteadyState	A Boolean value, 0/1 Option '1' is used for steady-state study only, where, there is no temperature dynamics. The fuel cell temperature is set by Tairinlet, meaning that Tout will not change Option '0' is used for dynamic analysis, where thermal dynamics are included in the model
♦ Output quantities	
Vout	Output voltage of the SOFC model (V)
Tout	Temperature of the SOFC model (K)

- *Steps to Run the Model*
  1. Start MATLAB/Simulink (Version 7.0.4.365 (R14) Service Pack 2).
  2. In the folder "\SOFC_Model," open the model file "SOFC_5kW.mdl" in MATLAB/Simulink.
  3. Set the proper model input quantities.
  4. Choose a proper solver and set the simulation time. The recommended solver for this model is "ode23tb." Select the solver by clicking on "Simulation" in the menu and then select "Simulation parameters."
  5. Click "Start simulation" to run the model.
  6. Observe any output by double-clicking on its corresponding scope block.
- *Examples*
  1. Steady-state characteristics of the SOFC model

     Set TSteadySate = 1. Set the proper Tairlet, which in this case is taken as the FC operating temperature. For example, set Tairlet = 1173K. Use the "Ramp" block for the load current input. Slope rate = 0.01. Simulation time = 15,800 s. Under this scenario, the load current will increase from 0 to 158 A at the rate 0.01 A/s. Other parameter values (for the parameters given in Table A.1) are as set in the model. The parameter values can be changed by the user. Run the model and check the results.
  2. Dynamic response of the SOFC model

     Set TSteadySate = 0. Set Tairinlet and Tfuelinlet accordingly. For example, set them both to 1173 K, as shown in Fig. A.3. Use "I_step" block given in the model file to simulate load transients. Simulation time = 15,800 s. Default values are used for other input quantities. Run the model and check its dynamic response.

III. FCDG SYSTEMS

The example FCDG systems based on both PEMFC and SOFC stack models are given in the CD. The FCDG system configurations have been discussed and shown in Fig. 7.1 of Chapter 7. The principle of operation and design of the power electronic circuits (the dc/dc converters and the inverter) in the systems are addressed in Chapter 6, and controller design

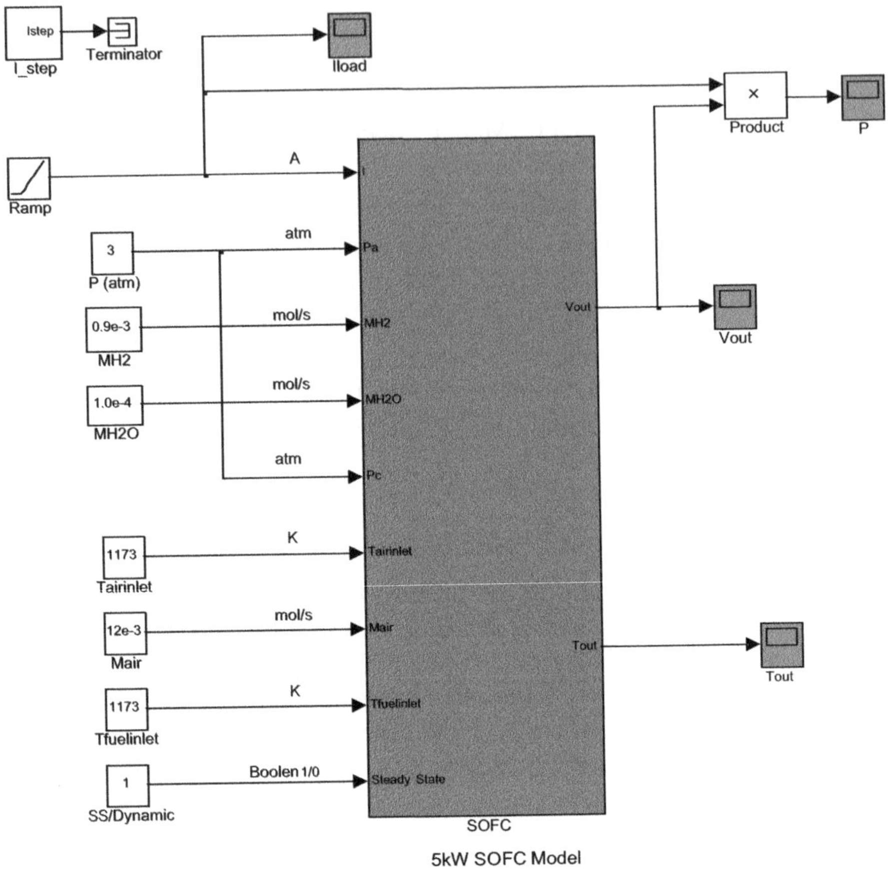

FIGURE A.3 Example for using the SOFC model in MATLAB/SIMULINK.

for the inverter and dc/dc converters is discussed in Chapter 7. A brief introduction for running the SOFCDG system simulation is given below. A similar procedure can be followed to run the PEMFC DG simulation system file.

- *Steps to Run the Model*
  1. Start MATLAB/Simulink (Version 7.0.4.365 (R14) Service Pack 2). SimPowerSystems toolbox (version 4.0.1) under Simulink was used to model the power electronic components in the system. Running the system on a different version of MATLAB/Simulink may cause incompatibility problems. Assistance from MATLAB technical support professionals may be required.

2. Look for the folder of "\FCDG" and open the model file with the name of "SOFC_DG.mdl". User should see the subsystem blocks of FC power plant, dc/dc converters and their controllers, dc/ac inverter and its controller, and the utility grid. The details of each subsystem can be explored by either double-clicking on it or by right-clicking on it and choosing "Look Under Mask" from the pop-up menu. A simple Delay Block with a small time constant has been used in the system to avoid any algebraic loop.
3. The default sampling time for the pulse width modulation (PWM) blocks used for the power switching devices is set to $T_s = 2\,\mu s$. This parameter can be changed as needed. However, any value larger than $10\,\mu s$ is not recommended.
4. Choose the proper solver and set the simulation time. The recommended solver for this model is "ode23tb". Select the solver by clicking on "Simulation" in the menu and then select "Simulation parameters."
5. In order to increase simulation speed, the use of Simulink Accelerator is recommended. Even with the Accelerator, *be aware that it may take quite some time (several hours) to finish the simulation.*
6. Click "Start simulation" to run the model.
7. When the simulation is done, the outputs can be observed by loading the output data files, saved as MAT type in the same folder, to MATLAB workspace. Note that the desired outputs are saved as their corresponding MAT files. For example, "P.mat" is the data file for real power data, P. The command "load P.mat" can be used to load the real power data when the simulation is finished. Then, the power curve can be plotted by using the command "plot(P(1,:), P(2,:))."

INDEX

abc/dq0, 156, 175, 178
Activation voltage drop, 60, 66, 67, 71, 74, 95, 96, 104, 286
Adiabatic, 31
Admittance, 175
AEDG, 219, 220, 221, 225
Aerospace, 263
AgriMet, 242, 263
Alloy, 21, 49
Alternative energy, 1, 2, 58, 126, 162, 219, 221, 222, 225, 278
Anode
  Anode channel, 59, 62, 88, 283
  Anode electrode, 61
  Anode supported, 48
Atmosphere, 18
Autonomous, 6, 283
Availability of fuel, 278
Average model, 127, 135, 144, 146, 149

Balance of plant (BOP), 268, 272
Balanced system, 175
Ballard, 11, 43
Battery discharge, 215
Biased
  Forward biased, 128, 137
  Reverse biased, 129
Biogas, 7, 266
Biomass, 7, 20
Bipolar junction transistor (BJT), 128, 130
Blade, 233, 236, 238
  Wind turbine blade, 236
Bode plot, 176, 179, 201
Boltzmann's constant, 234
Boost converter, 141, 144, 166
Breakdown voltage, 129, 151
Buck converter, 146, 148
Buffer, 203
Bundle, 46

Capacitance of double layer charge effect, 54, 75, 98

Modeling and Control of Fuel Cells, By M. Hashem Nehrir and Caisheng Wang

Capacitor, 82
  Super capacitor, 68, 98, 168, 285
Capacity
  Capacity factor, 20, 230
  Capacity range, 9
  Generation capacity, 4
Carbon capture and sequestration (CCS), 273
Carbon dioxide (CO2), 277, 278
Carbon monoxide (CO), 40, 267
Carbon sequestration, 17, 23
Carnot cycle, 33
Carrier
  Energy carrier, 14
Cartridge, 43
Catalyst(s), 16, 24, 116, 267, 274
Cathode
  Cathode channel, 63, 90, 108, 283
  Cathode electrode, 46, 61
  Cathode pressure, 71, 102
Circuit breaker, 183
Climate change, 279
CMPPT, 238, 243
Coal
  Coal-derived, 23, 277
  Coal fired, 17, 23
  Coal gasification, 17, 21
Cogeneration, 6, 43, 226, 247, 257
Cold startup, 11
Combined cycle, see also Cogeneration, 17, 226, 247
Combined heat and power (CHP), 8, 46, 226
Combustor, 226, 268
Compatibility
  Electromagnetic compatibility, 223
Compensator, 171, 201
Concentration
  Concentration region, 72
  Concentration voltage drop, 53, 60, 68, 75, 96, 286
Contaminants, 43
Controllability, 128
Crossover frequency, 172
Cryogenic storage, 21, 24, 273
Current control, 73, 145, 174, 181, 201
Cut-in speed, 243
Cut-out speed, 233

Damping, 204, 209, 212
Darlington, 131
DC/AC inverter, 150, 159, 222, 289
DC-coupled, DC coupling, 222, 225
DC/DC (DC-DC) converters, see boost and buck converters, rectifiers
Degradation, 40
Delphi, 49, 275, 276
Deoxidized, 49
Deregulated power market, 3, 7, 189
Deregulation, 3
Desulfurized, 275
Diaphragm, 117
Diffusion equations, 71, 91
Diffusivity, 21, 62, 86, 90
Diodes, 128, 137, 202
DISCO, 3
Dispatchable, 219
Distributed generation (DG), 3, 7, 58, 165, 227
Disturbance, 168, 193
Double-layer charge effect, see capacitance of double layer charge effect
dq/abc, 174, 181, 241
Droop, 271
DSP, 271
Dump load, 227
Durability, 24, 216, 269, 275
Duty ratio, 141, 148, 152, 201

Electric Power Research Institute (EPRI), 18
Electrical efficiency, 44, 250, 255, 258
Electrical fault, 164, 182, 192
Electrical power, 14, 251
Electrocatalyst, 267, 274
Electrochemical reaction, 35, 37, 40, 55
Electrode voltage, 38

Electrolyte, 38, 46, 54, 89
Electrolyzer, 134, 228
Electromagnetic compatibility, 223
Emission, 10, 23, 63, 272
Emissivity, 101
Emitter, 130
Endothermic reaction, 30
Energy balance in chemical reactions, 35, 56
Enthalpy of chemical reaction, 99
Enthalpy of formation, 36
Enthalpy of reaction, 253, 254
Entropy, 33
Equilibrium point, 34
Exothermic, 30

Faraday's Law, 62, 91
Fault protection, 195
Fault, see electrical fault
Federal Electric Regulatory Commission (FERC), 2, 37
Feedstock, 23
Flexibility, see fuel flexibility
Fossil energy, Office of Fossil Energy (FE), 227
Fossil fuel, 8, 16, 266, 273
Frequency
  Power Frequency, 175, 221, 225
Frequency modulation ratio, 152
Fuel cell DG, 9, 163, 193
Fuel cell energy, 50, 276
Fuel cell vehicle (FCV), 14, 57, 283
Fuel cell, see PEMFC, SOFC, PAFC, MCFC
Fuel flexibility, 48, 49, 267
Fuel oxidizer, 249, 253
Fuel-oxygen reaction, 253, 254
Fuel reformer, 99
FutureGen, 17, 277

Gas reformer, 275
Gate turn-off thyristor (GTO), 128, 132
GENCO, 3, 5
Geothermal, 7, 220, 278
Gibbs energy, 34, 273
Greenhouse gas, 21, 23
Grid-connected fuel cell, 163
Grid-connected system, 164, 225
Grove (William Grove), 13, 29

Harmonics, 136, 152, 160
High heating value (HHV), 254, 277
High voltage, 129, 130, 278
Humidified, 61
Hydride
  Metal hydride, 21, 24, 273
  Nickel-metal-hydride-battery, 228
  Sulfur hydride, 267
Hydrocarbon, 11, 16, 40, 267
Hydrogen economy, 13, 265, 274
Hydrogenics Corporation, 117, 118
Hydrogen rich gas, see also Synthesis gas, 11, 267
Hydroxide
  Hydroxide liquid, 51
  Potassium hydroxide, 19, 117

Independent power producer (IPP), 3, 7
Inductors, 225, 278
Insulated gate bipolar transistor (IGBT), 128, 133
Insulator, 42
Interaction, 269, 275
Interconnect, 22, 268, 274
Interfacing
  Power electronic interfacing circuit, 72, 126, 225
Intermittency, 278
Internal combustion engine (ICE), 8, 14, 274
Internal reforming, 46, 51, 85
Inversion, see also DC/AC, 127, 135
Ions
  Charged ions, 89
  Hydroxyle (OH) ions, 117
  Negative ions, 54, 89
  Oxygen ions, 19, 89
  Positive ions, 39

Irreversible process, 33, 34, 249
Island, see also power islands, 6, 221, 225
Islanding detection, 9
Isothermal reaction, 249

Kelvin, 285
Kinetic energy, 248

Latent heat, 70
Lead acid battery, 228, 202
Limitation current, 59
Lithium ion battery, 228
Load following, 189, 190
Load sharing, 2, 271
  Current sharing, 271
Load transient, 206
  Load transient mitigation, 209
Lower heating Value (LHV), 254

Master-slave, 271
Material conservation, 64, 92
MATLAB, 287
Maximum power point tracking (MPPT), 238
Mega rule, 2
Membrane, 23, 39, 42
Metal oxide semiconductor field effect transistor (MOSFET), 131
Methane, 7, 15, 267
Microgrid, 221, 189
Microturbine, 8, 9
Model, 12, 26
  Modeling, 57, 85
  Model validation, 77
Modulate, see also PWM, 152
Modulating signal, 174
Modulation index, 152
Modulator (PWM), 142
Molten carbonate fuel cells (MCFC), 10
MOS-controlled thyristor (MCT), 133
Multifuel capability, 46, 85
Multiobjective optimization, 271

Nanocrystalline, 278
Nanoscale, 26
Nanostructured, 24
National Aeronautic and Space Administration (NASA), 30
National Electric Reliability Council (NERC), 3
National Energy Technology Laboratory (NETL), 26
National Institute of Standards (NIST), 276
National Renewable Energy Laboratory (NREL), 24
National Research Council (NRC), 13
Nernst equation, 37
Nickel cadmium, 228
Nickel metal, 228
Nonlinearity, 182
Nonprecious, 267
Nonsinusoidal, 6
Normalized, 204
Nyquist, 142

Off grid, 266
Off peak, 19
Ohmic voltage drop, 53
On peak, 5
Oscillation, 204
Overshoot, 176, 205
Oxidant, 65, 66
Oxidation, 17
Oxidizer, 249

Palladium, 267
Peak demand, 19
Peak to peak value, 201
PEMFC (polymer electrolyte membrane fuel cell), 10, 57, 182, 209
Permanent magnet synchronous generator (PMSG), 226

Permittivity, 54
Petrochemicals, 274
Phase margin, 172
Phosphoric acid fuel cell (PAFC), 40, 9
Photocatalytic, 23
Photoelectrochemical, 19
Photoinduced, 23
Photosynthesis, 30
Photovoltaic (PV), 234, 7
Pitch angle control, 223
Platinum, 42, 51, 274
Plug and play, 221
Polarization, 98
Pollutant, 58
Pollution free, 20
Post loading, 210
Power coefficient, 233
Power conditioning, 12, 126
Power flow, 164, 180
Power islands, 6
Power outage, 6
Power plant, 12, 86, 166, 226
Power sharing, see also Load sharing, 222
Proportional integral (PI), 142, 236
Proton, 39, 19
Pspice, 74, 282
Pulse width modulation (PWM), 166, 289
Push-pull, 140

Quadrant
  Four quadrant operation, 181

Radiation, 88, 99
Reactant, 35, 60
Rechargeable battery, 228
Rectifier, 135
Reformer, see also gas reformer, fuel reformer, 99, 275
Regenerative fuel cell, 229
Regional transmission organization (RTO), 3
Regulation, see also Deregulation, 2, 276
ReliOn, 44
Renewable energy, see also Alternative energy, 7
Resonant, 201
Ripple, 135, 169, 199
Rotor efficiency, see also power coefficient, 233

Saw-tooth, 171
Scalable structure, 271
Seals, 268, 274
Self discharge, 202
Self excited induction generator (SEIG), 233
Semiconductor, 126
Sequestration, see also carbon sequestration, 18, 273
Short circuit current, 234
Siemens, 47
Silicon controlled rectifier (SCR), 130
Simpowersystems block-set, 139, 288
Simulink, see also MATLAB, 282
Sink, 258
Sinusoidal pulse-width-modulation (SPWM), 150
SOFC (solid oxide fuel cell), see also SOFC model, 10, 85,184, 212, 247
  Planar SOFC, 48
  Tubular SOFC, 98
SOFC model, 88
Soft switching, 270
Solar, 7, 19
Solid State Energy Conversion Alliance (SECA), 10
Spikes, 204
Spinning reserve, 4
Stability, 45, 182
Starvation
  Fuel starvation, 110
State space model, 148, 153, 157
Super capacitors, see under capacitor

Superconducting magnetic energy storage (SMES), 199
Survivability, 11
Sustainability, 274
Switched mode, 160
Switching frequency, 134
Synchronization, 222
Synchronous generator, 226
Syngas, 16

Tariffs, 2
Teflon like, 42
Thermodynamic time constant, 109
Thermodynamics
  The first law of thermodynamics, 31
  The second law of thermodynamics, 32
Thyristor, 129
Tip speed ratio (TSR), 233
Total harmonic distortion (THD), 136
TRANSCO, 3
Transformers
  Step-up transformer, 167, 269
  Zig-Zag transformer, 136
Transient mitigation, see also Load transient mitigation, 209
Transportation application, 10

Unitized regenerative fuel cell (URFC), 229
Utility, 1, 5, 164

Valve, 111
Versa Power Systems, 268
Voltage drop/loss, 120
Voltage source inverter (VSI), 150

Wind electrolysis, 20
Wind energy, 20, 232
Wind energy conversion system (WECS), 233
Wind farms, 20
Wind turbine generator (WTG), 20
Wind turbine, see also WTG, 20, 230

Zero current switching (ZCS), 271
Zero voltage switching (ZVS), 271
Zinc air, 228

Printed and bound by CPI Group (UK) Ltd, Croydon, CR0 4YY
16/09/2026
14963409-0002